T0310774

Percolation

Percolation theory was initiated some 50 years ago as a mathematical framework for the study of random physical processes such as flow through a disordered porous medium. It has proved to be a remarkably rich theory, with applications beyond natural phenomena to topics such as the theory of networks. Mathematically, it has many deep and difficult theorems, with a host of open problems remaining. The aims of this book are twofold. First, to present classical results, including the fundamental theorems of Harris, Kesten, Menshikov, Aizenman and Newman, in a way that is accessible to non-specialists. These results are presented with relatively simple proofs, making use of combinatorial techniques. Second, the authors describe, for the first time in a book, recent results of Smirnov on conformal invariance, and outline the proof that the critical probability for Voronoi percolation in the plane is 1/2.

Throughout, the presentation is streamlined, with elegant and straightforward proofs requiring minimal background in probability and graph theory, so that readers can quickly get up to speed. Numerous examples illustrate the important concepts and enrich the arguments. All in all, the book will be an essential purchase for mathematicians, probabilists, physicists, electrical engineers and computer scientists alike.

Percolation

BÉLA BOLLOBÁS
University of Cambridge and
University of Memphis

OLIVER RIORDAN
University of Cambridge

CAMBRIDGE UNIVERSITY PRESS
Cambridge, New York, Melbourne, Madrid, Cape Town,
Singapore, São Paulo, Delhi, Mexico City

Cambridge University Press
The Edinburgh Building, Cambridge CB2 8RU, UK

Published in the United States of America by Cambridge University Press, New York

www.cambridge.org
Information on this title: www.cambridge.org/9780521872324

First published 2006
Reprinted with corrections 2009

A catalogue record for this publication is available from the British Library

ISBN 978-0-521-87232-4 Hardback

To Gabriella and Gesine

Contents

Preface

Percolation theory was founded by Broadbent and Hammersley [1957] almost half a century ago; by now, thousands of papers and many books have been devoted to the subject. The original aim was to open up to mathematical analysis the study of random physical processes such as the flow of a fluid through a disordered porous medium. These *bona fide* problems in applied mathematics have attracted the attention of many physicists as well as pure mathematicians, and have led to the accumulation of much experimental and heuristic evidence for many remarkable phenomena. Mathematically, the subject has turned out to be much more difficult than might have been expected, with several deep results proved and many more conjectured.

The first spectacular mathematical result in percolation theory was proved by Kesten: in 1980 he complemented Harris's 1960 lower bound on the critical probability for bond percolation on the square lattice, and so proved that this critical probability is 1/2. To present this result, and numerous related results, Kesten [1982] published the first monograph devoted to the mathematical theory of percolation, concentrating on discrete two-dimensional percolation. A little later, Chayes and Chayes [1986b] came close to publishing the next book on the topic when they wrote an elegant and very long review article on percolation theory understood in a much broader sense.

For nearly two decades, Grimmett's 1989 book (with a second edition published in 1999) has been the standard reference for much of the basic theory of percolation on lattices. Other notable books on various aspects of percolation theory have been published by Smythe and Wierman [1978], Durrett [1988], Hughes [1995; 1996] and Meester and Roy [1996]; valuable survey articles have been written by Durrett [1984], Chayes, Puha and Sweet [1999], Kesten [2003] and Grimmett [2004], among others.

Our aims in this book are two-fold. First, we aim to present the 'classical' results of percolation in a way that is accessible to the non-specialist. To get straight to the point, we start with the best known result in the subject, the fundamental theorem of Harris and Kesten, even though this is a special case of later and more general results given in subsequent chapters.

The proof of the Harris–Kesten Theorem, in particular the upper bound due to Kesten, was a great achievement, and his proof not simple. Since then, however, especially with the advent of new tools in probabilistic combinatorics, many simple proofs have been found, a fact that most non-specialists are not aware of. For some of these arguments, all the pieces have been published some time ago, but perhaps not in one place, or only as comments that are easy to miss. Here, we bring together these various pieces and also more recent, very simple proofs of the Harris–Kesten Theorem.

In Chapters 4 and 5 we describe the very general results of Menshikov, and of Aizenman, Kesten and Newman; these results are again classical. Our aim here is to present them in the greatest generality that does not complicate the proofs.

Our second aim is to present recent results that have not yet appeared in book form. We give a complete proof of Smirnov's famous conformal invariance result; to our knowledge, no such account, with the i's dotted and t's crossed, has previously appeared. We finish by presenting an outline of the recent proof that the critical probability for random Voronoi percolation in the plane is $1/2$.

As is often the case, we have tried to write the kind of book we should like to read. By now, percolation theory is an immensely rich subject with enough material for a dozen books, so it is not surprising that the choice of topics strongly reflects our tastes and interests. We have striven to give streamlined proofs, and to bring out the elegance of the arguments. To make the book accessible to as wide a readership as possible, we have assumed very little mathematical background and have illustrated the important concepts and main arguments with numerous examples.

In writing this book we have received help from many people. Paul Balister, Gesine Grosche, Svante Janson, Henry Liu, Robert Morris, Amites Sarkar, Alex Scott and Mark Walters were kind enough to read parts of the manuscript and to correct many misprints; for the many that remain, we apologize.

1

Basic concepts and results

Percolation theory was founded by Broadbent and Hammersley [1957], in order to model the flow of fluid in a porous medium with randomly blocked channels. Interpreted narrowly, it is the study of the component structure of random subgraphs of graphs. Usually, the underlying graph is a lattice or a lattice-like graph, which may or may not be oriented, and to obtain our random subgraph we select vertices or edges independently with the same probability p. In the quintessential examples, the underlying graph is $\mathbb{Z}^d$.

The aim of this chapter is to introduce the basic concepts of percolation theory, and some easy fundamental results concerning them.

We shall use the definitions and notation of graph theory in a standard way, as in Bollobás [1998], for example. In particular, if Λ is a graph, then $V(\Lambda)$ and $E(\Lambda)$ denote the sets of vertices and edges of Λ, respectively. We write $x \in \Lambda$ for $x \in V(\Lambda)$. We also use standard notation for the limiting behaviour of functions: for $f = f(n)$ and $g = g(n)$, we write $f = o(g)$ if $f/g \to 0$ as $n \to \infty$, $f = O(g)$ if f/g is bounded, $f = \Omega(g)$ for $g = O(f)$, and $f = \Theta(g)$ if $f = O(g)$ and $g = O(f)$.

The standard terminology of percolation theory differs from that of graph theory: vertices and edges are called *sites* and *bonds*, and components are called *clusters*. When our random subgraph is obtained by selecting vertices, we speak of *site percolation*; when we select edges, *bond percolation*. In either case, the sites or bonds selected are called *open* and those not selected are called *closed*; the *state* of a site or bond is open if it is selected, and closed otherwise. (In some of the early papers, the term 'atom' is used instead of 'site', and 'dammed' and 'undammed' for 'closed' and 'open'.) In site percolation, the *open subgraph* is the subgraph induced by the open sites; in bond percolation, the *open subgraph* is formed by the open edges and all vertices; see Figure 1.

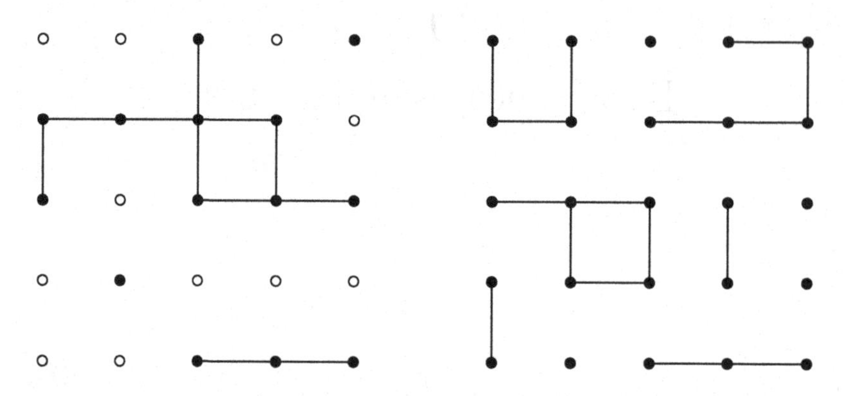

Figure 1. Parts of the open subgraphs in site percolation (left) and bond percolation (right) on the square lattice $\mathbb{Z}^2$. On the left, the filled circles are the open sites; the open subgraph is the subgraph of $\mathbb{Z}^2$ induced by these. For bond percolation, the open subgraph is the spanning subgraph containing all the open bonds.

To streamline the discussion, we shall concentrate on unoriented percolation, i.e., on (bond and site) percolation on an unoriented graph Λ. We assume that Λ is connected, infinite, and locally finite (i.e., every vertex has finite degree). In general, Λ is a *multi-graph*, so multiple edges between the same pair of vertices are allowed, but not loops. Most of the interesting examples will be simple graphs.

Often, we shall choose bonds or sites to be open with the same probability p, independently of each other. This gives us a probability measure on the set of subgraphs of Λ; in bond percolation we write $\mathbb{P}^{\mathrm{b}}_{\Lambda,p}$ for this measure, and in site percolation $\mathbb{P}^{\mathrm{s}}_{\Lambda,p}$. More often than not, we shall suppress the dependence of these measures on some or all parameters, and write simply $\mathbb{P}$ or $\mathbb{P}_p$. Similarly, Λ^{b}_p is the open subgraph in bond percolation, and Λ^{s}_p in site percolation.

Formally, given a graph Λ with edge-set E, a (bond) *configuration* is a function $\omega : E \to \{0,1\}$, $e \mapsto \omega_e$; we write $\Omega = \{0,1\}^E$ for the set of all (bond) configurations. A bond e is open in the configuration ω if and only if $\omega_e = 1$, so configurations correspond to open subgraphs. Let Σ be the σ-field on Ω generated by the cylindrical sets

$$C(F,\sigma) = \{\omega \in \Omega : \omega_f = \sigma_f \text{ for } f \in F\},$$

where F is a finite subset of E and $\sigma \in \{0,1\}^F$. Let $\mathbf{p} = (p_e)_{e\in E}$, with $0 \le p_e \le 1$ for every bond e. We denote by $\mathbb{P}^{\mathrm{b}}_{\Lambda,\mathbf{p}}$ the probability measure on (Ω, Σ) induced by

$$\mathbb{P}^{\mathrm{b}}_{\Lambda,\mathbf{p}}\big(C(F,\sigma)\big) = \prod_{\substack{f\in F\\ \sigma_f=1}} p_f \prod_{\substack{f\in F\\ \sigma_f=0}} (1-p_f). \tag{1}$$

When $p_e = p$ for every edge e, as before, we write $\mathbb{P}^{\mathrm{b}}_{\Lambda,p}$ for $\mathbb{P}^{\mathrm{b}}_{\Lambda,\mathbf{p}}$.

In the measure $\mathbb{P}^{\mathrm{b}}_{\Lambda,\mathbf{p}}$, the states of the bonds are independent, with the probability that e is open equal to p_e; thus, for two disjoint sets F_0 and F_1 of bonds,

$$\mathbb{P}^{\mathrm{b}}_{\Lambda,\mathbf{p}}\big(\text{the bonds in } F_1 \text{ are open and those in } F_0 \text{ are closed}\big) = \prod_{f\in F_1} p_f \prod_{f\in F_0} (1-p_f).$$

We call $\mathbb{P}^{\mathrm{b}}_{\Lambda,\mathbf{p}}$ an *independent bond percolation measure* on Λ. The special case where $p_e = p$ for every bond e is exactly the measure $\mathbb{P}^{\mathrm{b}}_{\Lambda,p}$ defined informally above. The formal definitions for *independent site percolation* are similar.

Let us remark that site percolation is more general, in the sense that bond percolation on a graph Λ is equivalent to site percolation on $L(\Lambda)$, the *line graph* of Λ. This is the graph whose vertices are the edges of Λ; two vertices of $L(\Lambda)$ are adjacent if the corresponding edges of Λ share a vertex; see Figure 2.

Although in this chapter we shall make some remarks about general infinite graphs, the main applications are always to 'lattice-like' graphs. These graphs have a finite number of 'types' of vertices and of edges. Occasionally, we may select vertices or edges of different types with different probabilities.

For a fixed underlying graph Λ, there is a natural coupling of the measures $\mathbb{P}^{\mathrm{b}}_{\Lambda,p}$, $0 \le p \le 1$: take independent random variables X_e for each bond e of Λ, with X_e uniformly distributed on $[0,1]$. We may realize Λ^{b}_p as the spanning subgraph of Λ containing all bonds e with $X_e \le p$. In this coupling, if $p_1 < p_2$, then $\Lambda^{\mathrm{b}}_{p_1}$ is a subgraph of $\Lambda^{\mathrm{b}}_{p_2}$. A similar coupling is possible for site percolation.

An *open path* is a path (i.e., a self-avoiding walk) in the open subgraph. For sites x and y, we write '$x \to y$' or $\{x \to y\}$ for the event that there is an open path from x to y, and $\mathbb{P}(x \to y)$ for the probability of this event in the measure under consideration. We also write '$x \to \infty$' for the event that there is an infinite open path starting at x.

An *open cluster* is a component of the open subgraph. As the graphs

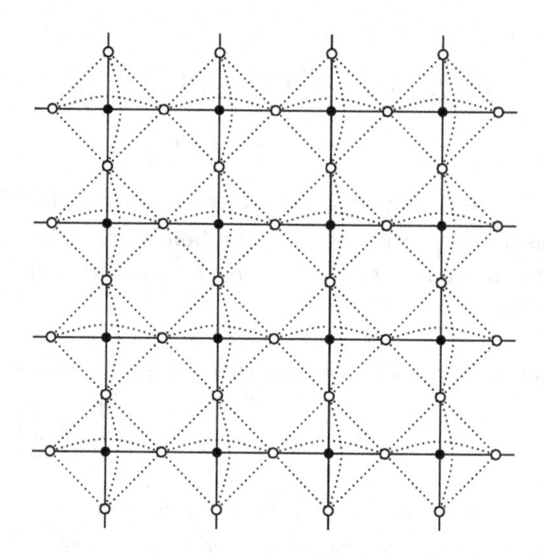

Figure 2. Part of the square lattice $\mathbb{Z}^2$ (solid circles and lines) and its line graph $L(\mathbb{Z}^2)$ (hollow circles and dotted lines). Note that $L(\mathbb{Z}^2)$ is isomorphic to the non-planar graph obtained from $\mathbb{Z}^2$ by adding both diagonals to every other face.

we consider are locally finite, an open cluster is infinite if and only if, for every site x in the cluster, the event $\{x \to \infty\}$ holds. Given a site x, we write C_x for the open cluster containing x, if there is one; otherwise, we take C_x to be empty. Thus $C_x = \{y \in \Lambda : x \to y\}$ is the set of sites y for which there is an open x–y path. Clearly, in bond percolation, C_x always contains x, and in site percolation, $C_x = \emptyset$ if and only if x is closed.

Let $\theta_x(p)$ be the probability that C_x is infinite, so $\theta_x(p) = \mathbb{P}_p(x \to \infty)$. Needless to say, $\theta_x(p)$ depends on the underlying graph Λ, and whether we take bond or site percolation. More formally, for bond percolation, for example,

$$\theta_x(p) = \theta_x(\Lambda_p^{\mathrm{b}}) = \theta_x^{\mathrm{b}}(\Lambda; p) = \mathbb{P}_{\Lambda,p}^{\mathrm{b}}(|C_x| = \infty),$$

where $|C_x| = |V(C_x)|$ is the number of sites in C_x. We shall use whichever form of the notation is clearest in any given context. In future, we shall introduce such self-explanatory variants of our notation without further comment; we believe that this will not lead to confusion. Two sites x and y of a graph Λ are *equivalent* if there is an automorphism of Λ mapping x to y. When all sites are equivalent (i.e., the symmetry group of the graph Λ acts transitively on the vertices), we write $\theta(p)$ for $\theta_x(p)$ for any site x. The quantity $\theta(p)$, or $\theta_x(p)$, is sometimes known as the *percolation probability*.

Clearly, if x and y are sites at distance d, then $\theta_x(p) \geq p^d\theta_y(p)$, so either $\theta_x(p) = 0$ for every site x, or $\theta_x(p) > 0$ for every x. Trivially, from the coupling described above, $\theta_x(p)$ is an increasing function of p. Thus there is a *critical probability* p_{H}, $0 \leq p_{\mathrm{H}} \leq 1$, such that if $p < p_{\mathrm{H}}$, then $\theta_x(p) = 0$ for every site x, and if $p > p_{\mathrm{H}}$, then $\theta_x(p) > 0$ for every x. The notation p_{H} is in honour of Hammersley. When the model under consideration is not clear from the context, we write $p_{\mathrm{H}}^{\mathrm{s}}(\Lambda)$ for site percolation on Λ and $p_{\mathrm{H}}^{\mathrm{b}}(\Lambda)$ for bond percolation.

The component structure of the open subgraph undergoes a dramatic change as p increases past p_{H}: if $p < p_{\mathrm{H}}$ then the probability of the event E that there is an infinite open cluster is 0, while for $p > p_{\mathrm{H}}$ this probability is 1. To see this, note that the event E is independent of the states of any finite set of bonds or sites, so Kolmogorov's 0-1 law (see Theorem 1 in Chapter 2) implies that $\mathbb{P}_p(E)$ is either 0 or 1. If $p < p_{\mathrm{H}}$, so that $\theta_x(p) = 0$ for every x, then

$$\mathbb{P}_p(E) \leq \sum_x \theta_x(p) = 0,$$

and if $p > p_{\mathrm{H}}$, then $\mathbb{P}_p(E) \geq \theta_x(p) > 0$ for some site x (and so for all sites), implying that $\mathbb{P}_p(E) = 1$. One says that *percolation occurs* in a certain model if $\theta_x(p) > 0$, so $\mathbb{P}_p(E) = 1$. With a slight abuse of terminology, we use the same word both for this particular event and for the measures studied; this is not ideal, but, as in so many subjects, the historical terminology is now entrenched.

To start with, the theory of percolation was concerned mostly with the study of critical probabilities, i.e., with the question of when percolation occurs. Now, however, it encompasses the study of much more detailed properties of the random graphs arising from percolation measures. In fact, great efforts are made to describe the structure of these random graphs at or near the critical probability, even when we cannot pin down the critical probability itself. In Chapter 7, we shall get a glimpse of the huge amount of work done in this area, although in a setting in which the critical probability is known.

The theory of percolation deals with infinite graphs, and many of the basic events studied (such as the occurrence of percolation) involve the states of infinitely many bonds. Nevertheless, it always suffices to consider events in *finite* probability spaces, since, for example,

$\theta_x(p) = \lim_{n\to\infty} \mathbb{P}_p(|C_x| \geq n)$. In this book, almost all the time, even the definition of the infinite product measure will be irrelevant.

For $p < p_{\mathrm{H}}$, the open cluster C_x is finite with probability 1, but its expected size need not be finite. This leads us to another critical probability, p_{T}, named in honour of Temperley. Again, we write $p_{\mathrm{T}}^{\mathrm{s}}(\Lambda)$ or $p_{\mathrm{T}}^{\mathrm{b}}(\Lambda)$ for site or bond percolation on Λ. For a site x, set

$$\chi_x(p) = \mathbb{E}_p(|C_x|),$$

where $\mathbb{E}_p$ is the expectation associated to $\mathbb{P}_p$. If all sites are equivalent, we write simply $\chi(p)$. Trivially, $\chi_x(p)$ is increasing with p, and, as before, $\chi_x(p)$ is finite for some site x if and only if it is finite for all sites. Hence there is a critical probability

$$p_{\mathrm{T}} = \sup\{p : \chi_x(p) < \infty\} = \inf\{p : \chi_x(p) = \infty\},$$

which does not depend on x. By definition, $p_{\mathrm{T}} \leq p_{\mathrm{H}}$. One of our aims will be to prove that $p_{\mathrm{T}} = p_{\mathrm{H}}$ for many of the most interesting ground graphs, including the lattices $\mathbb{Z}^d$, $d \geq 2$.

There are very few cases in which p_{H} and p_{T} are easy to calculate. The prime example is the *d-regular infinite tree*, otherwise known as the *Bethe lattice* (see Figure 3). For the purposes of calculation, it is more

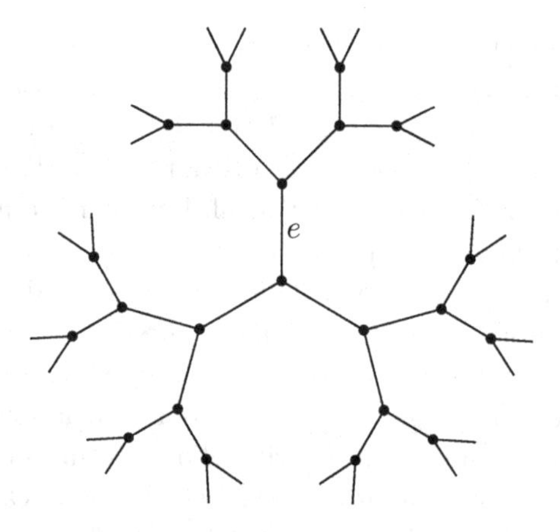

Figure 3. The 3-regular tree, for which $p_{\mathrm{T}}^{\mathrm{b}} = p_{\mathrm{H}}^{\mathrm{b}} = p_{\mathrm{T}}^{\mathrm{s}} = p_{\mathrm{H}}^{\mathrm{s}} = 1/2$. Deleting an edge ($e$, for example), this tree falls into two components, each of which is a 2-branching tree.

convenient to consider the *k-branching tree* T_k. This is the rooted tree in which each vertex has k children, so all sites but one have degree $k+1$. Writing v_0 for the root of T_k, let $T_{k,n}$ be the section of this tree up to

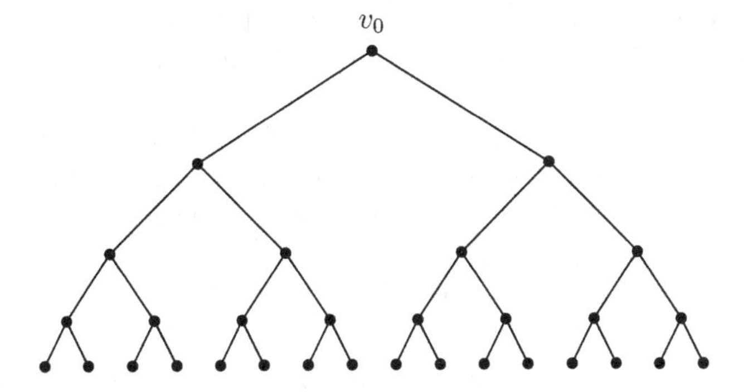

Figure 4. The tree $T_{2,4}$ with root v_0.

height (or, following the common mathematical convention of planting trees with the root at the top, depth) n, as in Figure 4. Taking the bonds to be open independently with probability p, let $\pi_n = \pi_{k,n}(p)$ be the probability that $T_{k,n}$ contains an open path of length n from the root to a leaf. Since such a path exists if and only if, for some child v_1 of v_0, the bond v_0v_1 is open and there is an open path of length $n-1$ from v_1 to a leaf, we have

$$\pi_n = 1 - (1 - p\pi_{n-1})^k = f_{k,p}(\pi_{n-1}). \tag{2}$$

On the interval $[0,1]$, the function $f_{k,p}(x)$ is increasing and concave, with $f_{k,p}(0) = 0$ and $f_{k,p}(1) < 1$, so $f_{k,p}(x_0) = x_0$ for some $0 < x_0 < 1$ if and only if $f'_{k,p}(0) = kp > 1$; furthermore, the fixed point x_0 is unique when it exists (see Figure 5). Thus, if $p > 1/k$, then, appealing to (2) we see

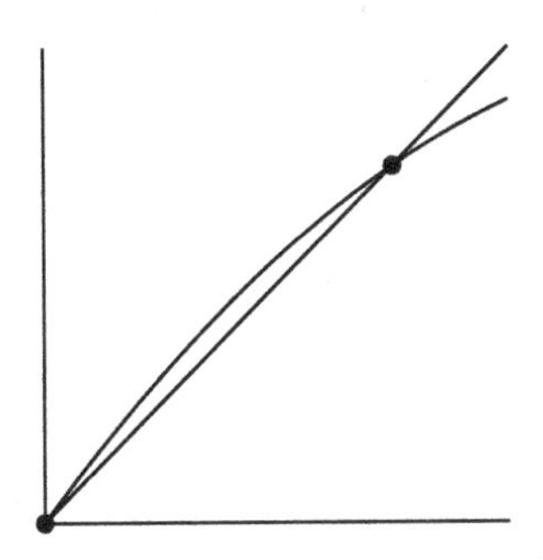

Figure 5. For $k = 2$ and $p = 2/3$, the increasing concave function $f(x) = f_{k,p}(x) = \frac{4}{3}x - \frac{4}{9}x^2$ satisfies $f' = \frac{4}{3} - \frac{8}{9}x$, $f(0) = 0$, $f(3/4) = 3/4$ and $f(1) = 8/9$.

that $\pi_{n-1} \geq x_0$ implies $\pi_n \geq x_0$. Since $\pi_0 = 1$, it follows that $\pi_n \geq x_0$ for every n, so $\theta_{v_0}^{\mathrm{b}}(T_k;p) \geq x_0 > 0$, implying that $p_{\mathrm{H}}^{\mathrm{b}}(T_k) \leq 1/k$. Also, if $p \leq 1/k$, then π_n converges to 0, the unique fixed point of $f_{k,p}(x)$, and so $\theta_{v_0}^{\mathrm{b}}(T_k;p) = 0$. Hence, the critical probability $p_{\mathrm{H}}^{\mathrm{b}}(T_k)$ is equal to $1/k$.

Turning to p_{T}, note that the probability that a site y at graph distance ℓ from the root v_0 belongs to C_{v_0} is exactly p^ℓ. Thus,

$$\chi_{v_0}^{\mathrm{b}}(T_k;p) = \mathbb{E}(|C_{v_0}|) = \sum_{y \in T_k} \mathbb{P}(y \in C_{v_0}) = \sum_{\ell=0}^{\infty} k^\ell p^\ell,$$

which is finite for $p < 1/k$ and infinite for $p \geq 1/k$. Thus the critical probability $p_{\mathrm{T}}^{\mathrm{b}}(T_k)$ is also equal to $1/k$.

For any infinite tree, after conditioning on the root x being open, the open clusters containing x in site and bond percolation have exactly the same distribution. Indeed, each child of a site in the open cluster lies in the open cluster with probability p. Thus, for the k-branching tree T_k, we have $p_{\mathrm{H}}^{\mathrm{s}} = p_{\mathrm{H}}^{\mathrm{b}} = p_{\mathrm{T}}^{\mathrm{s}} = p_{\mathrm{T}}^{\mathrm{b}} = 1/k$. It is easy to show similarly, or indeed to deduce, that the four critical probabilities associated to the $(k+1)$-regular tree are also equal to $1/k$.

The argument above amounts to a comparison between percolation on T_k and a certain branching process; we shall give a slightly less trivial example of such a comparison shortly. If Λ is any graph with maximum degree Δ, then a 'one-way' comparison with a branching process shows that all critical probabilities associated to Λ are at least $1/(\Delta-1)$. To see this more easily, note that for every $y \in C_x$ there is at least one open path in Λ from x to y. Thus $\chi_x(p) = \mathbb{E}_p(|C_x|)$ is at most the expected number of open (finite) paths in Λ starting at x. There are at most $\Delta(\Delta-1)^{\ell-1}$ paths in Λ of length ℓ starting at x, so

$$\chi_x^{\mathrm{b}}(p) \leq 1 + \sum_{\ell \geq 1} \Delta(\Delta-1)^{\ell-1} p^\ell$$

and

$$\chi_x^{\mathrm{s}}(p) \leq p + \sum_{\ell \geq 1} \Delta(\Delta-1)^{\ell-1} p^{\ell+1},$$

for bond and site percolation respectively. Both sums converge for any $p < 1/(\Delta-1)$, so $p_{\mathrm{T}}^{\mathrm{b}}(\Lambda)$, $p_{\mathrm{T}}^{\mathrm{s}}(\Lambda) \geq 1/(\Delta-1)$. As $p_{\mathrm{H}} \geq p_{\mathrm{T}}$, the corresponding inequalities for p_{H} follow. This shows that among all graphs with maximum degree Δ, the Δ-regular tree has the lowest critical probabilities.

There are various trivial changes we can make to a graph whose effect on the critical probability is easy to calculate. For example, if Λ is any graph and $\Lambda^{(\ell)}$ is obtained from Λ by subdividing each edge $\ell - 1$ times, then $p_c^b(\Lambda^{(\ell)}) = p_c^b(\Lambda)^{1/\ell}$, where p_c^b is p_H^b or p_T^b. Also, if $\Lambda^{[k]}$ is obtained from Λ by replacing each edge by k parallel edges, then $1 - p_c^b(\Lambda^{[k]}) = (1 - p_c^b(\Lambda))^{1/k}$, where p_c^b is p_H^b or p_T^b. Of course, $p_c^s(\Lambda^{[k]}) = p_c^s(\Lambda)$. Combining these operations, we may replace each bond of a graph by k independent paths of length ℓ to obtain a new graph. For bond percolation, the critical probabilities p_{old} and p_{new} satisfy

$$1 - (1 - p_{\mathrm{new}}^{\ell})^k = p_{\mathrm{old}}.$$

In this way, by a trivial operation on the graph, a critical probability in the interval $(0, 1)$ can be moved very close to any point of $(0, 1)$.

If we know the critical probability for a graph Λ, then we know instantly the critical probabilities for a family of graphs Λ' obtained by sequences of trivial operations from Λ, as in Figure 6.

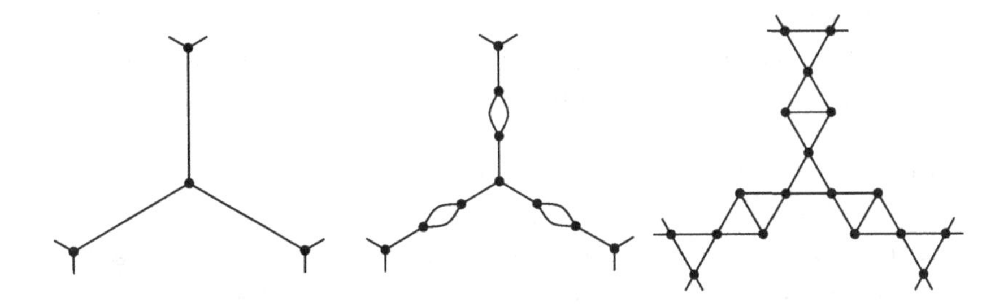

Figure 6. Transforming one bond percolation model into another, and then into a site percolation model. If the first (the hexagonal lattice) has critical probability p, then the second has critical probability r satisfying $r^3(2-r) = p$, which is also the critical probability for site percolation on the third graph.

It is easy to show that any $0 < \pi < 1$ is the critical probability for some graph, indeed, for some tree. Let T be a finite rooted tree with height (depth) h, with ℓ leaves. Let $T^1 = T$, and let T^n be the rooted tree of height hn formed from T^{n-1} by identifying each leaf with the root of a copy of T. For example, if T is a star with k edges, then T^n is the tree $T_{k,n}$ defined above. Let T^∞ be the 'limit' of the trees T^n, defined in the obvious way.

Taking the bonds of T to be open independently with probability p, the number of leaves of T joined to the root by open paths has a certain distribution X with expectation $p^h\ell$. Now suppose that the bonds of T^∞ are open independently with probability p, and let X_n be the number of sites of T^∞ at distance hn from the root joined to the root by open paths. Then the sequence $(X_0, X_1, X_2, \ldots)$ is a branching process: we have $X_0 = 1$, and each X_n is the sum of X_{n-1} independent copies of the distribution X. As X is bounded, excluding the trivial case $p = \ell = 1$, it is easy to show (arguing as above for T_k) that percolation occurs if and only if $\mathbb{E}(X) > 1$, i.e., if and only if $p^h\ell > 1$; this is a special case of the fundamental result of the theory of branching processes. In fact, one obtains

$$p_{\mathrm{T}}^{\mathrm{s}}(T^\infty) = p_{\mathrm{H}}^{\mathrm{s}}(T^\infty) = p_{\mathrm{T}}^{\mathrm{b}}(T^\infty) = p_{\mathrm{H}}^{\mathrm{b}}(T^\infty) = \ell^{-1/h}. \tag{3}$$

Suppose now that $k \geq 1$ and $1/(k+1) < \pi < 1/k$. Define $0 < \alpha < 1$ by $(k+1)^\alpha k^{1-\alpha} = 1/\pi$. Let $\mathbf{a} = (a_i)_{i=1}^\infty$ be the 0-1 sequence with density α constructed as follows: whenever 2^{j-1} divides i but 2^j does not, set $a_i = 1$ if and only if the jth bit in the binary expansion of α is 1. Let $T_{\mathbf{a}}$ be the rooted tree in which each site at distance i from the root has $k+a_{i+1}$ children. It is easy to check that, for each n, we can find trees T' and T'' of height $\ell = 2^n$ such that $(T')^\infty \subset T_{\mathbf{a}} \subset (T'')^\infty$, where T'' has $(k+1)/k$ times as many leaves as T'. Using (3), one can easily deduce that $p_{\mathrm{c}}(T_{\mathbf{a}}) = \pi$, where p_{c} denotes any of the four critical probabilities we have defined.

Alternatively, let T be the random rooted tree in which each site has $k+1$ children with probability r and k children with probability $1-r$, with the choices made independently for each site. It is easy to show that with probability 1 this random tree has $p_{\mathrm{c}}(T) = 1/(k+r)$.

In general, it is easy to calculate the various critical probabilities for a graph that is 'sufficiently tree-like'. For example, for $\ell \geq k \geq 3$, let $C_{k,\ell}$ be the *cactus* shown in Figure 7. This graph is formed by replacing each vertex of the k-regular tree T_k by a complete graph on ℓ vertices, and joining each pair of complete graphs corresponding to an edge of T_k by identifying a vertex of one with a vertex of the other, using no vertex in more than one identification. We call the vertices resulting from these identifications *attachment vertices*. Although $C_{k,\ell}$ contains many cycles, it still has the global structure of a tree, and percolation on $C_{k,\ell}$ may again be compared with a branching process.

Indeed, let K_ℓ be a complete graph with k distinguished (attachment) vertices $v_1, \ldots, v_k$. Taking the edges of K_ℓ to be open independently with probability p, let X_p be the random number of vertices among $v_2, \ldots, v_k$ that may be reached from v_1 by open paths. Let us explore the open cluster of a given initial site x of $C_{k,\ell}$ by working outwards from x. Except at the first step, from each attachment vertex that we

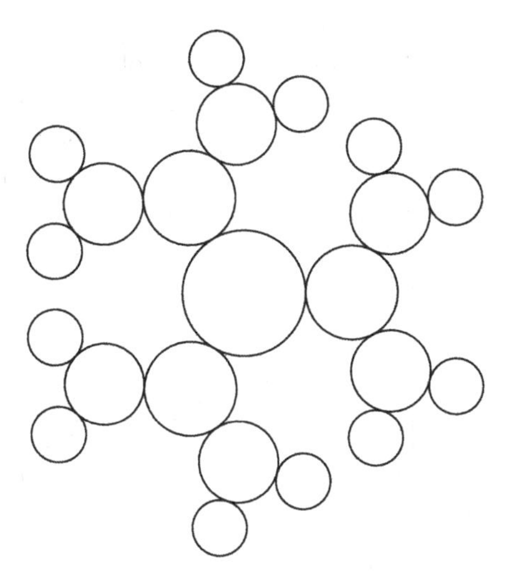

Figure 7. Part of the cactus $C_{3,\ell}$: each circle represents a complete graph on ℓ vertices. Where two circles touch, the corresponding complete graphs share an 'attachment' vertex.

reach at a given step, the number of further attachment vertices that we reach at the next step has the same distribution as X_p, and these numbers are independent. It follows, by considering a branching process as above, that

$$p_{\mathrm{H}}^{\mathrm{b}}(C_{k,\ell}) = p_{\mathrm{T}}^{\mathrm{b}}(C_{k,\ell}) = \inf\{p : \mathbb{E}(X_p) > 1\}.$$

This quantity may be easily calculated for given k and ℓ. The critical probabilities for site percolation on $C_{k,\ell}$ may be calculated in a corresponding way. Furthermore, there is nothing special about complete graphs: there are many similar constructions of 'tree-like' graphs for which the critical probabilities can be calculated using a branching process.

In the rest of this chapter, we shall prove some easy bounds on the various critical probabilities for $\mathbb{Z}^d$, $d \geq 2$. Of course, percolation on $\mathbb{Z}$ is trivial; all the critical probabilities we have defined are equal to 1.

In studying $\mathbb{Z}^2$, we shall make use of simple properties of plane graphs. As all such graphs we consider will be piecewise-linear (in fact, they will have subdivisions that are subgraphs of $\mathbb{Z}^2$, if we wish), there are no topological difficulties. In fact, all we shall need is that every polygon (in the graph $\mathbb{Z}^2$, say) separates the plane into two components, the *interior* and the *exterior*, with the interior bounded. (This is easily seen by considering the winding number of the polygon, which changes by 1 when we cross a side.) From this, Euler's formula follows easily by induction (see Chapter 1 of Bollobás [1998] for the details), which in turn implies that K_5 and $K_{3,3}$ are non-planar. Thus, for example, if C is a cycle in the plane and a, b, c, d are four vertices of C appearing in this order around C, then neither the interior nor the exterior of C can contain disjoint a–c and b–d paths; see Figure 8.

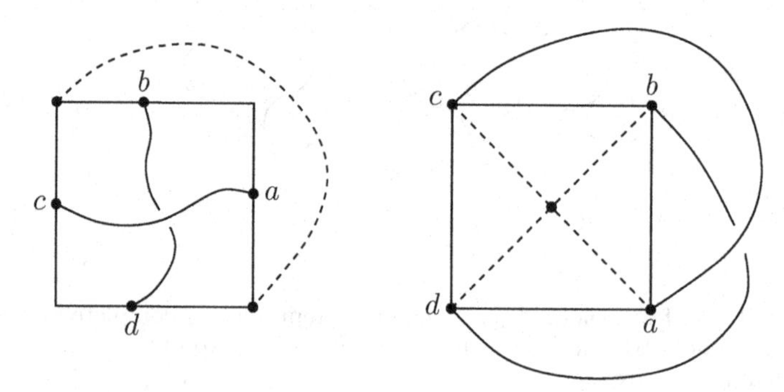

Figure 8. Neither the interior nor the exterior of a cycle visiting a, b, c, d in this order can contain disjoint a–c and b–d paths (solid lines). Both statements can be deduced either from the non-planarity of K_5 or from that of $K_{3,3}$, by considering the dashed lines.

Let us point out an important 'self-duality' property of $\mathbb{Z}^2$. This will be very important later; now we shall use it in a trivial way. The *dual* $\Lambda^\star$ of a graph Λ drawn in the plane has a vertex for each face of Λ, and an edge $e^\star$ for each edge e of Λ; this edge joins the two vertices of $\Lambda^\star$ corresponding to the faces of Λ in whose boundary e lies. When $\Lambda = \mathbb{Z}^2$, it is customary to take $v = (x+1/2, y+1/2)$ as the dual vertex corresponding to the face with vertices (x, y), $(x+1, y)$, $(x+1, y+1)$, $(x, y+1)$; see Figure 9. The dual lattice is then $\mathbb{Z}^2 + (1/2, 1/2)$, which is isomorphic to Λ. The dual graph $\Lambda^\star$ is important in the context of bond percolation on Λ. In this context, the sites and bonds of $\Lambda^\star$ are known as *dual sites* and *dual bonds*, and a dual bond $e^\star$ is usually taken to be open when e is closed and vice versa.

One trivial use of planar duality is to provide an alternative way to visualize site percolation. Let Λ be a plane graph; in *face percolation* on Λ, we assign a state, open or closed, to each face of Λ. The faces of Λ form the vertex set of a graph in which two faces are adjacent if they share an edge: this graph is precisely $\Lambda^\star$; see Figure 10. Thus, face

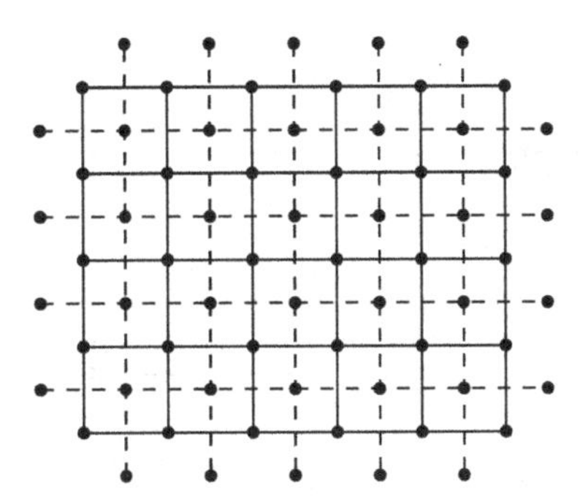

Figure 9. Portions of the lattice $\Lambda = \mathbb{Z}^2$ (solid lines) and the isomorphic dual lattice $\Lambda^\star$ (dashed lines).

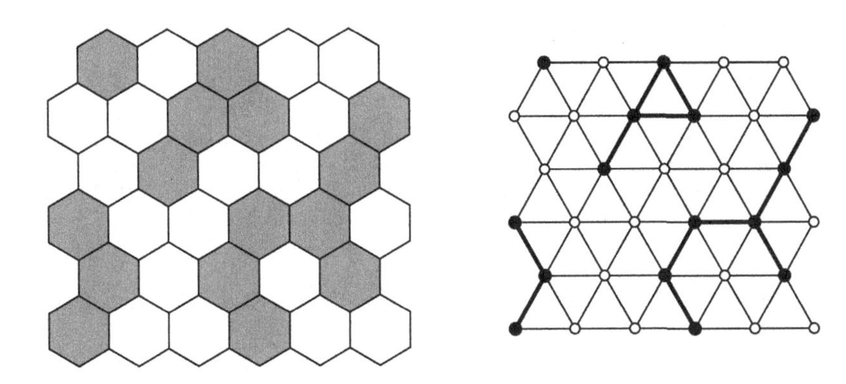

Figure 10. Face percolation on the hexagonal lattice (left): the open faces are shaded. The corresponding open subgraph in the site percolation on the triangular lattice is shown on the right (to the same scale!).

percolation on Λ is equivalent to site percolation on $\Lambda^\star$. Thinking of open faces or sites as coloured black, for example, and closed faces or sites as white, the face percolation picture is easier to visualize.

Returning to the study of percolation on $\mathbb{Z}^2$, if H is a finite connected subgraph of $\Lambda = \mathbb{Z}^2$ with vertex set C, then there is a unique infinite component C_∞ of $\Lambda - C$, the subgraph of Λ induced by the vertices outside H. By the *external boundary* $\partial^\infty C$ of C we mean the set of bonds of $\Lambda^\star$ dual to bonds of Λ joining C and C_∞; see Figure 11. Our first task is to show that $\partial^\infty C$ has the properties we expect of a boundary.

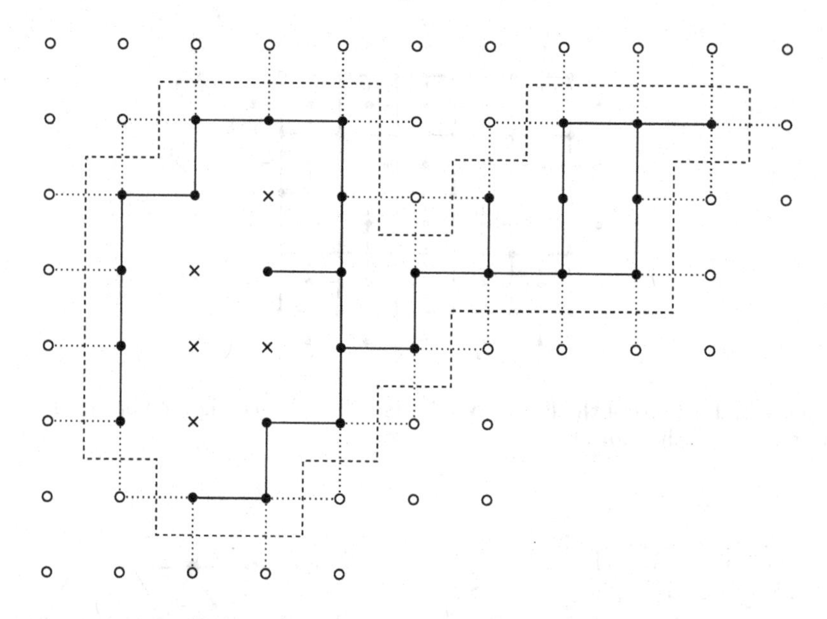

Figure 11. A finite connected subgraph H of $\mathbb{Z}^2$ with vertex set C (solid lines and circles). Vertices in finite components of $\mathbb{Z}^2 - C$ are shown with crosses, (some of) those in the infinite component with hollow circles. The dotted lines are the C–C_∞ bonds, and the dashed lines the external boundary of C.

Lemma 1. *If C is the vertex set of a finite connected subgraph of $\mathbb{Z}^2$, then $\partial^\infty C$ is a cycle with C in its interior.*

Proof. Let $\overrightarrow{F}$ be the set of C–C_∞ bonds, oriented from C to C_∞. For $\overrightarrow{f} = \overrightarrow{ab} \in \overrightarrow{F}$, orient the dual bond so that a is on its left, to obtain an oriented dual bond $\overrightarrow{f}^\star = \overrightarrow{uv}$. Equivalently, $\overrightarrow{f}^\star$ is $\overrightarrow{f}$ rotated counter-clockwise through $\pi/2$ about its midpoint. Let $\overrightarrow{\partial^\infty C} = \{\overrightarrow{f}^\star : \overrightarrow{f} \in \overrightarrow{F}\}$, so $\overrightarrow{\partial^\infty C}$ is an orientation of $\partial^\infty C$. We claim that if $\overrightarrow{f}^\star = \overrightarrow{uv} \in \overrightarrow{\partial^\infty C}$, then there is a unique bond of $\overrightarrow{\partial^\infty C}$ leaving v.

Let the vertices of $\mathbb{Z}^2$ in the face corresponding to v be a, b, c, d in cyclic order, as in Figure 12. Note that $\overrightarrow{f} = \overrightarrow{ab}$, so $a \in C$ and $b \in C_\infty$. Let $R = \mathbb{Z}^2 - C - C_\infty$ be the *rest* of $\mathbb{Z}^2$, i.e., the set of vertices of $\mathbb{Z}^2 - C$ in bounded components. Note that, as C_∞ is a component of $\mathbb{Z}^2 - C$, there is no C_∞–R bond in $\mathbb{Z}^2$.

Suppose first that $d \in C$. Then $c \in C$ or $c \in C_\infty$, since c is adjacent to $b \in C_\infty$, and so cannot lie in R. In the first case the bond $\overrightarrow{cb}^\star$ is the unique bond of $\overrightarrow{\partial^\infty C}$ leaving v; in the second, $\overrightarrow{dc}^\star$.

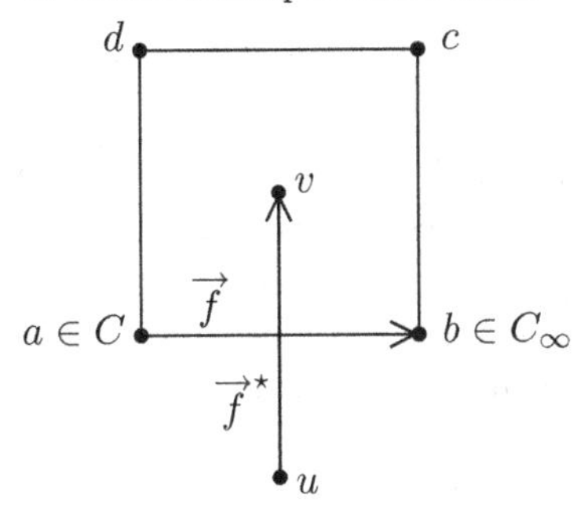

Figure 12. The oriented dual bond $\overrightarrow{f}^\star$ corresponding to an oriented bond $\overrightarrow{f}$ from C to C_∞.

Suppose next that $c \in C_\infty$. Then $d \notin R$, so either $d \in C$ or $d \in C_\infty$, and again the claim holds.

We may thus suppose that $c \notin C_\infty$, so c (which is adjacent to $b \in C_\infty$) lies in C, and that $d \notin C$. If $d \in R$ the claim again holds, leaving only the case a, $c \in C$, b, $d \in C_\infty$. But C and C_∞ are disjoint connected subgraphs of $\mathbb{Z}^2$, so there are disjoint paths in $\mathbb{Z}^2$, one joining a to c, and one joining b to d. As $abcd$ is a cycle in $\mathbb{Z}^2$ with no edges in its interior, both paths must lie in the exterior of this cycle, which is impossible; see Figure 8.

As every edge in $\overrightarrow{\partial^\infty C}$ has a unique successor, the underlying unoriented graph $\partial^\infty C$ contains a cycle S. This cycle separates the plane, and crosses only C–C_∞ bonds; thus C_∞ lies outside S, and C is inside S. If $f^\star \in \partial^\infty C$ with $f = ab$ then, since one of a, b is in C and the other in C_∞, the cycle S must cut the bond f. Hence $f^\star \in S$. In other words, every bond of $\partial^\infty C$ is in S, so $\partial^\infty C$ consists of a single cycle. □

In the rest of this chapter we present some basic results concerning critical probabilities. First, following Broadbent and Hammersley [1957] and Hammersley [1957a; 1959], we show that the phenomenon of bond percolation in $\mathbb{Z}^2$ is non-trivial: the critical probabilities are neither 0 nor 1. To do so, we consider the number $\mu_n(\Lambda; x)$ of *self-avoiding walks* in Λ starting at x, where in graph terminology, a self-avoiding walk is simply a path. All vertices of $\mathbb{Z}^2$ are equivalent, and, as $\mathbb{Z}^2$ is 4-regular,

$$\mu_n = \mu_n(\mathbb{Z}^2) = \mu_n(\mathbb{Z}^2; 0) \leq 4 \times 3^{n-1}.$$

Lemma 2. *For bond percolation in* $\mathbb{Z}^2$ *we have*

$$1/3 \leq p_{\mathrm{T}} \leq p_{\mathrm{H}} \leq 2/3.$$

Proof. As we noted earlier, $p_T \le p_H$ in any context, so we must show only the outer inequalities.

Suppose first that $p < 1/3$, and let C_0 be the open cluster of the origin in the independent bond percolation on $\mathbb{Z}^2$ where each bond is open with probability p. For every site $x \in C_0$, there is at least one open path from 0 to x. Hence $|C_0|$ is at most the number X of open paths starting at 0. As a path in $\mathbb{Z}^2$ of length n is open with probability p^n, we have

$$\chi(p) = \mathbb{E}_p(|C_0|) \le \mathbb{E}_p(X) = \sum_{n\ge 0} \mu_n p^n \le 1 + \frac{4}{3}\sum_{n\ge 1}(3p)^n < \infty.$$

Hence, $p_T \ge p$. Since $p < 1/3$ was arbitrary, it follows that $p_T \ge 1/3$.

For the upper bound we consider $\Lambda = \mathbb{Z}^2$ together with its dual $\Lambda^\star = \mathbb{Z}^2 + (1/2, 1/2)$ defined earlier, taking a dual bond $e^\star$ to be open if e is closed, and vice versa. An *open dual cycle* is a cycle in the dual graph consisting of dual bonds that are open.

Suppose now that $p > 2/3$. Let L_k be the line segment joining the origin to the point $(k, 0)$, and let S be a dual cycle surrounding L_k, of length 2ℓ. Then S must contain a dual bond $e^\star$ crossing the positive x-axis at some coordinate between $k + 1/2$ and $(2\ell - 3)/2$; thus we have fewer than ℓ choices for $e^\star$. As the rest of S is a path of length $2\ell - 1$ in the dual lattice, which is isomorphic to $\mathbb{Z}^2$, there are at most $\ell\mu_{2\ell-1}$ possibilities for S. Let Y_k be the number of open dual cycles surrounding L_k. Since dual bonds are open independently with probability $1 - p$,

$$\mathbb{E}_p(Y_k) \le \sum_{\ell\ge k+2} \ell\mu_{2\ell-1}(1-p)^{2\ell} \le \sum_{\ell\ge k+2} \frac{4\ell}{9}(3(1-p))^{2\ell}.$$

As $3(1-p) < 1$, the final sum is convergent, so $\mathbb{E}_p(Y_k) \to 0$ as $k \to \infty$, and there is some k with $\mathbb{E}_p(Y_k) < 1$. Let A_k be the event that $Y_k = 0$; since $\mathbb{E}_p(Y_k) < 1$, we have $\mathbb{P}_p(A_k) > 0$. Let B_k be the event that the k bonds in L_k are open. Note that A_k and B_k are independent. Also, if both hold, then there is no open dual cycle surrounding the origin so, by Lemma 1, the open cluster containing the origin is infinite. Hence,

$$\theta(p) = \theta_0(p) \ge \mathbb{P}_p(A_k \cap B_k) = \mathbb{P}_p(A_k)\mathbb{P}_p(B_k) = \mathbb{P}_p(A_k)p^k > 0.$$

This shows that $p_H \le p$. Since $p > 2/3$ was arbitrary, $p_H \le 2/3$ follows. □

The second part of the argument above, bounding the critical probability from above by estimating the number of separating cycles, is sometimes called a *Peierls argument*, after Peierls [1936]. In higher dimensions, we estimate the number of separating surfaces; even the easiest bounds on the number of these surfaces suffice to show that the critical probability is strictly less than 1.

Comparing $\mathbb{Z}^d$ with $\mathbb{Z}^2$ and with the $(2d)$-regular tree, we see that

$$1/(2d-1) \leq p_{\mathrm{T}}^{\mathrm{b}}(\mathbb{Z}^d) \leq p_{\mathrm{H}}^{\mathrm{b}}(\mathbb{Z}^d) \leq 2/3.$$

As we shall see later, of these two trivial inequalities, the first is close to an equality.

The bounds in Lemma 2 can be improved by bounding μ_n from above less crudely. As noted by Hammersley and Morton [1954], since $\mu_{m+n} \leq \mu_m \mu_n$, the sequence $\mu_n^{1/n}$ converges to a limit $\lambda = \lambda(\mathbb{Z}^2)$. This limit is known as the *connective constant* of $\mathbb{Z}^2$ (see Hammersley [1957a]). The proof of Lemma 2 shows that

$$1/\lambda \leq p_{\mathrm{T}}^{\mathrm{b}}(\mathbb{Z}^2) \leq p_{\mathrm{H}}^{\mathrm{b}}(\mathbb{Z}^2) \leq 1 - 1/\lambda. \tag{4}$$

This was what Broadbent and Hammersley [1957] and Hammersley [1959] actually proved.

In constructing a path step by step, at every step other than the first there are at most three possibilities, giving the bound $\lambda \leq 3$ used above. In fact, looking three steps ahead, two of the 27 combinations counted so far are always impossible, namely, those corresponding to turning left three times in a row, or right three times. It follows that $\lambda \leq 25^{1/3}$. Using a computer, it is easy to count paths of length 25, say, in a few minutes. As $\mu_{25} = 123,481,354,908$, this gives $\lambda \leq 2.736$.

For $p > 1/\lambda$, the expected number of open paths of length n in $\mathbb{Z}^2$ starting at the origin tends to infinity as $n \to \infty$. Thus, in analogy with numerous phenomena in probabilistic combinatorics, one might expect that, with probability bounded away from 0, there are arbitrarily long open paths starting at the origin. This would suggest that $p_{\mathrm{H}} = 1/\lambda$, as is indeed the case for the k-regular trees. The trouble is that, as there are $(\lambda + o(1))^{2n}$ dual cycles of length $2n$ surrounding the origin (see Hammersley [1961b]), the same intuition would indicate that for $p < 1 - 1/\lambda$ there are open dual cycles surrounding the origin, implying that $p_{\mathrm{H}} = 1 - 1/\lambda$. Thus, for both intuitions to be correct, λ would have to be 2. As we shall see now, this is not the case.

Any walk in which every step goes up or to the right is self-avoiding, so $\mu_n \geq 2^n$ and $\lambda \geq 2$. Let us say that a path P is a *building block* if P starts at 0, ends on the line $x + y = 2$, and every intermediate vertex of P lies in the region $0 < x + y \leq 2$; see Figure 13. Then any sequence of building blocks $P_1, P_2, \ldots,$ may be concatenated to make a self-avoiding

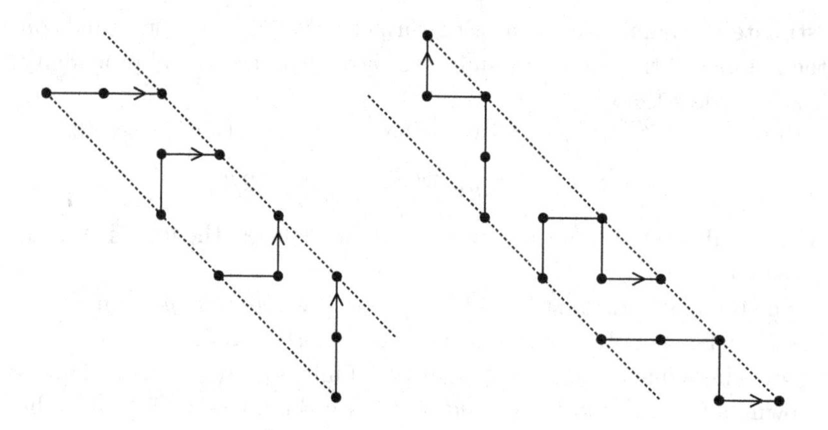

Figure 13. A selection of building blocks that may be put together in any order to create a self-avoiding walk.

walk W starting at 0. Starting with W, each P_i is the part of W lying in the region $2i-2 < x+y \leq 2i$ (plus the vertex on $x+y = 2i-2$ where W enters this region), translated to start at 0, so distinct sequences give distinct walks. Let w_n be the number of walks W of length n that may be obtained in this way. As there are four building blocks with two edges, and four with four edges, we have $w_4 \geq 4^2 + 4 = 20$, and $\mu_{4n} \geq w_{4n} \geq w_4^n = 20^n$, so $\lambda \geq 20^{1/4} = 2.114\ldots$. In fact, we have

$$w_n \geq 4w_{n-2} + 4w_{n-4}.$$

Solving the recurrence relation, it follows that λ is at least the positive root of $x^4 - 4x^2 - 4 = 0$, namely, $2.197\ldots$.

Much research has been done on calculating μ_n and bounding λ. Fisher and Sykes [1959] showed that $\mu_{16} = 17,245,332$. They also obtained the bounds $2.5767 \leq \lambda \leq 2.712$, by considering respectively a special sub-class of paths, and walks with no short cycles. More sophisticated algorithms and the use of computers have enabled these results to be greatly extended. For example, μ_{51} was calculated by Conway and Guttmann [1996] using a supercomputer (see also Guttmann and Conway [2001]), and Jensen [2004a] has found μ_{71}. The best published bounds on λ are $\lambda \leq 2.6792$, due to Pönitz and Tittmann [2000], and $\lambda \geq 2.6256$, obtained by Jensen [2004b] using the method of irreducible bridges introduced by Kesten [1963].

Connective constants of other lattices have also been studied. In particular, writing λ_d for the connective constant of $\mathbb{Z}^d$, Kesten [1964] showed that $\lambda_d = 2d - 1 - 1/(2d) + O(d^{-2})$. More recently, Hara and Slade [1995] showed that

$$\lambda_d = (2d) - 1 - (2d)^{-1} - 3(2d)^{-2} - 16(2d)^{-3} - 102(2d)^{-4} + O(d^{-5}),$$

and that a corresponding expansion exists to any order.

Returning to general graphs Λ, we next show that percolation is 'more likely' in the bond model than the site model. In the arguments to come, we shall consider step by step explorations of the states of the sites, say. Suppose that the states X_u of the sites are independent, and each site u is open with probability p_u. In each step, the next site v to be explored will depend only on the states of the previously explored sites. Hence, given the history of the exploration so far, the conditional probability that v is open is just p_v.

The reader may well feel that the observation above needs no justification. Note, however, that as a random variable, the history of the exploration up to step t depends on the states of all sites that *might* be explored in the first t steps. But the event that this history takes a particular value (i.e., the event that, for $1 \leq s \leq t$, at step s we tested site v_s and found that $X_{v_s} = x_s$) is independent of X_v for any $v \notin \{v_1, \ldots, v_t\}$. The next result is due to Hammersley [1961a]; certain special cases were proved by Fisher [1961].

Theorem 3. *Let Λ be a connected, infinite, locally finite multi-graph. Then*

$$p_{\mathrm{H}}^{\mathrm{s}}(\Lambda) \geq p_{\mathrm{H}}^{\mathrm{b}}(\Lambda) \text{ and } p_{\mathrm{T}}^{\mathrm{s}}(\Lambda) \geq p_{\mathrm{T}}^{\mathrm{b}}(\Lambda).$$

Proof. Both inequalities follow from the assertion that, for every site x of Λ, every integer $n \geq 1$, and every probability $0 < p < 1$, we have

$$\mathbb{P}_{\Lambda,p}^{\mathrm{s}}\big(|C_x| \geq n\big) \leq p\,\mathbb{P}_{\Lambda,p}^{\mathrm{b}}\big(|C_x| \geq n\big). \tag{5}$$

Indeed, letting $n \to \infty$, it follows that

$$\theta_x(\Lambda_p^{\mathrm{s}}) \leq p\theta_x(\Lambda_p^{\mathrm{b}}). \tag{6}$$

Thus, if $\theta_x(\Lambda_p^{\mathrm{b}}) = 0$, then $\theta_x(\Lambda_p^{\mathrm{s}}) = 0$, so $p_{\mathrm{H}}^{\mathrm{s}}(\Lambda) \geq p_{\mathrm{H}}^{\mathrm{b}}(\Lambda)$.

Also, if (5) holds, then

$$\chi_x(\Lambda_p^{\mathrm{s}}) = \sum_{n=1}^{\infty} n\mathbb{P}_{\Lambda,p}^{\mathrm{s}}(|C_x| = n) = \sum_{n=1}^{\infty} \mathbb{P}_{\Lambda,p}^{\mathrm{s}}(|C_x| \geq n)$$

$$\leq \sum_{n=1}^{\infty} p\mathbb{P}_{\Lambda,p}^{\mathrm{b}}(|C_x| \geq n) = p\chi_x(\Lambda_p^{\mathrm{b}}).$$

Hence, if $\chi_x(\Lambda_p^{\mathrm{b}}) < \infty$, then $\chi_x(\Lambda_p^{\mathrm{s}}) < \infty$, so $p_{\mathrm{T}}^{\mathrm{s}}(\Lambda) \geq p_{\mathrm{T}}^{\mathrm{b}}(\Lambda)$.

It remains to prove (5). As C_x is empty in Λ_p^{s} if x is closed, inequality (5) is equivalent to

$$\mathbb{P}_{\Lambda,p}^{\mathrm{s}}(|C_x| \geq n \mid x \text{ is open}) \leq \mathbb{P}_{\Lambda,p}^{\mathrm{b}}(|C_x| \geq n). \tag{7}$$

In proving this, we may and shall replace Λ by the finite subgraph Λ_n of Λ induced by the vertices within distance n of x, since the event that $|C_x| \geq n$ depends only on the states of sites or bonds in Λ_n.

Let us explore C_x^{s}, the open cluster in the site percolation on Λ_n containing x, conditioning throughout on x being open. We shall construct a random sequence $\mathcal{T} = (R_t, D_t, U_t)_{t=1}^{\ell}$ of tripartitions of the vertex set $V(\Lambda_n)$ of Λ_n; this sequence will be such that the final set R_ℓ, obtained after a random number ℓ of steps, will be the cluster C_x^{s}. The notation indicates that the sites in R_t have been 'reached' by step t, those in D_t are 'dead' (known to be closed), and those in U_t are 'untested'.

To define $\mathcal{T}$, set $R_1 = \{x\}$, $U_1 = V(\Lambda_n) \setminus \{x\}$, and $D_1 = \emptyset$. Given (R_t, D_t, U_t), if there is no R_t–U_t bond, set $\ell = t$ and stop the sequence. Otherwise, pick a bond $e_t = y_t z_t$ with $y_t \in R_t$, $z_t \in U_t$, and set $U_{t+1} = U_t \setminus \{z_t\}$. Now test whether the site z_t is open. If so, set $R_{t+1} = R_t \cup \{z_t\}$, $D_{t+1} = D_t$. Otherwise, set $D_{t+1} = D_t \cup \{z_t\}$, $R_{t+1} = R_t$. Note that, at each step, the conditional probability that z_t is open is p. The process terminates as Λ_n is finite.

By construction, for every t, R_t is a connected set of open sites, and all sites in D_t are closed. Hence, as no site in R_ℓ has a neighbour in $U_\ell = V(\Lambda_n) \setminus (R_\ell \cup D_\ell)$, the set R_ℓ is precisely the open cluster C_x^{s}.

To compare the distribution of $|C_x^{\mathrm{s}}|$ to that of $|C_x^{\mathrm{b}}|$, let us explore C_x^{b} in a similar manner, using a random sequence $\mathcal{T}' = (R_t', U_t', D_t')_{t=1}^{\ell'}$ as above. This sequence $\mathcal{T}'$ is constructed as $\mathcal{T}$, except that, having picked $e_t = y_t z_t$, we test whether the bond e_t is open. As this is the first (and only) time we test e_t, conditional on the sequence $\mathcal{T}'$ up to step t, the probability that e_t is open is just p. Consequently, the sequences $\mathcal{T}$ and $\mathcal{T}'$ have the same distribution. In particular, $|C_x^{\mathrm{s}}| = |R_\ell|$ and $|R_{\ell'}'|$ have the same distribution. This implies (7) since $R_{\ell'}'$ is contained in the open cluster C_x^{b} of x in Λ_p^{b}: it is the vertex set of the subgraph spanned by the set of bonds we have tested in the process $\mathcal{T}'$ and found open. $\square$

In the argument above we could have worked directly in the infinite graph, continuing the exploration indefinitely if C_x^{s} is infinite.

One can also consider *mixed percolation*, in which sites are open with probability p and bonds with probability p', independently of each other. Writing $\theta_x(\Lambda; p, p')$ for the probability that there is an infinite path starting at x all of whose sites and bonds are open, Hammersley [1980] noted that the more general inequality

$$\theta_x(\Lambda; \delta p, p') \leq \delta\theta_x(\Lambda; p, \delta p')$$

for $0 \leq p, p', \delta \leq 1$ is an immediate consequence of its special case (6) proved above.

The inequalities in Theorem 3 need not be strict, as shown by the k-regular tree. In most interesting examples, however, they are strict. Strict inequalities have been proved by Higuchi [1982], Kesten [1982], Menshikov [1987] and others. For a general result that includes the graphs $\mathbb{Z}^d$ as special cases, see Grimmett and Stacey [1998]. Aizenman and Grimmett [1991] proved that a certain 'essential enhancement' of a percolation model leads to strictly smaller critical values: under suitable conditions, adding edges to the graph strictly decreases the critical probability. This result has been extended by Bezuidenhout, Grimmett and Kesten [1993] and Grimmett [1994].

One might expect that, if $p_{\mathrm{H}}^{\mathrm{s}}(\Lambda_1) < p_{\mathrm{H}}^{\mathrm{s}}(\Lambda_2)$, then $p_{\mathrm{H}}^{\mathrm{b}}(\Lambda_1) < p_{\mathrm{H}}^{\mathrm{b}}(\Lambda_2)$. However, it is easy to construct examples to show that this is not the case, by modifying a graph in a way that decreases $p_{\mathrm{H}}^{\mathrm{b}}$ while leaving $p_{\mathrm{H}}^{\mathrm{s}}$ unchanged; see Wierman [2003a].

It is easy to bound $p_{\mathrm{H}}^{\mathrm{s}}$ above by a function of $p_{\mathrm{H}}^{\mathrm{b}}$. Indeed, let Λ be an infinite connected graph with maximal degree $\Delta < \infty$. Let $0 < r < 1$, and let $\{X_{xy}, X_{yx} : xy \in E(\Lambda)\}$ be a family of independent Bernoulli random variables with $\mathbb{P}(X_{xy} = 1) = \mathbb{P}(X_{yx} = 1) = r$ for every edge xy of Λ. (Note that for every edge there are two random variables.) Declaring a site $x \in V(\Lambda)$ open if there is at least one site y with $X_{xy} = 1$, we obtain an independent site percolation measure on Λ in which each site x is open with probability $1 - (1-r)^{d(x)} \leq 1 - (1-r)^{\Delta}$, where $d(x)$ is the degree of x. Furthermore, declaring a bond xy open if $X_{xy} = X_{yx} = 1$, we obtain an independent bond percolation measure in which each bond is open with probability r^2. Clearly, no matter what values the random variables X_{xy} take, every open cluster in the bond percolation is contained in an open cluster in the site percolation. Hence, if $r^2 > p_{\mathrm{H}}^{\mathrm{b}}(\Lambda)$ then $1-(1-r)^\Delta \geq p_{\mathrm{H}}^{\mathrm{s}}(\Lambda)$, showing that

$$p_{\mathrm{H}}^{\mathrm{s}}(\Lambda) \leq 1 - \left(1 - \sqrt{p_{\mathrm{H}}^{\mathrm{b}}(\Lambda)}\right)^{\Delta}.$$

There is a reason for expressing the bound above in terms of Δ: it is easy to see that there is a locally finite graph Λ with unbounded degrees in which $p_{\mathrm{H}}^{\mathrm{b}}(\Lambda) = 0$ and $p_{\mathrm{H}}^{\mathrm{s}}(\Lambda) = 1$. For example, take a one-way infinite path and replace the kth edge by a complete graph on 2^k vertices.

Our next result, due to Grimmett and Stacey [1998], shows that we can do better than the trivial argument above. The proof of this result will be a little less pleasant than that of Theorem 3, due to a minor complication introduced to improve the exponent below from Δ to $\Delta-1$. Later, we shall give a related result for oriented percolation, whose proof is much simpler.

Theorem 4. *Let Λ be a connected, infinite graph with maximum degree $\Delta < \infty$. Then*

$$p_{\mathrm{H}}^{\mathrm{s}}(\Lambda) \leq 1 - \left(1 - p_{\mathrm{H}}^{\mathrm{b}}(\Lambda)\right)^{\Delta-1}. \tag{8}$$

The same inequality holds for p_{T}.

Proof. It suffices to show that there is a constant $K = K(\Delta)$ such that, for every site x of Λ, every integer $n \geq 1$, and every $0 < p < 1$, we have

$$\mathbb{P}_{\Lambda,r}^{\mathrm{s}}\big(|C_x| \geq n \mid x \text{ is open}\big) \geq \mathbb{P}_{\Lambda,p}^{\mathrm{b}}\big(|C_x| \geq Kn\big), \tag{9}$$

where $r = 1-(1-p)^{\Delta-1}$. We shall prove this with $K = \Delta+1$. In doing so, we may replace Λ by the finite subgraph Λ_n induced by the sites within distance $(\Delta+1)n$ of x.

Let C_x^{b} be the open cluster of x in the bond percolation on Λ_n in which each edge is open independently with probability p, and let C_x^{s} be the open cluster in the site percolation in which x is open, and the other sites of Λ_n are open independently with probability r. In order to compare the distributions of $|C_x^{\mathrm{b}}|$ and $|C_x^{\mathrm{s}}|$, we first give an algorithm that finds a significant fraction of C_x^{b}, and then show that a slight variant of this algorithm finds a subset of C_x^{s}.

This time, we start by exploring C_x^{b}, using a random sequence $\mathcal{T} = (L_t, D_t, U_t)_{t=1}^{\ell}$ of tripartitions of the set $E(\Lambda_n)$ of *bonds*. The notation indicates that the bonds in L_t are 'live' (known to be open), those in D_t are 'dead' (known to be closed), and those in U_t are 'untested'. At each stage, the bonds of L_t will form the edge set of a tree containing x; we write R_t for the vertex set of this tree. (Thus R_t is the set of sites that may be reached from x along paths consisting of bonds in L_t.) The set U_t will shrink as we proceed, while the sets L_t and D_t will grow. We shall also work with a set P_t of *sites* growing with t, namely the set $P_t \subset V(\Lambda_n) \setminus R_t$ of 'peripheral' sites that will not contribute to the growth of R_t. The final sets R_ℓ and P_ℓ, obtained after a random number ℓ of steps, will be such that $R_\ell \subset C_x^{\mathrm{b}} \subset R_\ell \cup P_\ell$.

To define $\mathcal{T}$, set $U_1 = E(\Lambda_n)$, and $D_1 = L_1 = \emptyset$. Thus $R_1 = \{x\}$. Given (L_t, D_t, U_t), let P_t be the set of sites $v \notin R_t$ such that U_t contains exactly one bond incident with v. If U_t contains no bonds from R_t to $V(\Lambda_n) \setminus (R_t \cup P_t)$, set $\ell = t$ and stop the sequence. Otherwise, let $e_t = y_t z_t \in U_t$ be any such bond, with $y_t \in R_t$, $z_t \notin R_t \cup P_t$. Set $U_{t+1} = U_t \setminus \{e_t\}$, and test whether the bond e_t is open. If so, set $L_{t+1} = L_t \cup \{e_t\}$ and $D_{t+1} = D_t$, so $R_{t+1} = R_t \cup \{z_t\}$. Otherwise, set $D_{t+1} = D_t \cup \{e_t\}$ and $L_{t+1} = L_t$, noting that $R_{t+1} = R_t$. A possible state of this exploration is shown in Figure 14. Since we never test the

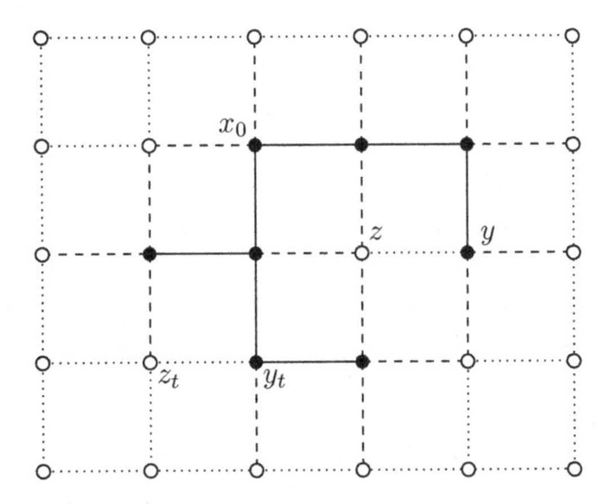

Figure 14. A possible value of (L_t, D_t, U_t) when exploring the open cluster of the site x_0 in bond percolation on $\mathbb{Z}^2$. Bonds in L_t, D_t and U_t are represented by solid, dashed and dotted lines, respectively. The set R_t is shown by filled circles. The site z lies in P_t, so the bond yz will never be tested; there is only one choice $y_t z_t$ for the next bond to test.

same bond twice, at each step the conditional probability that e_t is open is p. As Λ_n is finite, the process terminates.

By construction, the bonds in L_t are open, while those in D_t are closed. Thus R_ℓ, the set of sites that may be reached from x along paths consisting of bonds in L_ℓ, is a subset of C_x^{b}. Also, if $e = yz$ is an open bond with $y \in R_\ell$ and $z \notin R_\ell$, then $e \notin L_\ell$ and $e \notin D_\ell$, so $e \in U_\ell$. Since we stopped at step ℓ, we must have $z \in P_\ell$, so the site z is incident with no other bonds of U_ℓ. As $z \notin R_\ell$, the site z is incident with no bonds of L_ℓ, so all other bonds incident with z are in D_ℓ, and hence closed.

We have shown that all open bonds leaving R_ℓ terminate in sites of P_ℓ that are incident with no other open bonds. Thus, every site $z \in C_x^{\mathrm{b}} \setminus R_\ell$ is adjacent to a site in R_ℓ, and, crudely,

$$|C_x^{\mathrm{b}}| \le (\Delta + 1)|R_\ell|. \tag{10}$$

We now turn to the site percolation on Λ_n in which x is open, and the other sites are open independently with probability $r = 1 - (1-p)^{\Delta-1}$. To realize this probability measure, let $X_{y,i}$, $y \in V(\Lambda_n) \setminus \{x\}$, $1 \le i \le \Delta - 1$, be a set of i.i.d. random variables, with $\mathbb{P}(X_{y,i} = 1) = p$ and $\mathbb{P}(X_{y,i} = 0) = 1 - p$. We declare y to be open if and only if at least one of the $X_{y,i}$ is equal to 1.

Let us explore a subset of C_x^{s} as follows, using a random sequence $\mathcal{T}' = (L'_t, D'_t, U'_t)_{t=1}^{\ell'}$ of tripartitions of the set of bonds. This sequence $\mathcal{T}'$ is constructed as $\mathcal{T}$, except that, having picked $e_t = y_t z_t$, we test one of the variables $X_{z_t,i}$. More precisely, if this is the ith time that we have chosen the same site z, i.e., if there are $i-1$ values $1 \le s < t$ with $z_s = z$, then we test whether $X_{z_t,i} = 1$. Note that $i \le \Delta - 1$: initially, z is incident with at most Δ untested bonds, each time we choose $z_s = z$ we test an untested bond incident with z, and we cannot choose $z_s = z$ if there is only one such bond remaining. Since each variable $X_{z_t,i}$ is tested at most once, each test succeeds with conditional probability p, and the sequence $\mathcal{T}'$ has exactly the distribution of $\mathcal{T}$.

By the construction of $\mathcal{T}'$, if $e_t = y_t z_t$ was the bond chosen at step t and $e_t \in L_\ell$, then the test at step t succeeded, so one of the $X_{z_t,i}$ is equal to 1, and z_t is open. As $y_t \in R_{t-1}$, it follows by induction on t that $R_t \subset C_x^{\mathrm{s}}$. Using (10) and the fact that R_ℓ and $R'_{\ell'}$ have the same distribution, (9) follows, completing the proof. □

For graphs Λ which are not regular, the proof above gives a slightly stronger result than Theorem 4, showing that if $\theta_x > 0$ for the bond percolation on Λ in which each bond is open independently with probability p, then $\theta_x > 0$ for the site percolation in which the states of the sites are independent, and the probability p_v that a site v is open satisfies $1 - p_v = (1-p)^{d(v)-1}$, where $d(v)$ is the degree of v.

The bound in (8) can be improved when $p_{\mathrm{H}}^{\mathrm{b}}(\Lambda)$ is close to 1; in this case, the exponent $\Delta - 1$ may be replaced by an exponent a little larger than $\Delta/2$. This is shown by the much more difficult result of Chayes and Schonmann [2000] that, for any infinite connected graph Λ of maximum degree $\Delta \geq 3$,

$$p_{\mathrm{H}}^{\mathrm{s}}(\Lambda) \leq 1 - \frac{1}{2^{2\lfloor \Delta/2 \rfloor + 1}} \left(1 - p_{\mathrm{H}}^{\mathrm{b}}(\Lambda)\right)^{\lfloor \Delta/2 \rfloor}, \tag{11}$$

while there are such graphs with

$$p_{\mathrm{H}}^{\mathrm{s}}(\Lambda) \geq 1 - 2^{\lfloor \Delta/2 \rfloor + 1} \left(1 - p_{\mathrm{H}}^{\mathrm{b}}(\Lambda)\right)^{\lfloor \Delta/2 \rfloor}.$$

In proving (11), Chayes and Schonmann made use of the concept of mixed percolation mentioned above, where states are assigned to the sites and bonds of Λ.

So far, we have considered percolation on ordinary graphs, in which bonds are two-way. The basic concepts of percolation make just as good sense for *oriented graphs*, where each edge is oriented from one endpoint to the other. We could also consider *directed graphs*, where there may be an edge in each direction between the same pair of vertices, or indeed, undirected or directed *multi-graphs* where there may be several edges (in one or both directions) between the same pair of vertices.

In *oriented percolation*, the underlying (multi-)graph $\overrightarrow{\Lambda}$ is oriented. We assign states, open or closed, to the sites (vertices) or bonds (oriented edges) as usual, and define the open subgraph, $\overrightarrow{\Lambda}_p^{\mathrm{b}}$ or $\overrightarrow{\Lambda}_p^{\mathrm{s}}$, as before. Now, however, the open subgraph is of course oriented. An *oriented path* is a (finite or infinite) path (i.e., self-avoiding walk) $P = x_0 x_1 x_2 \ldots$ in $\overrightarrow{\Lambda}$ with edges $\overrightarrow{x_i x_{i+1}}$ oriented from x_i to x_{i+1}. An *open path* in the oriented percolation is an oriented path in the open subgraph, i.e., an oriented path all of whose sites (for oriented site percolation) or bonds (for oriented bond percolation) are open.

Given a site x, we write C_x^+ for the *open out-cluster* of x, namely, the set of sites y reachable from x by an open path:

$$C_x^+ = \{y \in \overrightarrow{\Lambda} : \text{there is an open path from } x \text{ to } y\}.$$

Similarly, C_x^-, the *open in-cluster* of x, is the set of sites y from which x can be reached, so $y \in C_x^+$ if and only if $x \in C_y^-$. Note that C_x^+ is a subgraph of the *out-subgraph* $\overrightarrow{\Lambda}_x^+$ of x, i.e., the set of all sites y reachable from x by oriented paths in $\overrightarrow{\Lambda}$. Since almost always we consider only C_x^+, we write C_x for C_x^+. We define $\theta_x = \theta_x^+$, and $\chi_x = \chi_x^+$ as before, using $C_x = C_x^+$.

As in the unoriented case, $\theta_x(p)$ and $\chi_x(p) = \mathbb{E}_p(|C_x|)$ are increasing functions of p, so we may define critical probabilities $p_{\mathrm{H}}(\overrightarrow{\Lambda};x)$ and $p_{\mathrm{T}}(\overrightarrow{\Lambda};x)$ as before. Of course, all these quantities depend on whether we are considering site or bond percolation. Sometimes we shall indicate this by writing $\theta_x^{\mathrm{s}}(p)$, or $p_{\mathrm{H}}^{\mathrm{b}}(\overrightarrow{\Lambda};x)$, and so on. In the oriented case, these critical probabilities may well depend on the site x: consider, for example, the rooted tree T shown in Figure 15, in which the root x_0 has

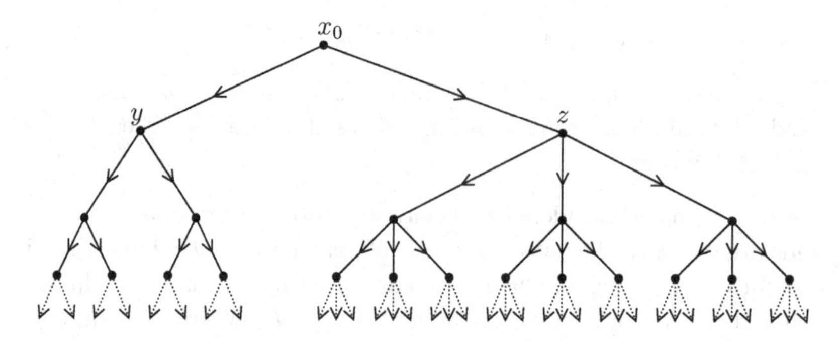

Figure 15. An oriented rooted tree $\overrightarrow{T}$ in which $\frac{1}{2} = p_{\mathrm{c}}(\overrightarrow{T};y) \neq p_{\mathrm{c}}(\overrightarrow{T};z) = \frac{1}{3}$.

two children, y and z, all descendants of y have two children, and all descendants of z have three children. Orienting the edges away from x_0 to obtain an oriented graph $\overrightarrow{T}$, we have $p_{\mathrm{c}}(\overrightarrow{T};x) = 1/2$ for y and all its descendants, and $p_{\mathrm{c}}(\overrightarrow{T};x) = 1/3$ for the remaining vertices, where p_{c} is any of $p_{\mathrm{H}}^{\mathrm{s}}$, $p_{\mathrm{T}}^{\mathrm{s}}$, $p_{\mathrm{H}}^{\mathrm{b}}$ or $p_{\mathrm{T}}^{\mathrm{b}}$.

For most graphs we shall consider, we do have $p_{\mathrm{H}}(\overrightarrow{\Lambda};x) = p_{\mathrm{H}}(\overrightarrow{\Lambda};y)$ for all sites x and y, and the same for p_{T}; when this holds, we write $p_{\mathrm{H}}(\overrightarrow{\Lambda})$ and $p_{\mathrm{T}}(\overrightarrow{\Lambda})$ for the common values. Indeed, if $\overrightarrow{\Lambda}$ is *strongly connected*, i.e., for every pair x and y of sites there is an oriented path from x to y, then, as in the unoriented case, it is immediate that $\theta_x(p) > 0$ if and only if $\theta_y(p) > 0$, and $\chi_x(p) < \infty$ if and only if $\chi_y(p) < \infty$, so p_{H} and p_{T} are well defined.

If $\overrightarrow{\Lambda}$ is any oriented graph obtained by orienting the edges of a simple graph Λ, then the paths in $\overrightarrow{\Lambda}$ are a subset of the paths in Λ. Thus percolation in Λ dominates that in $\overrightarrow{\Lambda}$, so, for example,

$$p_{\mathrm{H}}^{\mathrm{s}}(\overrightarrow{\Lambda};x) \geq p_{\mathrm{H}}^{\mathrm{s}}(\Lambda)$$

for all sites x.

One of the inequalities between site and bond critical probabilities presented above carries over to oriented percolation. The analogue of the other has a slightly different form, and a much simpler proof.

Theorem 5. *Let $\overrightarrow{\Lambda}$ be an infinite, connected, locally finite oriented multi-graph, and let x be any site of $\overrightarrow{\Lambda}$. Then*

$$p_{\mathrm{H}}^{\mathrm{b}}(\overrightarrow{\Lambda};x) \le p_{\mathrm{H}}^{\mathrm{s}}(\overrightarrow{\Lambda};x) \le 1 - \left(1 - p_{\mathrm{H}}^{\mathrm{b}}(\overrightarrow{\Lambda};x)\right)^{\Delta_{\mathrm{in}}}, \tag{12}$$

where Δ_{in} is the maximum in-degree of a vertex in $\overrightarrow{\Lambda}$. The same inequalities hold for p_{T}.

Proof. Let p_{c} denote either p_{H} or p_{T}, with all occurrences of p_{c} to be interpreted in the same way. The first inequality, $p_{\mathrm{c}}^{\mathrm{s}}(\overrightarrow{\Lambda};x) \ge p_{\mathrm{c}}^{\mathrm{b}}(\overrightarrow{\Lambda};x)$, is proved in the same way as Theorem 3, *mutatis mutandis*: it suffices to show that for all sites x, integers n, and $0 < p < 1$, we have

$$\mathbb{P}^{\mathrm{s}}_{\overrightarrow{\Lambda},p}\big(|C_x| \ge n\big) \le p\,\mathbb{P}^{\mathrm{b}}_{\overrightarrow{\Lambda},p}\big(|C_x| \ge n\big),$$

where here $C_x = C_x^+$. In showing this, we work in the finite subgraph $\overrightarrow{\Lambda}_n$ induced by the vertices reachable from x by oriented paths in $\overrightarrow{\Lambda}$ of length at most n. The construction of the sequences $\mathcal{T}$ and $\mathcal{T}'$ proceeds exactly as in the proof of Theorem 3, except that we take the orientation into account: at each step t we select an edge $\overrightarrow{e_t} = \overrightarrow{y_t z_t}$ with $y_t \in R_t$ and $z_t \in U_t$.

To prove the second inequality in (12), it suffices to show that for every site x of $\overrightarrow{\Lambda}$, every integer $n \ge 1$, and every $0 < p < 1$, we have

$$\mathbb{P}^{\mathrm{s}}_{\overrightarrow{\Lambda},r}\big(|C_x| \ge n \mid x \text{ is open}\big) \ge \mathbb{P}^{\mathrm{b}}_{\overrightarrow{\Lambda},p}\big(|C_x| \ge n\big), \tag{13}$$

where $r = 1 - (1-p)^{\Delta_{\mathrm{in}}}$.

Consider the independent bond percolation on $\overrightarrow{\Lambda}$ in which each bond is open with probability p. Declare a site $y \ne x$ to be open if there is at least one open bond $\overrightarrow{zy}$, and declare x to be always open. Since the sets of bonds coming into different sites are disjoint, we obtain an independent site percolation measure $\mathbb{P}^{\mathrm{s}}_{\overrightarrow{\Lambda},\mathbf{r}}$ in which each site y is open with some probabilty r_y. More precisely, $r_x = 1$ and, for $y \ne x$, $r_y = 1-(1-p)^{d_{\mathrm{in}}(y)}$, where $d_{\mathrm{in}}(y)$ is the in-degree of y. If an oriented path starting at x is open in the bond percolation, it is also open in the site percolation. Thus

$$\mathbb{P}^{\mathrm{s}}_{\overrightarrow{\Lambda},\mathbf{r}}\big(|C_x| \ge n\big) \ge \mathbb{P}^{\mathrm{b}}_{\overrightarrow{\Lambda},p}\big(|C_x| \ge n\big).$$

Since $r_x = 1$, while $r_y \le r$ for all $y \ne x$, relation (13) follows, completing the proof of the theorem. □

The bound (12) in Theorem 5 is tight for oriented multi-graphs. Indeed, let $\overrightarrow{\Lambda}$ be an infinite binary tree with each edge oriented away from the root. Then $p_{\mathrm{H}}^{\mathrm{s}}(\overrightarrow{\Lambda}) = p_{\mathrm{H}}^{\mathrm{b}}(\overrightarrow{\Lambda}) = 1/2$. (In other words, $p_{\mathrm{H}}^{\mathrm{s}}(\overrightarrow{\Lambda};x) = p_{\mathrm{H}}^{\mathrm{b}}(\overrightarrow{\Lambda};x) = 1/2$ for every site x.) Replacing each oriented edge by k parallel edges with the same orientation, we obtain a graph $\overrightarrow{\Lambda}^{[k]}$ with maximum in-degree k, with

$$1 - p_{\mathrm{H}}^{\mathrm{s}}(\overrightarrow{\Lambda}^{[k]}) = 1 - p_{\mathrm{H}}^{\mathrm{s}}(\overrightarrow{\Lambda}) = 1 - p_{\mathrm{H}}^{\mathrm{b}}(\overrightarrow{\Lambda}) = \left(1 - p_{\mathrm{H}}^{\mathrm{b}}(\overrightarrow{\Lambda}^{[k]})\right)^k,$$

and the same for p_{T}.

If we do not allow multiple edges, then (12) is still fairly tight when $p_{\mathrm{H}}^{\mathrm{s}}(\overrightarrow{\Lambda})$ is close to 1; in particular, the correct exponent is indeed Δ_{in}. For $p_{\mathrm{H}}^{\mathrm{s}}(\overrightarrow{\Lambda})$ very close to 1, this can be shown using the oriented analogue of the construction of Chayes and Schonmann [2000]. To get good bounds in a wider range seems to require a more complicated construction, which we now describe.

By a (d,k)-*unit* we mean the oriented graph shown in Figure 16; this has an *initial site* u, and a *final site* v. The site v is the root of a d-ary

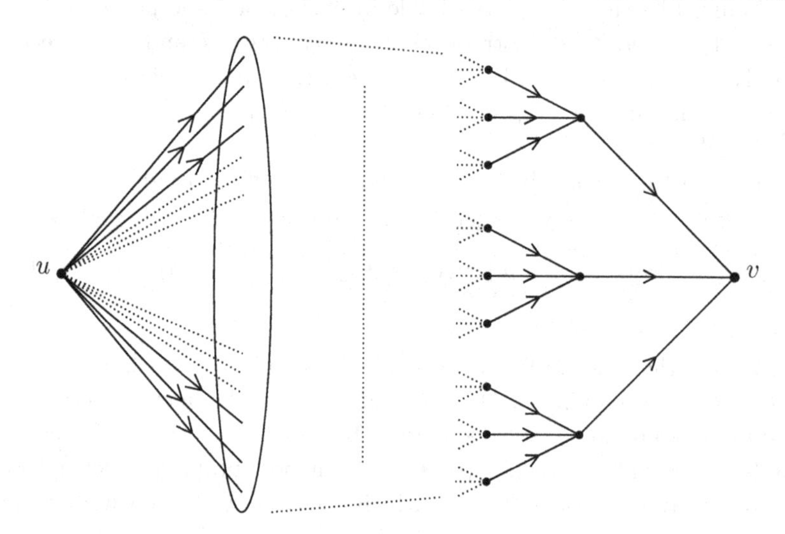

Figure 16. A $(3,k)$ unit. The initial site u is joined to all 3^k leaves of an oriented tree with root v.

tree of height k with every edge oriented towards v. (In the figure, the solid lines show such a tree with $k = 2$ and $d = 3$.) The initial site u sends an edge to all d^k leaves of this tree.

Suppose that the bonds of a (d,k)-unit are open independently with probability $p = C/d$. We shall take $C \geq 2$ to be a large constant, and let $d \to \infty$. Ignoring u for the moment, the numbers X_i of sites at distance i from v that are joined to v by an open path form a certain branching process: for $1 \leq i \leq k$, X_i is the sum of X_{i-1} independent copies of a binomial $\mathrm{Bi}(d,p)$ distribution. As $d \to \infty$ with C fixed, this binomial distribution converges to the Poisson distribution with mean C. When continued to infinity, the survival probability of this branching process is a certain function $\rho(C)$ of C.

We have $\rho(C) = 1 - e^{-C} - O\big(Ce^{-2C}\big)$ as $C \to \infty$: when C is large, the process is highly supercritical, so given that it survives at the first step, it is very likely to survive forever. In fact, as $k \to \infty$, with probability $\rho(C) - o(1)$ we have $X_k \geq C^{k/2}$, say. Let $f(d,k,p)$ be the probability that there is an open path from u to v. It follows that

$$\lim_{d\to\infty} \lim_{k\to\infty} f(d,k,C/d) = \rho(C),$$

so choosing d and then k large enough (depending on C), we have $f(d,k,C/d) = 1 - e^{-C} + O\big(Ce^{-2C}\big)$.

A similar but much simpler argument for site percolation shows that if u is open, and the other sites are open independently with probability p, then the probability $g(d,k,p)$ that there is an open path from u to v is between p and $p - O\big((1-p)^2\big)$ for p close to 1, $d \geq 2$, and any k. Let $0 < \pi < 1$, and let T be a tree with critical probability $p_{\mathrm{H}}(T) = p_{\mathrm{T}}(T) = \pi$. Choose an arbitrary site x_0 as the root, and orient each edge of T away from x_0 to obtain an oriented tree $\overrightarrow{T}$ with maximum in-degree 1 which also has critical probability $p_{\mathrm{H}}(\overrightarrow{T}; x_0) = p_{\mathrm{T}}(\overrightarrow{T}; x_0) = \pi$. (Depending on the choice of T, the critical probabilities $p_{\mathrm{c}}(\overrightarrow{T}; x)$ may or may not depend on the site x.)

Let us replace each bond $\overrightarrow{xy}$ of $\overrightarrow{T}$ by a (d,k)-unit, by identifying u with x and v with y and then deleting $\overrightarrow{xy}$. The resulting graph $\overrightarrow{\Lambda}$ has bond and site critical probabilities $p_{\mathrm{c}}^{\mathrm{b}} = p_{\mathrm{H}}^{\mathrm{b}}(\overrightarrow{\Lambda}; x_0) = p_{\mathrm{T}}^{\mathrm{b}}(\overrightarrow{\Lambda}; x_0)$, $p_{\mathrm{c}}^{\mathrm{s}} = p_{\mathrm{H}}^{\mathrm{s}}(\overrightarrow{\Lambda}; x_0) = p_{\mathrm{T}}^{\mathrm{s}}(\overrightarrow{\Lambda}; x_0)$ that satisfy

$$f(d,k,p_{\mathrm{c}}^{\mathrm{b}}) = \pi \text{ and } g(d,k,p_{\mathrm{c}}^{\mathrm{s}}) = \pi.$$

From the formulae above it follows that if $p_{\mathrm{c}}^{\mathrm{s}} \to 1$, then $1 - \pi \sim 1 - p_{\mathrm{c}}^{\mathrm{s}}$. Also, if $d \to \infty$ and $k \to \infty$ are chosen suitably, then

$$e^{-dp_{\mathrm{c}}^{\mathrm{b}}} + O\Big(dp_{\mathrm{c}}^{\mathrm{b}} e^{-2dp_{\mathrm{c}}^{\mathrm{b}}}\Big) = 1 - \pi \sim 1 - p_{\mathrm{c}}^{\mathrm{s}}.$$

If $p_{\mathrm{c}}^{\mathrm{b}} \to 0$ and $dp_{\mathrm{c}}^{\mathrm{b}} \to \infty$, this implies that

$$1 - p_{\mathrm{c}}^{\mathrm{s}} = (1 - p_{\mathrm{c}}^{\mathrm{b}})^{(1+o(1))d}.$$

It follows that, given any sequence $p_n \to 1$, we can construct graphs $\vec{\Lambda}_n$ with maximum in-degree $\Delta_{\rm in} = \Delta_{\rm in}(n) \to \infty$ such that $p_n^{\rm s} = p_{\rm H}^{\rm s}(\vec{\Lambda}_n; x_0) = p_{\rm T}^{\rm s}(\vec{\Lambda}_n; x_0)$ and $p_n^{\rm b} = p_{\rm H}^{\rm b}(\vec{\Lambda}_n; x_0) = p_{\rm T}^{\rm b}(\vec{\Lambda}_n; x_0)$ satisfy

$$p_n = p_n^{\rm s} = 1 - (1 - p_n^{\rm b})^{(1+o(1))\Delta_{\rm in}}.$$

This shows that the constant $\Delta_{\rm in}$ in the exponent in (12) is essentially best possible, at least when $p_{\rm H}^{\rm s} \to 1$.

If $\vec{\Lambda}$ is an oriented graph and x a site of $\vec{\Lambda}$, then we have defined eight associated critical probabilities: $p_{\rm H}$ and $p_{\rm T}$ for site and bond percolation on $\vec{\Lambda}$ and on the underlying simple graph Λ in any combination. We have eight trivial inequalities relating these probabilities; four of the form $p_{\rm T} \le p_{\rm H}$, and four of the form $p_{\rm c}(\Lambda) \le p_{\rm c}(\vec{\Lambda}; x)$. Also, we have that each critical probability for site percolation is at least that for the corresponding bond percolation. It follows that all eight critical probabilities are at least $p_{\rm T}^{\rm b}(\Lambda)$ and at most $p_{\rm H}^{\rm s}(\vec{\Lambda}; x)$.

For any d, let $\vec{\mathbb{Z}}^d$ denote the natural orientation of $\mathbb{Z}^d$, where each edge is oriented in the positive direction. Let $p_{\rm c}$ be one of $p_{\rm H}^{\rm s}, p_{\rm H}^{\rm b}, p_{\rm T}^{\rm s}, p_{\rm T}^{\rm b}$. Since all sites of $\vec{\mathbb{Z}}^d$ are equivalent, the critical probability $p_{\rm c}(\vec{\mathbb{Z}}^d; x)$ is independent of x, so we shall write $p_{\rm c}(\vec{\mathbb{Z}}^d)$ for the common value. Our next result shows that all eight critical probabilities associated to $\vec{\mathbb{Z}}^2$ are non-trivial.

Lemma 6.

$$1/3 \le p_{\rm T}^{\rm b}(\mathbb{Z}^2) \le p_{\rm H}^{\rm s}(\vec{\mathbb{Z}}^2) \le 80/81.$$

Proof. We have shown that $p_{\rm T}^{\rm b}(\mathbb{Z}^2) \ge 1/3$ in Lemma 2. It remains to show that $p_{\rm H}^{\rm s}(\vec{\mathbb{Z}}^2) \le 80/81$. The proof will be similar to the proof of the upper bound on $p_{\rm H}^{\rm b}(\mathbb{Z}^2)$ in Lemma 2. This time we shall use the concept of a blocking cycle. A cycle S in the lattice $\Lambda^\star$ which is the dual of the unoriented lattice $\Lambda = \mathbb{Z}^2$ is *blocking* if, for every *oriented* bond $\vec{e} = \vec{ab}$ of $\vec{\mathbb{Z}}^2$ with a inside S and b outside, the site b is closed; see Figure 17.

Fix $p > 80/81$, and let C_0 be the set of sites that may be reached from the origin by open paths in $\vec{\Lambda}_p^{\rm s}$, where $\vec{\Lambda} = \vec{\mathbb{Z}}^2$. Suppose that C_0 is finite. Since C_0 is a connected subgraph of $\mathbb{Z}^2$, by Lemma 1 the external boundary $\partial^\infty C_0$ of C_0 is a cycle in $\Lambda^\star$. If $\vec{ab}$ is a bond of $\vec{\mathbb{Z}}^2$ with a inside $\partial^\infty C_0$ and b outside, then, by the definition of $\partial^\infty C_0$, we have $a \in C_0$ and $b \notin C_0$. Hence, by the definition of C_0, the site b is closed.

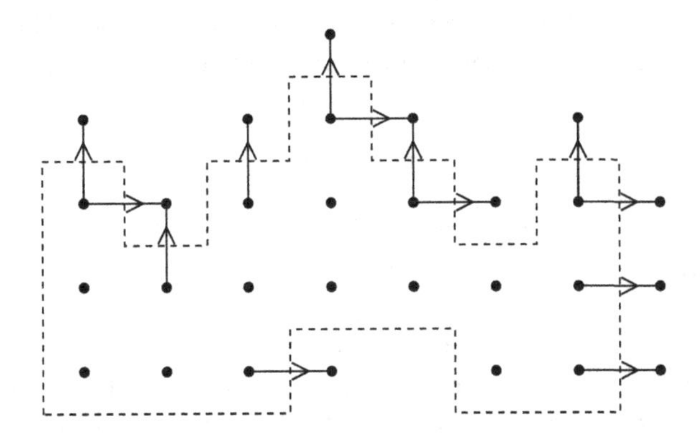

Figure 17. A dual cycle S (dashed lines), and the oriented bonds from sites inside S to sites outside S (arrows). S is blocking if every site pointed to is closed.

Therefore, whenever C_0 is finite, there is a blocking cycle surrounding C_0.

Let S be any cycle of length 2ℓ. As we walk around S anticlockwise, each time we take a step upwards or to the left, in order for S to be blocking, the site x on our right must be closed. As S ends where it starts, we take as many steps upwards as downwards, and as many steps to the left as to the right. Thus we take a total of ℓ steps that go up or to the left. Any site x is on the right of at most one upwards step and at most one leftwards step, so there is a set X of at least $\ell/2$ sites that must be closed for S to be blocking. Consequently

$$\mathbb{P}(S \text{ is blocking}) \leq (1-p)^{\ell/2}.$$

The rest of the argument is as in the unoriented case. Let L_k be the line segment joining the origin to the point $(k, 0)$, and let Y_k be the number of blocking cycles surrounding L_k. We have already shown that there are at most $\ell\mu_{2\ell-1}$ cycles of length 2ℓ surrounding L_k, and each is blocking with probability at most $(1-p)^{\ell/2}$, so

$$\mathbb{E}(Y_k) \leq \sum_{\ell \geq k+2} \ell\mu_{2\ell-1}(1-p)^{\ell/2} \leq \sum_{\ell \geq k+2} \frac{4\ell}{9}\left(3(1-p)^{1/4}\right)^{2\ell}.$$

As $p > 80/81$, the final sum is finite, so $\mathbb{E}(Y_k) \to 0$ as $k \to \infty$, and there is a k with $\mathbb{E}(Y_k) < 1$.

Continuing as in the proof of Lemma 2, let A_k be the event that $Y_k = 0$, so $\mathbb{P}(A_k) > 0$, and let B_k be the event that the $k+1$ sites in

L_k are open. Note that A_k and B_k are independent. Also, if both hold, then there is no blocking cycle surrounding the origin, so C_0 is infinite. Hence,

$$\theta(p) \geq \mathbb{P}(A_k \cap B_k) = \mathbb{P}(A_k)\mathbb{P}(B_k) = \mathbb{P}(A_k)p^{k+1} > 0.$$

This shows that $p_{\mathrm{H}}^{\mathrm{s}}(\vec{\mathbb{Z}}^2) \leq p$. Since $p > 80/81$ was arbitrary, $p_{\mathrm{H}}^{\mathrm{s}}(\vec{\mathbb{Z}}^2) \leq 80/81$ follows, as desired. □

For oriented bond percolation, a cycle S of length 2ℓ surrounding the origin is blocking if every bond crossing S from the inside to the outside is closed. As there are exactly ℓ such bonds, this event has probability $(1-p)^\ell$, and the argument above shows that $p_{\mathrm{H}}^{\mathrm{b}}(\vec{\mathbb{Z}}^2) \leq 8/9$. Just as in the unoriented case, one can obtain better bounds by more careful counting. Not every dual cycle surrounding the origin can arise as the external boundary $\partial^\infty C_0$ of the open cluster C_0: for example, the cycle S shown in Figure 17 cannot, as there is no way to reach the two sites at the bottom right along oriented bonds inside S. Balister, Bollobás and Stacey [1999] counted cycles that can arise as $\partial^\infty C_0$ more carefully, showing that there are $(K + o(1))^\ell$ such cycles of length 2ℓ, where K is a constant satisfying $5.1269 < K < 5.2623$. The constant $K = K(\vec{\mathbb{Z}}^2)$ plays one of the roles of the connective constant $\lambda = \lambda(\mathbb{Z}^2)$: the argument above shows that $p_{\mathrm{H}}^{\mathrm{b}}(\vec{\mathbb{Z}}^2) \leq 1 - 1/K \leq 0.81$, which is analogous to the upper bound on $p_{\mathrm{H}}^{\mathrm{b}}(\mathbb{Z}^2)$ given in (4). As we shall see in later chapters, much better bounds on this critical probability may be obtained in other ways.

We now turn to the asymptotic behaviour of the various critical probabilities associated to percolation on $\mathbb{Z}^d$ and $\vec{\mathbb{Z}}^d$. As noted earlier, for each fixed d, we have defined eight critical probabilities, of which $p_{\mathrm{T}}^{\mathrm{b}}(\mathbb{Z}^d)$ is the smallest and $p_{\mathrm{H}}^{\mathrm{s}}(\vec{\mathbb{Z}}^d)$ the largest. Since $\vec{\mathbb{Z}}^{d+1}$ includes $\vec{\mathbb{Z}}^d$ as a subgraph, each of these eight functions decreases with d.

Theorem 7. *For any $d \geq 2$ we have*

$$\frac{1}{2d-1} \leq p_{\mathrm{T}}^{\mathrm{b}}(\mathbb{Z}^d) \leq p_{\mathrm{H}}^{\mathrm{s}}(\vec{\mathbb{Z}}^d) = O(1/d).$$

Proof. The first inequality is just as easy as the special case $d = 2$ proved in Lemma 2. Indeed, as $\mathbb{Z}^d$ is $2d$-regular, the number $\mu_n = \mu_n(\mathbb{Z}^d)$ of paths in $\mathbb{Z}^d$ starting at 0 and having length n is at most $2d(2d-1)^{n-1}$. For bond percolation with $p < 1/(2d-1)$, the expected number of open paths starting at 0 is $\sum_n \mu_n p^n < \infty$, so $\chi(p)$ is finite and $p_T \geq p$. As $p < 1/(2d-1)$ was arbitrary, $p_T \geq 1/(2d-1)$ follows.

For the upper bound on $p_H^s(\vec{\mathbb{Z}}^d)$ we shall in fact prove that

$$p_H^s(\vec{\mathbb{Z}}^d) \leq 1 - \left(1 - p_H^s(\vec{\mathbb{Z}}^2)\right)^{1/\lfloor d/2 \rfloor}.$$

Note that the right-hand side is indeed $O(1/d)$, as $p_H^s(\vec{\mathbb{Z}}^2) < 1$. Since $p_H^s(\vec{\mathbb{Z}}^d)$ is decreasing, we may assume that d is even. Fix any p_2 with $p_H^s(\vec{\mathbb{Z}}^2) < p_2 < 1$, and set $p_d = 1 - (1-p_2)^{2/d}$, so $(1-p_d)^{d/2} = 1 - p_2$. We shall compare oriented site percolation on $\vec{\mathbb{Z}}^d$ with $p = p_d$ to that on $\vec{\mathbb{Z}}^2$ with $p = p_2$. It will be convenient to take the origin to be always open in both cases; this multiplies $\theta_0(p)$ and $\chi_0(p)$ by a constant factor $1/p$.

By *layer* t of $\mathbb{Z}^d$, we shall mean the set of points whose coordinates sum to t, with each coordinate non-negative. (In considering oriented site percolation starting at 0, we may of course ignore sites with negative coordinates.) Taking the origin open, let $R_t = C_0 \cap \{x + y = t\}$ be the set of points of $\mathbb{Z}^2$ in layer t that are reachable from 0 in the oriented site percolation on $\vec{\mathbb{Z}}^2$. Clearly, given R_{t-1}, every site (x, y) in layer t with at least one neighbour in R_{t-1} is in R_t with probability p, independently of the other sites in layer t. A site in layer t with no neighbours in R_{t-1} cannot be in R_t. By choice of p_2, we have $\theta_0(p_2) > 0$, so with positive probability every R_t is non-empty.

Let $\varphi : \mathbb{Z}^d \to \mathbb{Z}^2$ be the projection

$$\varphi : (x_1, x_2, \ldots, x_d) \mapsto \left(\sum_{i=1}^{d/2} x_i, \sum_{i=d/2+1}^{d} x_i \right).$$

Thus φ is the linear map sending the first $d/2$ coordinate vectors to $(1,0)$, and the last $d/2$ to $(0,1)$. Let the origin of $\mathbb{Z}^d$ be open, and each other site be open independently with probability p_d. Let C_0 be the open cluster of the oriented site percolation on $\vec{\mathbb{Z}}^d$. We will construct sequences $(R'_0, R'_1, R'_2, \ldots)$ and $(S_0, S_1, S_2, \ldots)$ with the following properties: for each point $(x, y) \in R'_t$ there is a unique $\mathbf{v} \in S_t$ such that $\varphi(\mathbf{v}) = (x, y)$, each S_t is a subset of C_0, and $(R'_0, R'_1, \ldots)$ has exactly the distribution of $(R_0, R_1, \ldots)$.

We start the construction by taking $S_0 = R'_0 = \{0\}$. At each step in the construction, R'_t and S_t will depend only on the states of sites in layers up to t of $\mathbb{Z}^d$. Given R'_{t-1} and S'_{t-1}, let us say that a point (x, y) is *eligible* if it is in layer t of $\mathbb{Z}^2$ and has at least one neighbour in R'_{t-1}. Let

(x, y) be an eligible point, and suppose without loss of generality that $(x-1, y) \in R'_{t-1}$. Then there is a $\mathbf{v} \in S_{t-1}$ with $\varphi(\mathbf{v}) = (x-1, y)$. The point $\mathbf{v}$ has exactly $d/2$ neighbours $\mathbf{w}$ in $\mathbb{Z}^d$ with $\varphi(\mathbf{w}) = (x, y)$. Since each is open independently with probability p_d, the probability that at least one is open is exactly p_2. If at least one $\mathbf{w}$ is open, include (x, y) in R'_t, and (one of the possible) $\mathbf{w}$ in S_t. Note that we decide whether to include (x, y) by looking at the states of its preimages under φ. As the sets of preimages of distinct points are disjoint, each eligible (x, y) is included in R'_t independently. Thus, the distribution of R'_t conditional on R'_{t-1} is exactly that of R_t conditional on R_{t-1}, and our construction has the required properties.

As the distribution of $(R'_0, R'_1, \ldots)$ is the same as that of $(R_0, R_1, \ldots)$, there is a positive probability that every R'_t is non-empty. But then every S_t is non-empty, and C_0 is infinite, so $p_{\mathrm{H}}^{\mathrm{s}}(\vec{\mathbb{Z}}^d) \le p_d$, completing the proof. □

Numerous considerably stronger bounds have been proved about critical probabilities in high dimensions. In particular, Cox and Durrett [1983] showed that

$$d^{-1} + \frac{1}{2}d^{-3} + o(d^{-3}) \le p_{\mathrm{H}}^{\mathrm{b}}(\vec{\mathbb{Z}}^d) \le d^{-1} + d^{-3} + O(d^{-4}).$$

As we shall see in Chapter 4, Menshikov [1986] proved that p_{H} and p_{T} are equal in a very general context, including all the cases considered here, so we may write p_{c} for their common value. Kesten [1990; 1991] showed that $p_{\mathrm{c}}^{\mathrm{s}}(\mathbb{Z}^d)$, $p_{\mathrm{c}}^{\mathrm{b}}(\mathbb{Z}^d) \sim (2d)^{-1}$, and, independently, Hara and Slade [1990] and Gordon [1991] gave stronger results for $p_{\mathrm{c}}^{\mathrm{b}}(\mathbb{Z}^d)$. Concerning site percolation, Bollobás and Kohayakawa [1994] gave a simple combinatorial argument showing that $p_{\mathrm{c}}^{\mathrm{s}}(\mathbb{Z}^d) = (1 + d^{o(1)-1/3})/(2d)$. In fact, there are very precise formulae for both bond and site percolation:

$$p_{\mathrm{c}}^{\mathrm{b}}(\mathbb{Z}^d) = \frac{1}{2d} + \frac{1}{4d^2} + \frac{7}{16d^3} + O(d^{-4})$$

was given by Hara and Slade [1995], confirming the first few terms in an expansion reported in Gaunt and Ruskin [1978] without rigorous error bounds. A similar expansion for site percolation,

$$p_{\mathrm{c}}^{\mathrm{s}}(\mathbb{Z}^d) = \frac{1}{2d} + \frac{5}{8d^2} + \frac{31}{32d^3} + \frac{75}{32d^4} + \cdots,$$

was given by Gaunt, Sykes and Ruskin [1976], again without rigorous error bounds.

Throughout this chapter, we have concentrated on the critical probabilities of independent percolation models. As we remarked earlier, although these were the first graph invariants associated with percolation, there is likely to be only so much that can be said about them. Except in certain special cases, it seems that the exact critical probabilities are in some sense rather arbitrary; for example, there may well be no formula for $p_c^s(\mathbb{Z}^2)$, say, in which case this quantity will never be known exactly.

The quintessential example of a known critical probability is the critical probability for bond percolation on $\mathbb{Z}^2$; we shall present this celebrated result of Harris and Kesten in Chapter 3. Using the same method, Kesten determined the critical probability for site percolation on the triangular lattice. Applying tricks of changing one graph into another, as in Figure 6, and the star-delta transformation to be discussed in Chapter 5, one may obtain certain other critical probabilities in a similar way, including those for bond percolation on the triangular and hexagonal lattices. Nevertheless, these are only sporadic examples.

There is another, very different, percolation model whose critical probability is known: face percolation on a random Voronoi tessellation in the plane. This result is considerably harder than the results concerning exact critical probabilities on lattices: we shall sketch a proof in Chapter 8.

There are several other invariants defined from the component structure that are likely to be considerably more significant than the critical probability in the long term. Rather little is known about these; we shall say a few words about them in Chapter 7.

2
Probabilistic tools

In this chapter we present some fundamental tools from combinatorial probability that we shall use when we come to study percolation. These results have many applications in other areas, for example, the study of random graphs.

One of the most basic results in probability theory is Kolmogorov's 0-1 law.

Theorem 1. *Let $X = (X_1, X_2, \ldots)$ be a sequence of independent random variables, and let A be an event in the σ-field generated by X. Suppose that, for every n, the event A is independent of $X_1, \ldots, X_n$. Then $\mathbb{P}(A)$ is 0 or 1.* □

Note that different X_i are not assumed to have the same distribution. An event A with the property described above is known as a *tail event*: Theorem 1 states that any tail event in a product probability space has probability 0 or 1.

In fact, Theorem 1 was not Kolmogorov's original formulation of this result. What he showed was that, if $X = (X_1, X_2, \dots)$ is a sequence of real-valued random variables, $f : \mathbb{R}^{\mathbb{N}} \to \mathbb{R}$ is Baire function, and

$$\mathbb{P}\big(f(X) = 0 \mid X_1, \ldots, X_n\big) = \mathbb{P}\big(f(X) = 0\big),$$

then $\mathbb{P}\big(f(X) = 0\big)$ is 0 or 1; see Kolmogorov [1956, pp. 69–70]. As Kolmogorov noted, these assumptions are satisfied if the X_i are independent and the value of the function $f(X)$ remains unchanged when only a finite number of variables are changed.

The following observation, known as Fekete's Lemma [1923], is frequently used to prove the convergence of various sequences.

Lemma 2. *Let (a_n) be a sequence of non-negative reals such that $a_{n+m} \leq a_n + a_m$ for $n, m \geq 1$. Then $\lim_{n\to\infty} a_n/n$ exists.*

Proof. Let $c = \inf_n a_n/n$, so $0 \leq c \leq a_1$. Given $\varepsilon > 0$, there is a k such that $a_k/k < c + \varepsilon$. If $n = kq + r$, $0 \leq r < k$, then

$$a_n \leq qa_k + a_r \leq qa_k + b_k,$$

where $b_k = \max_{i<k} a_k < \infty$. Hence

$$a_n/n \leq a_k/k + b_k/n < c + 2\varepsilon,$$

if n is large enough, so $\limsup a_n/n \leq c + 2\varepsilon$. Since $\varepsilon > 0$ was arbitrary, the result follows. □

If we do not assume that $a_n \geq 0$, then the conclusion is again that $\lim_{n\to\infty} a_n/n$ exists, but we must allow the limit to take the value $-\infty$. Taking logarithms, it follows that if (a_n) is a submultiplicative sequence of positive real numbers, i.e., if $a_{n+m} \leq a_n a_m$ for $n, m \geq 1$, then $\lim_{n\to\infty} a_n^{1/n}$ exists.

Many sequences occurring in percolation theory do not satisfy the conditions of subadditivity or submultiplicativity cleanly, and one needs the following extension of Fekete's Lemma proved by de Bruijn and Erdős [1952].

Lemma 2′. *Let $\varphi : \mathbb{R}^+ \to \mathbb{R}^+$ be an increasing function with*

$$\int_1^\infty \varphi(t)t^{-2}\,\mathrm{d}t < \infty,$$

and let (a_n) be a sequence of reals such that

$$a_{n+m} \leq a_n + a_m + \varphi(n+m) \tag{1}$$

whenever $n/2 \leq m \leq n$. Then $a_n/n \to L$ for some L with $-\infty \leq L < \infty$. □

This result is essentially best possible. For example, if $\varepsilon > 0$, then it does not suffice to impose condition (1) for $(1+\varepsilon)n/2 \leq m \leq n$.

The framework for much of combinatorial probability (including percolation theory), is the *weighted hypercube* Q_p^n. We shall take our time over introducing this important concept.

For any set S, we may identify the *power set* $\mathcal{P}(S)$, i.e., the set of all subsets of S, with the set $\{0,1\}^S$ of all functions $f\colon S \to \{0,1\}$. In this identification, a set $A \subset S$ corresponds to its characteristic function $\mathbf{1}_A : S \to \{0,1\}$, defined by

$$\mathbf{1}_A(x) = \begin{cases} 1 & \text{if } x \in A, \\ 0 & \text{if } x \in S \setminus A. \end{cases}$$

(We follow the standard convention in combinatorics that $S \subset S$.)

If S is finite (and this is the case we are mostly interested in) then S is usually taken to be $[n] = \{1, 2, \ldots, n\}$, and then $\mathcal{P}(n) = \mathcal{P}([n]) = \mathcal{P}(S)$ is identified with the *hypercube* or, simply, the *cube* $Q^n = 2^n = \{0,1\}^n$, i.e., the set of all 0-1 sequences of length n. Under this identification, a set $A \subset [n]$ corresponds to the 0-1 sequence $(a_i)_{i=1}^n$ in which $a_i = 1$ if $i \in A$ and $a_i = 0$ if $i \in [n] \setminus A$.

It is natural to consider Q^n as a graph whose vertex set is $\mathcal{P}(n)$, in which two sets A and B are joined if their symmetric difference $A \triangle B$ consists of one element. Equivalently, Q^n has vertex set $\{0,1\}^n \subset \mathbb{R}^n$, and two vertices $\mathbf{a}$ and $\mathbf{b}$ are joined if $\mathbf{a} - \mathbf{b} = \pm \mathbf{e}_i$ for some i, where $(\mathbf{e}_1, \mathbf{e}_2, \ldots, \mathbf{e}_n)$ is the canonical basis of $\mathbb{R}^n$. Thus the graph Q^n has 2^n vertices and $n2^{n-1}$ edges.

For $A, B \in \mathcal{P}(n)$, put $A \le B$ if $A \subset B$. This turns the power set $\mathcal{P}(n)$ into a *partially ordered set*, a *poset*. Equivalently, the graph Q^n has a natural orientation: an edge $\mathbf{ab}$ is oriented from $\mathbf{a}$ to $\mathbf{b}$ if $\mathbf{b} - \mathbf{a} = \mathbf{e}_i$ for some i, $1 \le i \le n$. Then $\mathbf{a} \le \mathbf{b}$ if there is an oriented path from $\mathbf{a}$ to $\mathbf{b}$, i.e., if $a_i \le b_i$ for every i.

A point (vertex) of Q^n is naturally identified with the outcome of a sequence of n coin tosses: for $\mathbf{a} = (a_i)_{i=1}^n \in Q^n$ we have $a_i = 1$ if the ith toss lands as heads. This puts a probability measure on (the vertex set of) Q^n. If we use fair coins then we get the uniform measure on Q^n, i.e., the normalized counting measure: for $A \subset Q^n$ the probability of A is $\mathbb{P}(A) = |A|/2^n$. If we use biased coins then we get a 'more interesting' measure: if the probability that the ith coin lands as heads is p_i, then

$$\mathbb{P}(A) = \sum_{\mathbf{a} \in A} \prod_{a_i = 1} p_i \prod_{a_i = 0} (1 - p_i) = \sum_{\mathbf{a} \in A} \prod_{a_i = 1} (p_i/(1-p_i)) \prod_{i=1}^{n} (1 - p_i). \quad (2)$$

The cube Q^n endowed with this probability measure is denoted by $Q^n_{\mathbf{p}}$, and called the *weighted cube with probability* $\mathbf{p} = (p_i)_{i=1}^n$.

Putting it slightly more formally, let $x_1, \ldots, x_n$ be independent Bernoulli random variables with $\mathbb{P}(x_i = 1) = p_i$ and $\mathbb{P}(x_i = 0) = 1 - p_i$, and let $\mathbf{x} = (x_i)$ be the random sequence obtained in this way. Then $Q^n_{\mathbf{p}}$ is the space of these random sequences. Also, $X = \{i : x_i = 1\}$ is a random element of $\mathcal{P}(n)$, so $Q^n_{\mathbf{p}}$ is the *space of random subsets* of $[n]$, in which the elements $1, 2, \ldots, n$ are chosen independently of each other, and i is chosen with probability p_i.

If $p_i = 1/2$ for every i, then the weighted cube is just the unweighted cube mentioned above; in this case the probability measure is the normalized counting measure. In the next simplest case we have $p_i = p$ for every i, so that $\mathbb{P}(X = A) = p^{|A|}(1-p)^{n-|A|}$; in this case we write Q^n_p instead of $Q^n_{\mathbf{p}}$. The standard percolation models correspond to Q^n_p.

The events in our probability space $Q^n_{\mathbf{p}}$ are the subsets of $Q^n_{\mathbf{p}}$, i.e., the subsets of Q^n. A subset $U \subset Q^n$ is a *(monotone) increasing* event, or simply an *up-set*, if $a, b \in Q^n$, $a \in U$ and $a \le b$ imply that $b \in U$. Replacing Q^n by $\mathcal{P}(S)$, a set system $\mathcal{U} \subset \mathcal{P}(S)$ is *increasing*, or an *up-set*, if $A \subset B \subset S$ and $A \in \mathcal{U}$ imply that $B \in \mathcal{U}$. Similarly, $D \subset Q^n$ is a *(monotone) decreasing* event or a *down-set* if $a, b \in Q^n$, $a \in D$ and $a \ge b$ imply that $b \in D$. Equivalently, $\mathcal{D} \subset \mathcal{P}(S)$ is a *decreasing* set system or a *down-set* if $A \subset B \in \mathcal{D}$ implies that $A \in \mathcal{D}$. Clearly, $U \subset Q^n$ is monotone increasing if and only if its complement, $D = Q^n \setminus U$, is monotone decreasing.

The fundamental correlation inequality in the cube, proved by Harris [1960] in the context of percolation and rediscovered by Kleitman [1966], states the intuitively obvious fact that increasing events are positively correlated.

Given a set $A \subset Q^n$, for $t = 0, 1$, define

$$A_t = \left\{(a_i)_{i=1}^{n-1} \colon (a_1, \ldots, a_{n-1}, t) \in A\right\} \subset Q^{n-1}.$$

If A is monotone increasing then $A_0 \subset A_1$, and if A is monotone decreasing then $A_1 \subset A_0$. Let $Q^{n-1}_{\mathbf{p}'}$ be the weighted cube whose probability measure is induced by the sequence $\mathbf{p}' = (p_i)_{i=1}^{n-1}$. With a slight (and usual) abuse of notation, let us write $\mathbb{P}$ for two different measures, namely the probability measures in $Q^n_{\mathbf{p}}$ and $Q^{n-1}_{\mathbf{p}'}$. Note that

$$\mathbb{P}(A) = (1 - p_n)\mathbb{P}(A_0) + p_n \mathbb{P}(A_1) \tag{3}$$

for every set $A \subset Q^n$.

We are ready to state and prove Harris's Lemma.

Lemma 3. *Let A and B be subsets of $Q^n_{\mathbf{p}}$. If both are up-sets or both are down-sets then*

$$\mathbb{P}(A \cap B) \ge \mathbb{P}(A)\mathbb{P}(B). \tag{4}$$

If A is an up-set and B is a down-set then

$$\mathbb{P}(A \cap B) \leq \mathbb{P}(A)\mathbb{P}(B). \tag{5}$$

Proof. Let us prove (4) by induction on n. For $n = 1$ (or, indeed, $n = 0$) the inequality is trivial, so suppose that $n \geq 2$ and that (4) holds for $n - 1$.

Suppose that A and B are both up-sets or both down-sets. Then either $A_0 \subset A_1$ *and* $B_0 \subset B_1$, or else $A_1 \subset A_0$ *and* $B_1 \subset B_0$. In particular,

$$\big(\mathbb{P}(A_0) - \mathbb{P}(A_1)\big)\big(\mathbb{P}(B_0) - \mathbb{P}(B_1)\big) \geq 0. \tag{6}$$

Also, by (3), the induction hypothesis and (6) we have

$$\begin{aligned}
\mathbb{P}(A \cap B) &= (1 - p_n)\mathbb{P}(A_0 \cap B_0) + p_n\mathbb{P}(A_1 \cap B_1) \\
&\geq (1 - p_n)\mathbb{P}(A_0)\mathbb{P}(B_0) + p_n\mathbb{P}(A_1)\mathbb{P}(B_1) \\
&\geq \{(1 - p_n)\mathbb{P}(A_0) + p_n\mathbb{P}(A_1)\}\{(1 - p_n)\mathbb{P}(B_0) + p_n\mathbb{P}(B_1)\} \\
&= \mathbb{P}(A)\mathbb{P}(B),
\end{aligned}$$

with the second inequality following from (6), and the last equality a consequence of (3).

For A an up-set and B a down-set, inequality (5) follows by applying (4) to the up-sets A and $B^c = Q^n \setminus B$:

$$\begin{aligned}
\mathbb{P}(A \cap B) &= \mathbb{P}(A) - \mathbb{P}(A \cap B^c) \\
&\leq \mathbb{P}(A) - \mathbb{P}(A)\mathbb{P}(B^c) \\
&= \mathbb{P}(A) - \mathbb{P}(A)\{1 - \mathbb{P}(B)\} \\
&= \mathbb{P}(A)\mathbb{P}(B).
\end{aligned}$$

□

The extension to infinite product spaces is immediate; we shall not need it here.

As the intersection of two up-sets is again an up-set, Lemma 3 implies that

$$\mathbb{P}(A_1 \cap A_2 \cap \cdots \cap A_t) \geq \mathbb{P}(A_1)\mathbb{P}(A_2)\cdots\mathbb{P}(A_t) \tag{7}$$

whenever $A_1, \ldots, A_t$ are all up-sets, or all down-sets.

A simple consequence of Harris's Lemma is that if $A_1, \ldots, A_t$ are increasing events in $Q^n_{\mathbf{p}}$ whose union A has very high probability, then one of the A_i must have high probability. Indeed, the complements A_i^c

are decreasing, or down-sets, so from (7) we have

$$\prod_{i=1}^{t} \mathbb{P}(A_i^c) \le \mathbb{P}(A^c).$$

It follows that for some i we have

$$\mathbb{P}(A_i^c) \le \Big(\mathbb{P}(A^c)\Big)^{1/t},$$

i.e.,

$$\mathbb{P}(A_i) \ge 1 - \Big(1 - \mathbb{P}(A_1 \cup \cdots \cup A_t)\Big)^{1/t}. \tag{8}$$

In the case where each A_i has the same probability, inequality (8) holds for every i. For $t = 2$, this observation is sometimes known as the 'square-root trick'; for $t = n$, this is the 'nth-root trick'.

If $\mu : Q^n \to \mathbb{R}$ is any function, then we may think of μ as a signed measure on Q^n. Thus, for $E \subset Q^n$ we have

$$\mu(E) = \sum_{x \in E} \mu(x),$$

and the integral of a function $h : Q^n \to \mathbb{R}$ with respect to μ is

$$\int_{Q^n} h \, \mathrm{d}\mu = \sum_{x \in Q^n} h(x)\mu(x) = (h\mu)(Q^n).$$

In particular, writing $\mathbf{1}_E$ for the characteristic function of a set $E \subset Q^n$, we have

$$\int \mathbf{1}_E \, \mathrm{d}\mu = \mu(E).$$

In this notation, Harris's Lemma states that for the probability measure $\mathbb{P}_{\mathbf{p}}$ on the cube $Q^n_{\mathbf{p}}$ we have

$$\int \mathbf{1}_A \mathbf{1}_B \, \mathrm{d}\mathbb{P}_{\mathbf{p}} \ge \int \mathbf{1}_A \, \mathrm{d}\mathbb{P}_{\mathbf{p}} \int \mathbf{1}_B \, \mathrm{d}\mathbb{P}_{\mathbf{p}} \tag{9}$$

whenever A, $B \subset Q^n$ are up-sets.

Inequality (9) yields a more high-brow formulation of Harris's Lemma. Although the terminology is self-explanatory, we note that a function $h : Q^n \to \mathbb{R}$ is *(monotone) increasing* if $h(x) \le h(y)$ whenever $x \le y$.

Lemma 4. *Let f and g be increasing functions on Q^n. Then*

$$\int fg \, \mathrm{d}\mathbb{P}_{\mathbf{p}} \ge \int f \, \mathrm{d}\mathbb{P}_{\mathbf{p}} \int g \, \mathrm{d}\mathbb{P}_{\mathbf{p}}. \tag{10}$$

Proof. Adding a constant C to f increases both sides of (10) by the same amount, namely, $C\int g$. As $C+f$ is positive for some C, we may thus assume that $f \geq 0$. Similarly, we may assume that $g \geq 0$. Then $f = \sum_{i=1}^{k} c_i f_i$ and $g = \sum_{j=1}^{\ell} d_j g_j$, where the c_i and d_j are positive constants and the functions f_i, g_j are characteristic functions of up-sets. Thus,

$$\int fg = \sum_{i,j} c_i d_j \int f_i g_j \geq \sum_{i,j} c_i d_j \int f_i \int g_j = \int f \int g,$$

where the inequality is Lemma 3. □

The van den Berg–Kesten inequality [1985] for monotone events is a partial converse of Harris's Lemma: it also states an intuitively obvious fact. Let $A \subset Q_{\mathbf{p}}^n$ be the increasing event of having at least one run of three heads, and let $B \subset Q_{\mathbf{p}}^n$ be the increasing event that there are at least five heads. Let $A \square B$ be the event that there is a run of three heads *and* there are at least five *other* heads. It is hard not to be convinced that $\mathbb{P}(A \square B) \leq \mathbb{P}(A)\mathbb{P}(B)$.

Let us give two formal definitions for the 'square' or 'box' operation for set systems. First, we define $\mathcal{A} \square \mathcal{B}$ for $\mathcal{A}, \mathcal{B} \subset \mathcal{P}(S)$:

$$\begin{aligned}\mathcal{A} \square \mathcal{B} \;=\; \big\{C \subset S : & \text{ there are disjoint sets } Y, Z \subset S \text{ such that} \\ & D \cap Y = C \cap Y \text{ implies } D \in \mathcal{A}, \text{ and} \\ & D \cap Z = C \cap Z \text{ implies } D \in \mathcal{B}, \text{ for any } D \subset S\big\}.\end{aligned}$$

Let us write $f_{|X}$ for the restriction of a function to a domain X. If $c = (c_i)_{i=1}^n$, $d = (d_i)_{i=1}^n$, and $I \subset [n]$, then the condition $c_{|I} = d_{|I}$ means exactly that $c_i = d_i$ for all $i \in I$. For $A, B \subset Q^n = \{0,1\}^n$, we may define $A \square B$ by

$$\begin{aligned}A \square B \;=\; \big\{c \in Q^n : & \text{ there are disjoint sets } I,\, J \subset [n] \text{ such that} \\ & d_{|I} = c_{|I} \text{ implies } d \in A, \text{ and} \\ & d_{|J} = c_{|J} \text{ implies } d \in B, \text{ for any } d \in Q^n\big\}.\end{aligned}$$

This definition is equivalent to that for set systems given above.

Note that $A \square B$ is a subset of $A \cap B$. In fact, it may be a rather small subset of $A \cap B$: for example, if neither A nor B contains a subcube of dimension at least $n/2$, then $A \square B = \emptyset$.

If $d_{|I} = c_{|I}$ implies that $d \in A$, then we call $c_{|I}$ a *witness* or a *certificate* for A, with *support* I. Thus $A \square B$ is the set of points c for which there are disjoint sets I and J such that $c_{|I}$ is a witness for A and $c_{|J}$ is a witness for B. Note that the square operation, which is obviously commutative, need not be associative, since a witness for $A \square B$ need not contain disjoint witnesses for A and for B.

For increasing events (and for decreasing events), the definition of the square operation can be simplified considerably. Indeed, if $\mathcal{A}$, $\mathcal{B} \subset \mathcal{P}(S)$ are increasing set systems, then

$$\mathcal{A} \square \mathcal{B} = \{A \cup B \colon A \cap B = \emptyset,\ A \in \mathcal{A},\ B \in \mathcal{B}\},$$

which is itself increasing. In the context of increasing set systems, we may take a *witness* for $\mathcal{A}$ to be simply a set $A \in \mathcal{A}$, rather than the function that is 1 on A. Then $\mathcal{A} \square \mathcal{B}$ is the set of sets C such that C contains disjoint witnesses for $\mathcal{A}$ and for $\mathcal{B}$. Since a witness for $\mathcal{A} \square \mathcal{B}$ is a disjoint union of witnesses for $\mathcal{A}$ and for $\mathcal{B}$, it follows that on increasing set systems, the square operation is associative.

Similarly, identifying $Q^n = \{0,1\}^n$ with the algebra $\mathbb{Z}_2^n$, so that for $a = (a_i)_{i=1}^n$ and $b = (b_i)_{i=1}^n$ we have $a + b = (a_i + b_i)_{i=1}^n$ and $ab = (a_i b_i)_{i=1}^n$, with $1+1$ defined to be 0, we have the following simple description of $A \square B$ for *increasing subsets* A, B of Q^n:

$$A \square B = \{a + b \colon\ ab = 0,\ a \in A,\ b \in B\}.$$

In other words, for increasing events A and $B \subset Q^n$, the event $A \square B$ happens if some (minimal) elements of A and B occur *disjointly*: $A \square B$ is the up-set generated by disjointly supported elements of A and B, as in the example we started out with.

Here, then, is the van den Berg–Kesten inequality. The simple proof below is due to Bollobás and Leader.

Theorem 5. *Let $A, B \subset Q_{\mathbf{p}}^n$ be increasing events. Then*

$$\mathbb{P}(A \square B) \le \mathbb{P}(A)\mathbb{P}(B). \tag{11}$$

Proof. We shall prove (11) by induction on n; the case $n = 1$ (or, indeed, the case $n = 0$, which makes perfect sense) is trivial.

As before, let $Q_{\mathbf{p}'}^{n-1}$ be the weighted $(n-1)$-dimensional cube with probability measure defined by $\mathbf{p}' = (p_1, \ldots, p_{n-1})$. Let $C = A \square B$. It is easily checked that, as A and B are up-sets,

$$C_0 = A_0 \square B_0$$

and

$$C_1 = (A_0 \square B_1) \cup (A_1 \square B_0). \tag{12}$$

As $A_0 \subset A_1$ and $B_0 \subset B_1$, it follows that

$$C_0 \subset (A_0 \,\square\, B_1) \cap (A_1 \,\square\, B_0) \tag{13}$$

and that

$$C_1 \subset A_1 \,\square\, B_1.$$

Thus, by induction,

$$\mathbb{P}(C_0) = \mathbb{P}(A_0 \,\square\, B_0) \le \mathbb{P}(A_0)\mathbb{P}(B_0),$$

and

$$\mathbb{P}(C_1) \le \mathbb{P}(A_1 \,\square\, B_1) \le \mathbb{P}(A_1)\mathbb{P}(B_1).$$

Furthermore, from (12) and (13),

$$\begin{aligned} \mathbb{P}(C_0) + \mathbb{P}(C_1) &\le \mathbb{P}\big((A_0 \,\square\, B_1) \cap (A_1 \,\square\, B_0)\big) \\ &\qquad\qquad + \mathbb{P}\big((A_0 \,\square\, B_1) \cup (A_1 \,\square\, B_0)\big) \\ &= \mathbb{P}(A_0 \,\square\, B_1) + \mathbb{P}(A_1 \,\square\, B_0) \\ &\le \mathbb{P}(A_0)\mathbb{P}(B_1) + \mathbb{P}(A_1)\mathbb{P}(B_0). \end{aligned}$$

Multiplying the last three inequalities by $(1-p_n)^2$, p_n^2 and $p_n(1-p_n)$, respectively, and summing, we find that

$$\begin{aligned} &(1-p_n)\mathbb{P}(C_0) + p_n\mathbb{P}(C_1) \\ &\qquad \le \big\{(1-p_n)\mathbb{P}(A_0) + p_n\mathbb{P}(A_1)\big\}\big\{(1-p_n)\mathbb{P}(B_0) + p_n\mathbb{P}(B_1)\big\}. \end{aligned}$$

Using (3) three times, this is just $\mathbb{P}(C) \le \mathbb{P}(A)\mathbb{P}(B)$, as required. □

Van den Berg and Kesten [1985] conjectured that their inequality holds for all events, not only monotone events. Van den Berg and Fiebig [1987] proved that the inequality holds for *convex* events, that is, intersections of increasing and decreasing events. The full conjecture resisted all attempts until Reimer [2000] proved it.

Theorem 6. *Let $A, B \subset Q_{\mathbf{p}}^n$. Then $\mathbb{P}(A \,\square\, B) \le \mathbb{P}(A)\mathbb{P}(B)$.* □

The proof is *much* harder than that of the van den Berg–Kesten inequality.

Harris's Lemma has many extensions, including some correlation inequalities of Griffiths [1967a; 1967b] concerning Ising ferromagnets, and extensions by Kelly and Sherman [1968]. The FKG inequality of Fortuin, Kasteleyn and Ginibre [1971] extends these to a correlation inequality on partially ordered sets, with applications to statistical mechanics;

the FKG inequality is the most often quoted correlation inequality in physics: even Harris's Lemma tends to be called the 'FKG inequality'.

Ahlswede and Daykin [1978] extended the FKG inequality to a very general correlation inequality on lattices. What is amazing is that such an inequality could be true: its proof, although far from trivial, is not very difficult.

Given points $a = (a_i)_{i=1}^n$ and $b = (b_i)_{i=1}^n$ of the cube $Q^n = \{0,1\}^n$, their *join* is $a \vee b = (c_i)_{i=1}^n$, where $c_i = a_i \vee b_i = \max\{a_i, b_i\}$, and their *meet* is $a \wedge b = (d_i)_{i=1}^n$, where $d_i = a_i \wedge b_i = \min\{a_i, b_i\}$. (Thus, identifying Q^n with the power set $\mathcal{P}([n])$, so that a and b become subsets of $[n]$, we have $a \vee b = a \cup b$ and $a \wedge b = a \cap b$.) For $A, B \subset Q^n$, define the *join* of A and B as

$$A \vee B = \{a \vee b\colon\ a \in A,\ b \in B\}$$

and their *meet* as

$$A \wedge B = \{a \wedge b\colon\ a \in A,\ b \in B\}.$$

As usual, we shall identify a function $\varphi\colon Q^n = \{0,1\}^n \to \mathbb{R}$ with the signed measure it defines on Q^n, so that if $E \subset Q^n$ then

$$\varphi(E) = \sum_{e \in E} \varphi(e).$$

Here, then, is the Four Functions Theorem of Ahlswede and Daykin, stating that a certain trivial necessary condition for an inequality is also sufficient.

Theorem 7. *Let $\alpha, \beta, \gamma, \delta\colon Q^n \to \mathbb{R}^+ = [0,\infty)$ be such that*

$$\alpha(a)\beta(b) \le \gamma(a \vee b)\delta(a \wedge b)$$

for all $a, b \in Q^n$. Then

$$\alpha(A)\beta(B) \le \gamma(A \vee B)\delta(A \wedge B)$$

for all subsets $A, B \subset Q^n$. □

Choosing appropriate functions α, β, γ and δ, Theorem 7 implies a host of inequalities. In most of these results, α, β, γ and δ are chosen to be the *same* function (measure) $\mu\colon Q^n \to \mathbb{R}^+$. We may choose *any* function μ, provided it is *log-supermodular*, i.e. satisfies

$$\mu(a)\mu(b) \le \mu(a \vee b)\mu(a \wedge b)$$

for all $a, b \in Q^n$. (Occasionally, such a function μ is said to be *log-monotone.*) By the Ahlswede–Daykin Four Functions Theorem, if μ is log-supermodular, then

$$\mu(A)\mu(B) \leq \mu(A \vee B)\mu(A \wedge B) \tag{14}$$

for all A, $B \subset Q^n$. The very special case where μ is the normalized counting measure implies Harris's Lemma, since, when A and B are increasing, $A \vee B$ is just $A \cap B$. The simple proof of Lemma 4 shows that (14) implies the FKG inequality, which may be stated as follows.

Theorem 8. *Let $\mu : Q^n \to \mathbb{R}^+$ be a log-supermodular probability measure, and let f, $g : Q^n \to \mathbb{R}^+$ be increasing functions. Then*

$$\int fg \, \mathrm{d}\mu \geq \int f \, \mathrm{d}\mu \int g \, \mathrm{d}\mu.$$

□

Our final aim in this chapter is to present some fundamental results concerning thresholds for events, stating that under rather general conditions, the probability $\mathbb{P}_p(A)$ of an increasing event $A \subset Q_p^n$ undergoes a sharp transition as p passes through a critical value. These sharp-threshold results will be of fundamental importance in several of the major results on percolation.

Let A be an event in the weighted cube $Q_{\mathbf{p}}^n$. Given a point $\omega \in Q^n = \{0,1\}^n$, the ith variable ω_i is *pivotal* for A if precisely one of $\omega = (\omega_1, \ldots, \omega_n)$ and $r_i(\omega) = (\omega_1, \ldots, \omega_{i-1}, 1-\omega_i, \omega_{i+1}, \ldots, \omega_n)$ is in A. Note that whether the ith coordinate is pivotal depends on the point ω and on the event A. The *influence* of the ith variable on A is

$$\beta_i(A) = \beta_{\mathbf{p},i}(A) = \mathbb{P}_{\mathbf{p}}\big(\{\omega : \omega_i \text{ is pivotal for } A\}\big),$$

or, in the usual notation of probability theory, simply

$$\beta_i(A) = \mathbb{P}_{\mathbf{p}}\big(\omega_i \text{ is pivotal for } A\big).$$

The fundamental lemma of Margulis [1974] connects the derivatives of $\mathbb{P}_{\mathbf{p}}(A)$ with the influences of the variables; this lemma was rediscovered some years later by Russo [1981].

Lemma 9. *Let A be an increasing event in the weighted cube $Q_{\mathbf{p}}^n$, where $\mathbf{p} = (p_1, \ldots, p_n)$. Then*

$$\frac{\partial}{\partial p_i}\mathbb{P}_{\mathbf{p}}(A) = \beta_i(A).$$

In particular,

$$\frac{\mathrm{d}}{\mathrm{d}p}\mathbb{P}_p(A) = \sum_{i=1}^{n} \beta_i(A).$$

Proof. It suffices to prove the first statement with $i = n$. Given a point $\mathbf{x} = (x_1, \ldots, x_k) \in Q^k$, where $k \leq n$, write

$$\mathbf{p}^{\mathbf{x}} = \prod_{i:\, x_i=1} p_i \prod_{i:\, x_i=0} (1 - p_i)$$

so that if $k = n$ then $\mathbb{P}_{\mathbf{p}}(\{\mathbf{x}\}) = \mathbf{p}^{\mathbf{x}}$. Also, for $\mathbf{x} \in Q^{n-1}$, set $\mathbf{x}_+ = (x_1, \ldots, x_{n-1}, 1)$ and $\mathbf{x}_- = (x_1, \ldots, x_{n-1}, 0)$, so that $\mathbf{x}_+, \mathbf{x}_- \in Q^n$. Note that $\mathbf{p}^{\mathbf{x}_+} + \mathbf{p}^{\mathbf{x}_-} = \mathbf{p}^{\mathbf{x}}$.

For an up-set $A \subset Q^n$, let

$$A_a = \{\mathbf{x} \in Q^{n-1} : \ \ \mathbf{x}_+ \in A,\ \mathbf{x}_- \in A\}$$

and

$$A_b = \{\mathbf{x} \in Q^{n-1} : \ \ \mathbf{x}_+ \in A,\ \mathbf{x}_- \notin A\},$$

and note that

$$\begin{aligned} \mathbb{P}_{\mathbf{p}}(A) &= \sum_{\mathbf{x}\in A_a} (\mathbf{p}^{\mathbf{x}_+} + \mathbf{p}^{\mathbf{x}_-}) + \sum_{\mathbf{x}\in A_b} \mathbf{p}^{\mathbf{x}_+} \\ &= \sum_{\mathbf{x}\in A_a} \mathbf{p}^{\mathbf{x}} + p_n \sum_{\mathbf{x}\in A_b} \mathbf{p}^{\mathbf{x}}. \end{aligned} \tag{15}$$

Hence

$$\frac{\partial}{\partial p_n}\mathbb{P}_{\mathbf{p}}(A) = \sum_{\mathbf{x}\in A_b} \mathbf{p}^{\mathbf{x}}.$$

At the point $\mathbf{x}' = (\mathbf{x}, x_n)$, the nth coordinate is pivotal if and only if $\mathbf{x} \in A_b$, so this last expression is exactly $\beta_n(A)$. □

To prove sharp-threshold results, we wish to find lower bounds on $\sum_i \beta_i(A)$. In the unweighted cube Q_n, $\sum_{i=1}^n \beta_i(A)2^{n-1}$ is precisely the edge-boundary of A. Thus the edge-isoperimetric inequality in Q_n tells us that if $\mathbb{P}_{1/2}(A) = t$, then

$$\sum_{i=1}^{n} \beta_i(A) \geq t2^n(n - \log_2(t2^n))/2^{n-1} = 2t\log_2(1/t),$$

so

$$\max_i \beta_i(A) \geq 2t\log_2(1/t)/n.$$

Ben-Or and Linial [1985; 1990] conjectured that this last inequality can be improved substantially: up to a constant factor, $\log_2(1/t)$ can be replaced by $\log n$. This conjecture was proved by Kahn, Kalai and Linial [1988] with the aid of the Bonami–Beckner inequality from harmonic analysis. For a combinatorial proof, see Falik and Samorodnitsky [2005].

Theorem 10. *Let A be a subset of the n-dimensional discrete cube $Q_n = \{0,1\}^n$ with probability $t = |A|/2^n$. Then*

$$\sum_{i=1}^{n} \beta_i(A)^2 \geq ct^2(1-t)^2 \frac{(\log n)^2}{n}, \tag{16}$$

where $c > 0$ is an absolute constant. □

The bound above is often written with $\min\{t^2, (1-t)^2\}$ instead of $t^2(1-t)^2$; apart from a change in the constant, this makes no difference. A similar comment applies to the bounds below.

Relation (16) immediately implies that

$$\max_i \beta_i(A) \geq \sqrt{c}t(1-t)\,(\log n)/n.$$

Ben-Or and Linial gave an example showing that this is best possible up to an absolute constant.

In order to apply Lemma 9, we need bounds on influences in weighted cubes. Such an extension of Theorem 10 was first proved by Bourgain, Kahn, Kalai, Katznelson and Linial [1992]; a simpler proof has since been given by Friedgut [2004].

Theorem 11. *Let A be a subset of the weighted cube $Q_{\mathbf{p}}^n$ with probability $\mathbb{P}_{\mathbf{p}}(A) = t$. Then*

$$\max_i \beta_i(A) \geq ct(1-t)\frac{\log n}{n},$$

where $c > 0$ is an absolute constant. □

Friedgut and Kalai [1996] noticed that a slight variant of the proof of Theorem 11 gives a stronger result (apart from the constant): if the maximal influence is not much larger than the bound in Theorem 11 then there are many variables of comparably large influence, so the sum of the influences is large.

Theorem 12. *Let A be a subset of the weighted cube $Q^n_{\mathbf{p}}$ with $\mathbb{P}_{\mathbf{p}}(A) = t$. If $\beta_i(A) \leq \delta$ for every i, then*

$$\sum_{i=1}^{n} \beta_i(A) \geq ct(1-t)\log(1/\delta),$$

where $c > 0$ is an absolute constant. □

There is a simple condition under which one large influence guarantees that the sum of the influences is large. A set $A \subset Q^n$ is *symmetric* if it is invariant under the action of some group acting transitively on $[n]$. Thus, $A \subset Q^n$ is symmetric if, for all $1 \leq j, k \leq n$, there is a permutation π of $[n]$ such that $\pi(j) = k$ and if $\mathbf{x} = (x_i)_1^n$ is in A then so is $\pi(\mathbf{x}) = (x_{\pi(i)})_1^n$. All we shall need about a symmetric event is that every variable has the same influence on it.

Using this observation and Lemma 9, Friedgut and Kalai [1996] deduced from Theorem 11 that every symmetric increasing property in Q^n_p has a *sharp threshold*: an $O(1/\log n)$ increase in the probability p suffices to increase the probability of the property from close to 0 to close to 1.

Theorem 13. *There is an absolute constant c_1 such that if $A \subset Q^n$ is symmetric and increasing, $0 < \varepsilon < 1/2$, and $\mathbb{P}_p(A) > \varepsilon$, then $\mathbb{P}_q(A) > 1 - \varepsilon$ whenever*

$$q - p \geq c_1 \frac{\log(1/(2\varepsilon))}{\log n}.$$ □

Friedgut and Kalai [1996] showed that, for events whose threshold occurs at very small values of p, this result can be strengthened considerably.

Theorem 14. *There is an absolute constant c_2 such that if $A \subset Q^n$ is symmetric and increasing, $0 < \varepsilon < 1/2$, and $\mathbb{P}_p(A) > \varepsilon$, then $\mathbb{P}_q(A) > 1 - \varepsilon$ whenever*

$$q - p \geq c_2 p \log(1/p) \frac{\log(1/(2\varepsilon))}{\log n}.$$ □

It is not hard to adapt the proof of this result to powers of probability spaces with 3, 4, 5... elements, where all elements but one have small probability.

3

Bond percolation on $\mathbb{Z}^2$ – the Harris–Kesten Theorem

From the publication of the first papers on percolation theory in the late 1950s, for over two decades one of the main challenges of the theory was the rigorous determination of $p_{\mathrm{H}} = p_{\mathrm{H}}^{\mathrm{b}}(\mathbb{Z}^2)$, the critical probability for bond percolation on the square lattice. Hammersley's Monte Carlo experiments suggested that the value of p_{H} might be $1/2$; further evidence for this was given by Domb [1959], Elliott, Heap, Morgan and Rushbrooke [1960], and Domb and Sykes [1961].

The first major result on this topic was due to Harris [1960], who proved that $p_{\mathrm{H}} \geq 1/2$. In the light of this result and the Monte Carlo evidence, Hammersley conjectured that the critical probability is indeed $1/2$. Sykes and Essam [1964] gave a non-rigorous justification of $p_{\mathrm{T}}^{\mathrm{b}}(\mathbb{Z}^2) = 1/2$, further supporting this conjecture.

The next important step towards proving the conjecture was taken by Russo [1978] and by Seymour and Welsh [1978] who, independently, proved that $p_{\mathrm{T}}^{\mathrm{b}}(\mathbb{Z}^2) + p_{\mathrm{H}}^{\mathrm{b}}(\mathbb{Z}^2) = 1$. Kesten [1980] finally settled the conjecture: building on the Russo–Seymour–Welsh result, he gave an ingenious and intricate proof that $p_{\mathrm{H}} = 1/2$.

By now, there are many proofs of this famous Harris–Kesten Theorem; in fact, there are a number of global strategies based on various different ingredients, and frequently even the same ingredients have several different proofs. Here we shall give several variants of what is essentially one proof, using some of the basic probabilistic results presented in Chapter 2. What follows will be heavily based on the presentation in Bollobás and Riordan [2006c]. Later, in Chapter 5, we shall indicate another proof, based on Menshikov's exponential decay theorem and the uniqueness theorem of Aizenman, Kesten and Newman.

Intuitively, the main 'reason why' $p_{\mathrm{H}} = 1/2$ is the fact that, for $p = 1/2$, the probability that there is an open path crossing an n by $n-1$

rectangle the 'long way' is $1/2$. As we shall now see, this is an immediate consequence of self-duality. However, having proved this simple fact, we are still rather far from proving that $p_H = 1/2$: in some sense, the real problem starts only then.

Recall that the dual of $\Lambda = \mathbb{Z}^2$ is the lattice $\Lambda^\star$ with vertex set $\{(a + 1/2, b + 1/2) : (a, b) \in \mathbb{Z}^2\}$, in which sites at distance 1 are adjacent. Thus there is one dual bond $e^\star$ for each bond e of $\mathbb{Z}^2$; this bond is the bond of $\Lambda^\star$ that crosses e.

A *rectangle* R in $\mathbb{Z}^2$ is a subgraph induced by a set of sites of the form $[a, b] \times [c, d]$, where $a \leq b$ and $c \leq d$ are integers. We shall use the same notation for the vertex set $[a, b] \times [c, d]$ and for the rectangle it induces. If $k = b - a + 1$ and $\ell = d - c + 1$ then we call R a *k by ℓ rectangle*; note that such a rectangle has $k\ell$ sites and $2k\ell - k - \ell$ bonds. A rectangle in $\Lambda^\star$ is defined as in $\Lambda = \mathbb{Z}^2$; equivalently, $R^\star$ is a rectangle in $\Lambda^\star$ if $R^\star + (1/2, 1/2)$ is a rectangle in Λ.

Although we have defined rectangles as subgraphs of $\mathbb{Z}^2$ and its dual, it is natural to define an 'abstract' k by ℓ rectangle as a 'grid graph' with $k\ell$ vertices and $2k\ell - k - \ell$ edges. Clearly, a subgraph of $\mathbb{Z}^2$ isomorphic to a k by ℓ abstract rectangle with k, $\ell \geq 2$ is a k by ℓ rectangle as defined above.

Turning to bond percolation on $\Lambda = \mathbb{Z}^2$, let us consider a configuration $\omega \in \{0, 1\}^{E(\Lambda)}$ on Λ. For the moment, the measure on the space of configurations will be irrelevant, but we shall still call a set of configurations an event. As in Chapter 1, we define a bond $e^\star$ in $\Lambda^\star$ to be open if e is closed and vice versa; thus the configuration ω specifies the states of the bonds in Λ *and* in $\Lambda^\star$.

The *horizontal dual* of a rectangle $R = [a, b] \times [c, d]$ in Λ or $\Lambda^\star$ is the rectangle $R^h = [a + 1/2, b - 1/2] \times [c - 1/2, d + 1/2]$ in the dual lattice. Analogously, the *vertical dual* of R is the rectangle $R^v = [a - 1/2, b + 1/2] \times [c + 1/2, d - 1/2]$; see Figure 1. Somewhat artificially, the horizontal dual of a 1 by ℓ rectangle is the 'empty rectangle', as is the vertical dual of a k by 1 rectangle. For k, $\ell \geq 2$, the horizontal dual of a k by ℓ rectangle is a $k - 1$ by $\ell + 1$ rectangle, and the vertical dual is a $k + 1$ by $\ell - 1$ rectangle. Also, $(R^h)^v = (R^v)^h = R$.

Given a configuration ω, an *open horizontal crossing* of a rectangle $R = [a, b] \times [c, d]$ in Λ or $\Lambda^\star$ is an open path $P \subset R$ joining a site (a, y) to a site (b, z), i.e., a path in the appropriate lattice, all of whose edges are open, joining the left-hand side of R to the right-hand side. We write $H(R)$ for the event that R has such a crossing. Similarly, we write $V(R)$ for the event that R has an *open vertical crossing*, defined analogously.

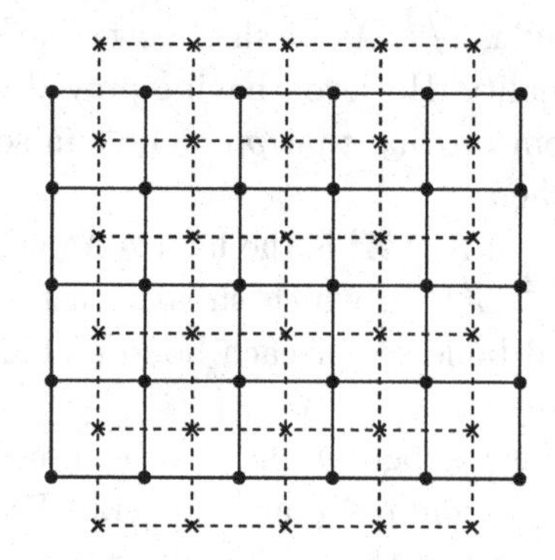

Figure 1. A rectangle R (solid lines) and its horizontal dual R^h (dashed lines). R^h is the vertical dual of R.

As a minimal open horizontal crossing of a rectangle R does not contain any bond joining two vertices on the same vertical side (left or right) of R, the event $H(R)$ depends only on the states of those bonds in R whose duals appear in R^h. Similarly $V(R)$ depends only on the states of bonds with duals in R^v.

The next lemma is the promised 'reason why' $p_{\mathrm{H}}(\mathbb{Z}^2) = 1/2$. The result is 'obvious', and it is tempting to state it without a proof. However, it is not entirely trivial to prove, and it usually takes quite a while to dot the i's; for a proof of a closely related result see Kesten [1982, pp. 386–392]. Here, we shall give a simple proof needing no topology. Logically, the proof is equivalent to that given in Bollobás and Riordan [2006c], but the presentation is more like that of a related result in Bollobás and Riordan [2006b].

Lemma 1. *Let R be a rectangle in $\mathbb{Z}^2$ or its dual. Whatever the states of the bonds in R, exactly one of the events $H(R)$ and $V(R^h)$ holds.*

Proof. Consider the partial tiling of the plane by octagons and squares shown in Figure 2. (This is, in fact, part of the Archimedean lattice $(4,8^2)$; see Figure 18 of Chapter 5.) We take a black octagon for each site of R, and a white one for each site of R^h. The bonds of R and of R^h are represented by squares, with the same square representing a bond e and its dual $e^\star$. A square representing a bond e of R and its dual $e^\star$ in R^h is coloured black if e is open, so $e^\star$ is closed, and white if e is closed and $e^\star$ is open. In the first case, the black square joins

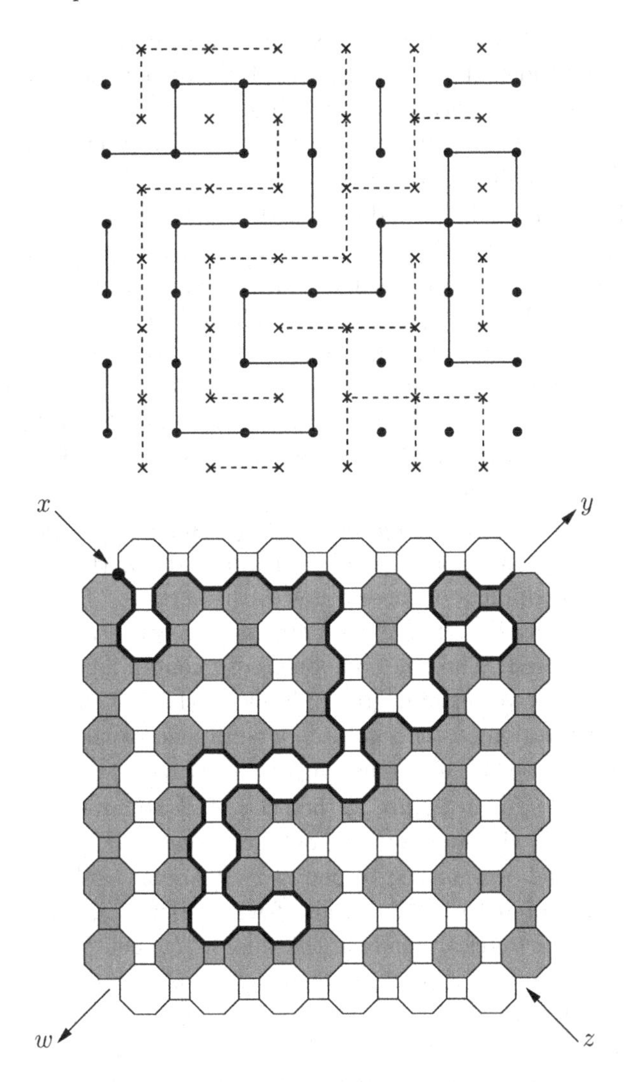

Figure 2. The upper figure shows the open bonds of a rectangle R (solid) and its horizontal dual R^h (dashed). In the lower figure each site of R is drawn as a black (shaded) octagon, and each site of R^h as a white octagon. The central squares correspond to bonds $e \in R$ with duals $e^* \in R^h$; such a square is black if e is open and white if e^* is open. There are additional black/white squares around the edges to connect the sides of the rectangles.

the two black octagons corresponding to the sites e joins. In the second, this white square joins the two white octagons corresponding to sites of the dual lattice that $e^\star$ joins. The squares corresponding to the bonds in the vertical sides of R are coloured black, to 'join up' each side of R; the states of the corresponding bonds are irrelevant to the event $H(R)$. Similarly, the squares corresponding to the (dual) bonds in the horizontal sides of R^h are coloured white.

Note that $H(R)$ holds if and only if there is a black path of squares and octagons from the left of the figure to the right, and $V(R^h)$ if and only if there is a white path of squares and octagons from top to bottom. In particular, $H(R)$ and $V(R^h)$ cannot both hold: otherwise, these black and white paths contain disjoint (piecewise linear) curves in the plane lying in the interior of a cycle C (the boundary of the partial tiling), and joining two pairs of points ac and bd, with a, b, c and d appearing in this cyclic order around C. As noted in Chapter 1 (see Figure 8), K_5 could then be drawn in the plane.

Let I be the *interface graph*, formed by taking those edges of octagons/squares that separate a black region from a (bounded) white one, with the endpoints of these edges as the vertices. Then every vertex of I has degree exactly 2 except for the four vertices x, y, z and w, which have degree 1. Thus the component of I containing x is a path W, ending at another vertex of degree 1. Walking along W from x, there is always a black region on the right and a white one on the left. Thus W cannot end at z, so W ends either at y or at w.

If W ends at y, as in Figure 2, then the black squares and octagons on the right of W give an open horizontal crossing of R. More precisely, these squares and octagons correspond to a connected subgraph S of R joining the left of R to the right, all of whose bonds are open, except possibly for some vertical bonds in the sides of R. Let P be a minimal connected subgraph of S connecting the left of R to the right. Then P is a path, and P uses no vertical bonds in the sides of R, so P is an open horizontal crossing of R, and $H(R)$ holds. Similarly, if W ends at w, then the white squares and octagons on the left of W give an open vertical crossing of R^h in the dual lattice. Thus at least one of $H(R)$ and $V(R^h)$ holds. $\square$

The path W in the interface graph I described above may be found step by step: we enter the tiling at x, and at each vertex we 'test' the two edges leaving this vertex and follow one of them. If $H(R)$ holds, so that W leaves the tiling at y, then the path P we find is the *the top-most open horizontal crossing* of R. This has the very useful property that it can be found without examining the states of bonds below P.

Algorithms such as this, that test for crossings by examining 'interfaces', are sometimes of practical significance, as they can be much faster than exploring, say, the set of all vertices of R reachable from the left, and testing whether this set contains a vertex on the right.

Lemma 1 may be worded as a statement about a plane graph and its dual: contract each vertical side of R to a single vertex, and add a single edge f joining the two resulting vertices to form a graph G. Also, contract each horizontal side of R^h to a single vertex, and add an edge $f^\star$ joining these vertices, obtaining a graph $G^\star$. Then G and $G^\star$ are planar duals; see Figure 3.

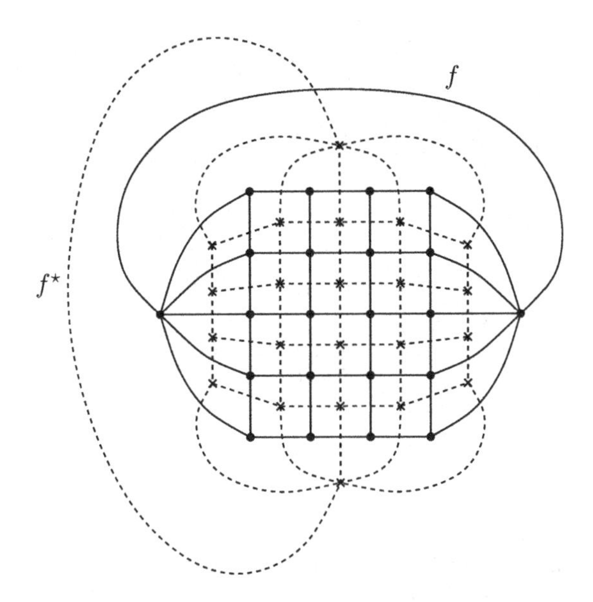

Figure 3. A rectangle R with its left and right sides pinched to single vertices (solid lines), and the rectangle R^h with its top and bottom sides pinched (dashed lines). With additional edges f, $f^\star$ as shown, the solid and dashed plane graphs are dual to each other.

Figure 3 shows that Lemma 1 is a special case of the following result for a general plane graph and its dual.

Lemma 2. *Let G be a graph drawn in the plane, and let $G^\star$ be its dual, with edge set $\{e^\star : e \in E(G)\}$. Let f be any edge of G, and suppose that each edge $e \neq f$ of G is assigned a state, open or closed. Taking an edge $e^\star$ to be open if e is closed and vice versa, either there is a path in $G - f$ consisting of open edges and joining the endpoints of f, or there is a path* *in $G^\star - f^\star$ consisting of open (dual) edges and joining the endpoints of $f^\star$. Paths of both types cannot exist simultaneously.*

Proof. The proof is the same as that of Lemma 1: replace each vertex of G with degree d by a black (topological) $2d$-gon, and each degree d vertex of $G^\star$ by a white $2d$-gon. Replace each pair $\{e, e^\star\}$ of dual edges with a 4-gon, sharing edges with the polygons corresponding to the endpoints of e and $e^\star$, respecting the cyclic order of the edges and faces around each vertex of G and $G^\star$, as in Figure 4. Colour all 4-gons except that corresponding to $\{f, f^\star\}$ black or white in any way. Then in the interface graph formed by the sides of polygons separating a black region from a white one, every vertex has degree 2 except the four vertices of the 4-gon corresponding to $\{f, f^\star\}$; the rest of the proof is as before. □

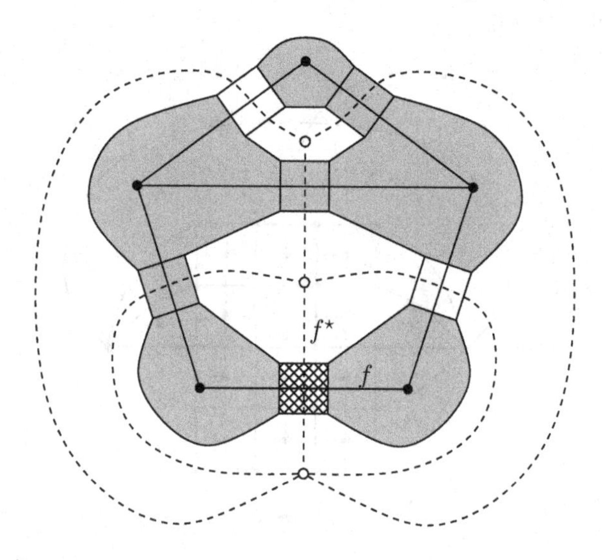

Figure 4. A plane graph G (solid circles and lines) drawn with its dual $G^\star$ (hollow circles and dashed lines). There is a black $2d$-gon for each degree d vertex of G, and a white $2d$-gon for each degree d vertex of $G^\star$. There is a 4-gon for each edge/dual-edge pair $\{e, e^\star\}$; this is black or white according to whether e is open or closed, apart from the hatched 4-gon, corresponding to the distinguished edge f and its dual $f^\star$.

Turning to the probability measure $\mathbb{P}_p = \mathbb{P}^{\mathrm{b}}_{\mathbb{Z}^2,p}$ in which each bond of $\mathbb{Z}^2$ is open with probability p independently of the other bonds, Lemma 1 has the following immediate consequence.

Corollary 3. *(i) If R and R' are k by $\ell - 1$ and $k - 1$ by ℓ rectangles in $\mathbb{Z}^2$, respectively, then*

$$\mathbb{P}_p(H(R)) + \mathbb{P}_{1-p}(V(R')) = 1.$$

(ii) If R is an $n+1$ by n rectangle, then

$$\mathbb{P}_{1/2}(H(R)) = 1/2. \tag{1}$$

(iii) If S is an n by n square, then

$$\mathbb{P}_{1/2}(V(S)) = \mathbb{P}_{1/2}(H(S)) \geq 1/2.$$

Proof. For part (i), recall that each bond of $\Lambda = \mathbb{Z}^2$ is open independently with probability p, and the dual $e^\star$ of e is open if and only if e is closed. By Lemma 1, every configuration ω lies in exactly one of $H(R)$ and $V(R^h)$, so $\mathbb{P}_p(H(R)) + \mathbb{P}_p(V(R^h)) = 1$. But R^h is a $k-1$ by ℓ rectangle in $\Lambda^\star$, where bonds are open independently with probability $1-p$, so $\mathbb{P}_p(V(R^h)) = \mathbb{P}_{1-p}(V(R'))$.

Parts (ii) and (iii) follow immediately from part (i). □

It is easy to deceive oneself into thinking that (1) shows that $p_{\mathrm{H}} = 1/2$. Although self-duality is of course the reason 'why' $p_{\mathrm{H}} = 1/2$, a rigorous deduction is far from easy, and took twenty years to accomplish.

3.1 The Russo–Seymour–Welsh method

The next ingredient we shall need in our proof of the Harris–Kesten Theorem is some form of the Russo–Seymour–Welsh Theorem relating crossings of squares to crossings of rectangles. The proof we present is from Bollobás and Riordan [2006c].

Lemma 4. *Let $R = [m] \times [2n]$, $m \geq n$, be an m by $2n$ rectangle. Let $X(R)$ be the event that there are paths P_1 and P_2 of open bonds, such that P_1 crosses the n by n square $S = [n] \times [n]$ from top to bottom, and P_2 lies within R and joins some site on P_1 to some site on the right-hand side of R. Then $\mathbb{P}_p(X(R)) \geq \mathbb{P}_p(H(R))\mathbb{P}_p(V(S))/2$.*

Proof. Suppose that $V(S)$ holds, so there is a path P_0 of open bonds crossing S from top to bottom. Note that any such P_0 separates S into two pieces, one to the left of P_0 and one to the right. Let $LV(S)$ be the left-most open vertical crossing, when one exists, defined analogously to the top-most open horizontal crossing discussed in the remark after

Lemma 1. By that remark, for any possible value P_1 of $LV(S)$, the event $\{LV(S) = P_1\}$ does not depend on the states of bonds of S to the right of P_1.

We claim that, for any possible value P_1 of $LV(S)$, we have

$$\mathbb{P}_p(X(R) \mid LV(S) = P_1) \geq \mathbb{P}_p(H(R))/2.$$

To see this, let P be the (not necessarily open) path formed by the union of P_1 and its reflection P_1' in the horizontal symmetry axis of R, with one additional bond joining P_1 to P_1'; see Figure 5. This path

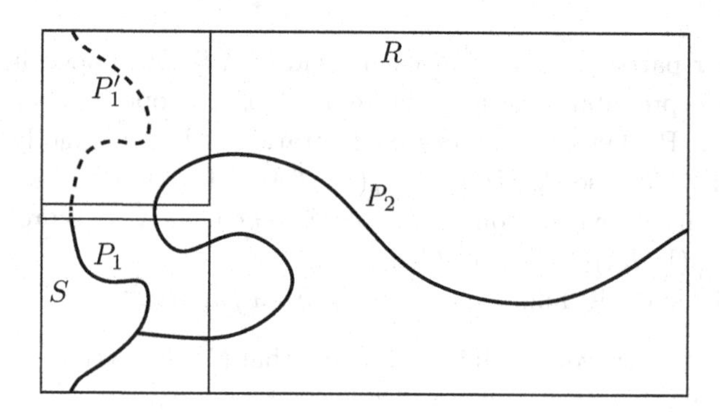

Figure 5. A rectangle R and square S inside it, drawn with paths (solid curves) whose presence as open paths would imply $X(R)$. The path P formed by P_1, its reflection P_1', and the single bond joining them crosses R from top to bottom.

crosses $[n] \times [2n]$ from top to bottom. With (unconditional) probability $\mathbb{P}_p(H(R))$ there is a path P_3 of open bonds crossing R from right to left – this path must meet P. By symmetry, the (unconditional) probability that some such path first meets P at a site of P_1 is at least $\mathbb{P}_p(H(R))/2$. Hence the event $Y(P_1)$ – that there is an open path P_2 in R to the right of P joining some site on P_1 to the right-hand side of R – has probability at least $\mathbb{P}_p(H(R))/2$. But $Y(P_1)$ depends only on the states of bonds to the right of P. All such bonds in S are to the right of P_1 in S. As the states of these bonds are independent of $\{LV(S) = P_1\}$, we have

$$\mathbb{P}_p(Y(P_1) \mid LV(S) = P_1) = \mathbb{P}_p(Y(P_1)) \geq \mathbb{P}_p(H(R))/2.$$

If $Y(P_1)$ holds and $LV(S) = P_1$, then $X(R)$ holds. Thus

$$\mathbb{P}_p(X(R) \mid LV(S) = P_1) \geq \mathbb{P}_p(H(R))/2.$$

As the event $V(S)$ is a disjoint union of events of the form $\{LV(S) = P_1\}$, we thus have $\mathbb{P}_p(X(R) \mid V(S)) \geq \mathbb{P}_p(H(R))/2$, and the result follows. □

Lemma 4 allows us to bound from below the crossing probability of some non-square rectangle; in particular, of a $3n$ by $2n$ rectangle. Let $h_p(m, n) = \mathbb{P}_p(H(R))$, where R is any m by n rectangle in $\mathbb{Z}^2$, and let $h(m, n) = h_{1/2}(m, n)$.

Corollary 5. *For all $n \geq 1$ we have $h(3n, 2n) \geq 2^{-7}$.*

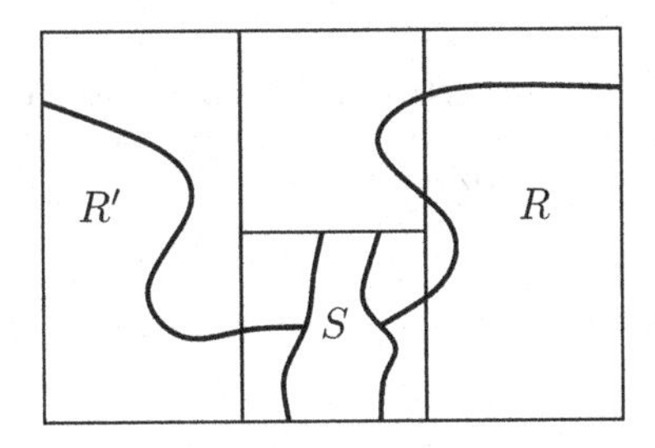

Figure 6. Two $2n$ by $2n$ (square) rectangles R, R', and an n by n square S in their intersection. The solid lines indicate paths witnessing the event $X(R)$, and the corresponding reflected event for R'.

Proof. Consider two $2n$ by $2n$ (square) rectangles R, R' arranged as in Figure 6, and the n by n square S in their intersection. Let $X'(R')$ be the event defined analogously to $X(R)$ but reflected horizontally. Applying Lemma 4 to the rectangle R (which happens to be square),

$$\mathbb{P}_p(X'(R')) = \mathbb{P}_p(X(R)) \geq \mathbb{P}_p(H(R))\mathbb{P}_p(V(S))/2.$$

The events $X(R)$, $X'(R')$ and $H(S)$ are increasing, and hence positively correlated by Harris's Lemma (Lemma 3 of Chapter 2). If all three events hold, so does $H(R \cup R')$. Thus,

$$\begin{aligned} h(3n, 2n) &= \mathbb{P}_{1/2}(H(R \cup R')) \\ &\geq \mathbb{P}_{1/2}(X'(R'))\mathbb{P}_{1/2}(X(R))\mathbb{P}_{1/2}(H(S)) \\ &\geq \mathbb{P}_{1/2}(H(R))^2\mathbb{P}_{1/2}(V(S))^2\mathbb{P}_{1/2}(H(S))/4. \end{aligned}$$

But R and S are squares, so, by Corollary 3,

$$h(3n, 2n) \geq (1/2)^2(1/2)^2(1/2)/4 = 2^{-7}.$$

□

It turns out that the difficult step is getting from squares to elongated rectangles: from Corollary 5 it is very easy to deduce Harris's Theorem. Indeed, considering m_1 by $2n$ and m_2 by $2n$ rectangles intersecting in a $2n$ by $2n$ square as in Figure 7, by Harris's Lemma we have

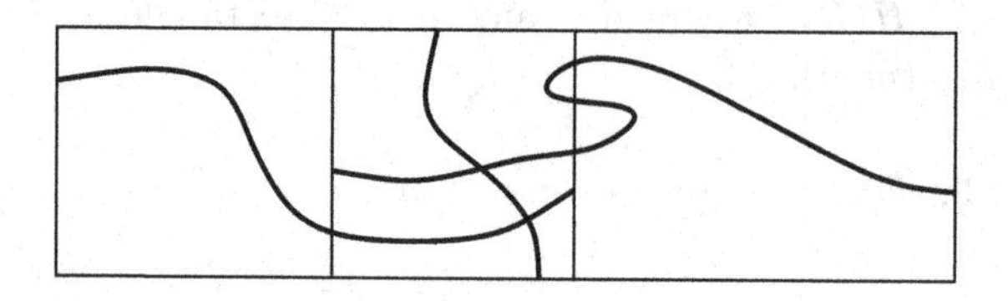

Figure 7. Two rectangles intersecting in a square. If the rectangles have open horizontal crossings, and the square an open vertical one, then the union of the rectangles has an open horizontal crossing.

$$h(m_1 + m_2 - 2n, 2n) \geq h(m_1, 2n)h(m_2, 2n)/2 \tag{2}$$

for all $m_1, m_2 \geq 2n$. In particular,

$$h(m+n, 2n) \geq h(m, 2n)h(3n, 2n)/2 \geq 2^{-8}h(m, 2n),$$

so $h(kn, 2n) \geq 2^{17-8k}$ for all $k \geq 3$ and $n \geq 1$. As $h(m, 2n+1) \geq h(m, 2n)$, it follows that there are constants $h_k > 0$ such that

$$h(kn, n) \geq h_k \tag{3}$$

for all $k \geq 2$ and $n \geq 1$.

Alternatively, starting with an m_1 by $2n$ rectangle R and an m_2 by $2n$ rectangle R', the proof of Lemma 4 actually shows that

$$h(m_1 + m_2 - n, 2n) \geq h(m_1, 2n)h(m_2, 2n)/2^5$$

for $m_1, m_2 \geq 2n$. Thus

$$h(5n, 2n) \geq h(3n, 2n)^2/2^5 \geq 2^{-19},$$

and

$$h(6n, 2n) \geq h(5n, 2n)h(2n, 2n)/2^5 \geq 2^{-19-1-5} = 2^{-25}. \tag{4}$$

As $2^{-7} \leq h(3n, 2n) \leq 1/2$ for every n, it is natural to expect that $h(3n, 2n)$ converges to a limit as $n \to \infty$. More generally, one would expect $\lim_{n\to\infty} h(an, bn) = f(a/b)$ for some function $f(x)$ with $0 < f(x) < 1$. Indeed, it would be astonishing if this were not the case. Surprisingly, this conjecture is still open. In fact, this is a very special case of the general conformal invariance conjecture of Aizenman and Langlands, Pouliot and Saint-Aubin [1994]. This conjecture was proved by Smirnov [2001a] for site percolation in the triangular lattice; see Chapter 7.

3.2 Harris's Theorem

From crossings of long, thin rectangles to Harris's Theorem is a very short step. Let us write $r(C_0)$ for the *radius* of the open cluster containing the origin, so

$$r(C_0) = \sup\{d(x, 0) : x \in C_0\},$$

where $d(x, y)$ denotes the graph distance between two vertices of $\mathbb{Z}^2$.

Theorem 6. *For bond percolation in* $\mathbb{Z}^2$, $\theta(1/2) = 0$.

Proof. In fact, we shall prove slightly more, namely that

$$\mathbb{P}_{1/2}\big(r(C_0) \geq n\big) \leq n^{-c} \tag{5}$$

for all $n \geq 1$, where $c > 0$ is a constant.

At $p = 1/2$, the bonds of the dual lattice $\Lambda^\star$ are open independently with probability $1/2$, so, from (4), the probability that a $6n$ by $2n$ rectangle in $\Lambda^\star$ has an open crossing is at least 2^{-25}. Consider two $6n$ by $2n$ and two $2n$ by $6n$ rectangles in $\Lambda^\star$, arranged to form a 'square annulus' as in Figure 8. By Harris's Lemma, with probability at least

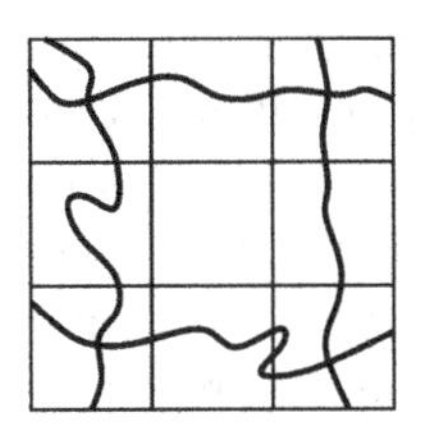

Figure 8. Four rectangles forming a square annulus.

$\varepsilon = 2^{-100} > 0$ each rectangle is crossed the long way by an open (dual) path. If this happens, then the union of these paths contains an open dual cycle surrounding the centre of the annulus; see Figure 8.

For $k \geq 1$, let A_k be the square annulus centred on the origin made

up of two 3×4^k by 4^k and two 4^k by 3×4^k dual rectangles arranged as in Figure 8, and let E_k be the event that A_k contains an open dual cycle surrounding the interior of A_k, and hence the origin. Then $\mathbb{P}(E_k) \geq \varepsilon$ for every k. As the A_k are disjoint, the events E_k are independent. If E_k holds, then no point inside A_k can be joined to a point outside A_k by an open path in $\mathbb{Z}^2$, so $r(C_0) \leq 3 \times 4^k/2 < 4^{k+1}$. Thus,

$$\mathbb{P}_{1/2}\Big(r(C_0) \geq 4^{\ell+1}\Big) \leq \mathbb{P}_{1/2}\left(\bigcap_{k=1}^{\ell} E_k^c\right) = \prod_{k=1}^{\ell} \mathbb{P}_{1/2}(E_k^c) \leq (1-\varepsilon)^{\ell},$$

and (5) follows. Also, for any n,

$$\theta(1/2) = \mathbb{P}_{1/2}\big(r(C_0) = \infty\big) \leq \mathbb{P}_{1/2}\big(r(C_0) \geq n\big) \leq n^{-c},$$

so $\theta(1/2) = 0$. □

Let S be an n by n square. If $H(S)$ holds, then at least one of the n sites v on the left of S is joined by an open path to a site at graph distance at least n from v. It follows that $\mathbb{P}_{1/2}(H(S)) \leq n\mathbb{P}_{1/2}\big(r(C_0) \geq n\big)$. Hence, $\mathbb{P}_{1/2}\big(r(C_0) \geq n\big) \geq 1/(2n)$ for all $n \geq 1$. It is natural to expect that $\mathbb{P}_{1/2}\big(r(C_0) \geq n\big)$ decays as a power of n, i.e., that the limit

$$\lim_{n\to\infty} -\frac{\log \mathbb{P}_{1/2}\big(r(C_0) \geq n\big)}{\log n}$$

exists. Once again, this natural conjecture is still open, although a corresponding result for site percolation on the triangular lattice is known; see Chapter 7. The limit above, if it does exist, is often denoted by $1/\rho$, or $1/\delta_r$. It is one of the *critical exponents* associated to percolation; see Chapter 7.

Looking back, the most difficult step in the proof of Theorem 6 is Lemma 4, or the equivalent results of Russo [1978] and Seymour and Welsh [1978]. Although Harris's proof is very different from that presented here, a key step is similar to a step in the proof of Lemma 4: roughly speaking, given an 'outermost open semi-circle S around the origin' within a certain region, one reflects this to form a cycle $S \cup S'$. Then a path meeting this cycle from inside is as likely to meet the 'real' part S (all of whose bonds are open) as the reflected part S'.

It is possible to prove Harris's Theorem using Lemma 1, Harris's Lemma, and the independence of disjoint regions, without ever considering left-most crossings or their equivalent; such a proof has applications in other contexts, where the proof of Lemma 4 does not work. Indeed, as we shall see in Chapter 8, a (long) proof of this kind was given for random

Voronoi percolation by Bollobás and Riordan [2006a]. In that setting, no result directly equivalent to Lemma 4 or to the Russo–Seymour–Welsh Theorem is known.

3.3 A sharp transition

Fairly early on it was recognized that there is an absolute constant $c < 1$ for which the following statement easily implies Kesten's Theorem: for any $p > 1/2$, there is an $n = n(p)$ such that $\mathbb{P}_p(H(R)) \geq c$ for a $2n$ by n rectangle R. For example, Chayes and Chayes [1986b] gave an argument showing that $c = 0.921$ will do. The factor 2 here is not important: recalling that $h_p(m, n)$ is the $\mathbb{P}_p$-probability that an m by n rectangle has an open horizontal crossing, it follows from Harris's Lemma that if $h_p((1 + \varepsilon)n, n) \to 1$ as $n \to \infty$ for some $\varepsilon > 0$, then $h_p(Cn, n) \to 1$ as $n \to \infty$ for every $C > 0$.

We shall give an explicit lower bound on $h_p(m, n)$, using a sharp-threshold result of Friedgut and Kalai, Theorem 12 of Chapter 2, and the Margulis–Russo formula, Lemma 9 of Chapter 2. We start by bounding the influence of a bond in a rectangle R on the event $H(R)$.

In the context of percolation, a bond e is *pivotal* for an event E in a configuration ω if precisely one of ω^+ and ω^- is in E, where $\omega^\pm$ are the configurations that agree with ω on all bonds other than e, with e open in ω^+ and closed in ω^-. In other words, e is pivotal if changing the state of e changes whether E holds or not. The *influence* of e on E is

$$I_p(e, E) = \mathbb{P}_p(\text{ } e \text{ is pivotal for } E \text{ }).$$

If E is increasing, then

$$I_p(e, E) = \mathbb{P}_p(\omega^+ \in E,\ \omega^- \notin E).$$

Lemma 7. *Let R be an m by n rectangle in $\mathbb{Z}^2$, and let e be a bond in R. Then*

$$I_p(e, H(R)) \leq 2\mathbb{P}_{1/2}\big(r(C_0) \geq \min\{m/2 - 1, (n-1)/2\}\big) \tag{6}$$

for all $0 < p < 1$.

Proof. Throughout the proof we work entirely within R, considering only the states of bonds e in R. Suppose that a bond e in R is pivotal for the increasing event $H(R)$ in the configuration ω. As $\omega^+ \in H(R)$, in the configuration ω^+, there is an open horizontal crossing of R. Since

$\omega^- \notin H(R)$, any such crossing must use e. Hence, in the configuration ω, one endpoint of e is joined by an open path to the left of R, and the other to the right; see Figure 9. Thus, at least one endpoint of e is the

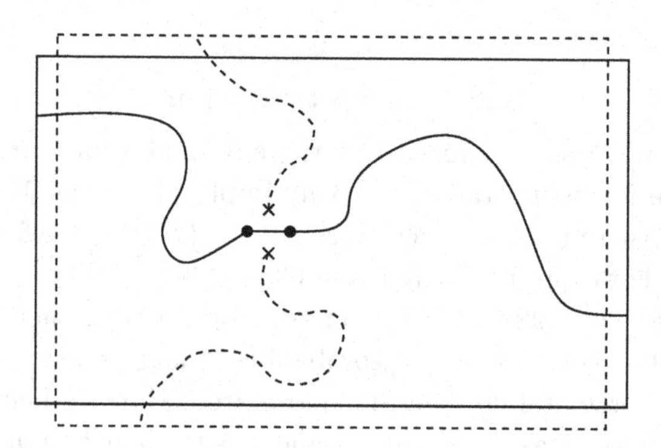

Figure 9. A bond e (joining the solid discs) that is pivotal for $H(R)$; the dual bond $e^\star$ (endpoints shown as crosses) is pivotal for $V(R^h)$.

start of an open path of length at least $m/2-1$, so

$$I_p(e, H(R)) \le 2\mathbb{P}_p\big(r(C_0) \ge m/2-1\big). \tag{7}$$

As $\omega^- \notin H(R)$, by Lemma 1 we have $\omega^- \in V(R^h)$. Similarly, $\omega^+ \notin V(R^h)$, so in ω^- there is an open dual path crossing R^h vertically and using the edge $e^\star$ dual to e. Hence, in ω, one endpoint of $e^\star$ is the start of an open dual path of length at least $(n-1)/2$. Since dual edges are open with probability $1-p$, it follows that

$$I_p(e, H(R)) \le 2\mathbb{P}_{1-p}\big(r(C_0) \ge (n-1)/2\big). \tag{8}$$

For any a the event $r(C_0) \ge a$ is increasing, so $\mathbb{P}_p\big(r(C_0) \ge a\big)$ is an increasing function of p. Thus (6) follows immediately from (7), for $p \le 1/2$, and from (8), for $p \ge 1/2$. □

Lemma 8. *Let $p > 1/2$ and an integer $\rho > 1$ be fixed. There are constants $\gamma = \gamma(p) > 0$ and $n_0 = n_0(p, \rho)$ such that*

$$h_p(\rho n, n) \ge 1 - n^{-\gamma} \tag{9}$$

for all $n \ge n_0$.

Proof. Let R be a ρn by n rectangle. From (3), we have

$$\mathbb{P}_{1/2}(H(R)) \ge h_\rho, \tag{10}$$

for some constant $h_\rho > 0$ depending only on ρ. From Lemma 7 and (5), for $n \geq 2$ we have

$$I_{p'}(e, H(R)) \leq n^{-a} = \delta$$

for every bond e of R and all $p' \in [1/2, p]$, where $a > 0$ is an absolute constant. Writing $f(p')$ for $\mathbb{P}_{p'}(H(R))$, from the Friedgut–Kalai result Theorem 12 of Chapter 2 it follows that

$$\sum_{e \in H(R)} I_{p'}(e, H(R)) \geq cf(p')(1 - f(p')) \log(1/\delta)$$

for all $p' \in [1/2, p]$, where $c > 0$ is an absolute constant.

By the Margulis–Russo formula (Lemma 9 of Chapter 2), the sum above is exactly the derivative of $f(p')$ with respect to p'. Thus, writing $g(p')$ for $\log\big(f(p')/(1 - f(p'))\big)$,

$$\frac{\mathrm{d}}{\mathrm{d}p'} g(p') = \frac{1}{f(p')(1 - f(p'))} \frac{\mathrm{d}}{\mathrm{d}p'} f(p') \geq c \log(1/\delta) = ac \log n.$$

From (10), $g(1/2)$ is bounded below by a constant that depends on ρ. Hence, taking $n_0(p, \rho)$ large enough, for $n \geq n_0(p, \rho)$ we have $g(p) \geq ac(p - 1/2) \log n + g(1/2) \geq ac(p - 1/2)(\log n)/2$, and (9) follows. □

Using his weaker 'approximate 0-1 law' instead of the more recent Friedgut–Kalai result, Russo [1982] proved a weaker form of Lemma 8; this weak form is more than enough to deduce Kesten's Theorem.

Following Bollobás and Riordan [2006c], we give an alternative proof of Lemma 8, using a version of the Friedgut–Kalai result for symmetric events, Theorem 13 of Chapter 2. The other ingredients of this proof are Harris's Lemma and (3), say, rather than (5). The idea is that, if $H(R)$ were a symmetric event, then Lemma 8 would follow immediately from (3) and the Friedgut–Kalai result. Of course, $H(R)$ is not symmetric, but it is very easy to convert it into a suitable symmetric event.

Alternative proof of Lemma 8. Fix $p > 1/2$. From (3), there is an absolute constant $0 < c_2 < 1/2$ such that $\mathbb{P}_{1/2}(H(R)) \geq c_2$ for any $4n$ by n rectangle R.

For $n \geq 3$, let $\mathbb{T}_n^2$ be the graph $C_n \times C_n$, obtained from $\mathbb{Z}^2$ by identifying all pairs of vertices for which the corresponding coordinates are congruent modulo n. This graph is often known as the *n by n discrete torus.* Note that $\mathbb{T}_n^2$ has n^2 vertices and $2n^2$ edges.

For $1 \leq k, \ell \leq n-2$, a *k by ℓ rectangle* R in $\mathbb{T}_n^2$ is an induced subgraph of $\mathbb{T}_n^2$ corresponding to a k by ℓ rectangle $R' = [a+1, a+k] \times [b+1, b+\ell]$

in $\mathbb{Z}^2$. Our rectangles in the torus are always too small to 'wrap around', so, as in the plane, any such subgraph is an abstract k by ℓ rectangle.

We shall work in $\mathbb{T}^2 = \mathbb{T}^2_{5n}$, taking the bonds to be open independently with probability p. We write $\mathbb{P}_p^{\mathbb{T}^2} = \mathbb{P}_p^{\mathbb{T}^2_{5n}}$ for the corresponding probability measure.

Let E_n be the event that $\mathbb{T}^2 = \mathbb{T}^2_{5n}$ contains *some* $4n$ by n rectangle with an open horizontal crossing, or some n by $4n$ rectangle with an open vertical crossing. Then E_n is symmetric as a subset of $\mathcal{P}(X)$, where X is the set of all $50n^2$ edges of $\mathbb{T}^2_{5n}$.

Considering one fixed $4n$ by n rectangle R in $\mathbb{T}^2$, we have

$$\mathbb{P}_{1/2}^{\mathbb{T}^2}(E_n) \geq \mathbb{P}_{1/2}^{\mathbb{T}^2}(H(R)) = \mathbb{P}_{1/2}(H(R)) \geq c_2.$$

Let $\delta = (p - 1/2)/(25c_1)$, where c_1 is the constant in Theorem 13 of Chapter 2, and set $\varepsilon = n^{-50\delta}$. As δ depends only on p, there is an $n_0 = n_0(p)$ such that $\varepsilon < c_2 \leq 1/2$ for all $n \geq n_0$. Now

$$p - \frac{1}{2} = 25c_1\delta = c_1\frac{\log(1/\varepsilon)}{\log(n^2)} > c_1\frac{\log(1/(2\varepsilon))}{\log(50n^2)}.$$

Hence, by Theorem 13 of Chapter 2, as $\mathbb{P}_{1/2}^{\mathbb{T}^2}(E_n) \geq c_2 > \varepsilon$, we have

$$\mathbb{P}_p^{\mathbb{T}^2}(E_n) \geq 1 - \varepsilon = 1 - n^{-50\delta} \tag{11}$$

for all $n \geq n_0$.

Let $R_1, \ldots, R_{25}$ be the $3n$ by $2n$ rectangles in $\mathbb{T}^2$ whose bottom-left coordinates are all possible multiples of n. Then any $4n$ by n rectangle R in $\mathbb{T}^2$ crosses one of the R_i, in the sense that the intersection of R and R_i is a $3n$ by n subrectangle of R_i. Similarly, there are $2n$ by $3n$ rectangles $R_{26}, \ldots, R_{50}$ so that any n by $4n$ rectangle in $\mathbb{T}^2$ crosses one of these from top to bottom.

It follows that if E_n holds, then so does one of the events $E_{n,i}$, $i = 1, \ldots, 50$, that R_i is crossed the long way by an open path. Thus E_n^c, the complement of E_n, contains the intersection of the $E_{n,i}^c$.

Applying Harris's Lemma (Lemma 3 of Chapter 2) to the product measure $\mathbb{P}_p^{\mathbb{T}^2}$, for each i the decreasing events $E_{n,i}^c$ and $\bigcap_{j<i} E_{n,j}^c$ are positively correlated. Hence,

$$\mathbb{P}_p^{\mathbb{T}^2}(E_n^c) \geq \mathbb{P}_p^{\mathbb{T}^2}\left(\bigcap_{i=1}^{50} E_{n,i}^c\right) \geq \prod_{i=1}^{50} \mathbb{P}_p^{\mathbb{T}^2}(E_{n,i}^c) = \mathbb{P}_p^{\mathbb{T}^2}(E_{n,1}^c)^{50}.$$

Thus, from (11), for $n \geq n_0$ we have

$$\mathbb{P}_p^{\mathbb{T}^2}(E_{n,1}^c) \leq \mathbb{P}_p^{\mathbb{T}^2}(E_n^c)^{1/50} \leq n^{-\delta},$$

so $\mathbb{P}_p^{\mathbb{T}^2}(E_{n,1}) \geq 1 - n^{-\delta}$.

Now $E_{n,1}$ is the event that there is an open horizontal crossing of a fixed $3n$ by $2n$ rectangle R in the torus, which we may identify with a corresponding rectangle in $\mathbb{Z}^2$. Thus

$$\mathbb{P}_p(H(R)) = \mathbb{P}_p^{\mathbb{T}^2}(H(R)) \geq 1 - n^{-\delta}$$

whenever R is a $3n$ by $2n$ rectangle in $\mathbb{Z}^2$ and n is large enough. Using (2), it is easy to deduce (9). □

The proofs above involved various explicit bounds. These are not really relevant. As noted earlier, the following much weaker form of Lemma 8 is enough to deduce Kesten's Theorem.

Lemma 9. *Let $p > 1/2$ be fixed. If R_n is a $3n$ by n rectangle in $\mathbb{Z}^2$, then $\mathbb{P}_p(H(R_n)) \to 1$ as $n \to \infty$.*

This may be proved by either of the methods above. Indeed, it follows from Harris's Theorem that the influences of the bonds e on the event $H(R)$ tend uniformly to zero as the shorter side of R tends to infinity. Using a qualitative form of Theorem 12 of Chapter 2 (that the sum of the influences is at least $f(\delta)$ if all are at most δ, with $f(\delta) \to \infty$ as $\delta \to 0$), or Russo's approximate 0-1 law, it follows that the sum of the influences tends to infinity. Using $\inf_n h_{1/2}(3n, n) > 0$ (from (3)), Lemma 9 follows. Alternatively, one may use (3) and a qualitative form of the symmetric Friedgut–Kalai result, working in the torus as above. Either argument shows that, unlike Lemma 8, Lemma 9 does not depend on the particular form of the Friedgut–Kalai bound.

3.4 Kesten's Theorem

As noted earlier, it is well known that Kesten's Theorem follows easily from Lemma 9; in fact, from the quantitative form of this lemma, Lemma 8, Kesten's Theorem is more or less immediate.

Let E_∞ denote the event that there is an infinite open cluster. Recall that $\mathbb{P}_p(E_\infty) > 0$ implies that $\mathbb{P}_p(E_\infty) = 1$ and $\theta(p) > 0$.

Theorem 10. *For bond percolation in $\mathbb{Z}^2$, if $p > 1/2$ then $\mathbb{P}_p(E_\infty) = 1$.*

Proof. Fix $p > 1/2$, and let $\gamma = \gamma(p)$ and $n_0 = n_0(p, 2)$ be as in Lemma 8. Let $n \geq n_0$ be an integer to be chosen below. For $k = 0, 1, 2, \ldots$, let R_k be a rectangle whose bottom-left corner is at the origin, with side-lengths $2^k n$ and $2^{k+1}n$, where the longer side is vertical if k is even and horizontal if k is odd; see Figure 10.

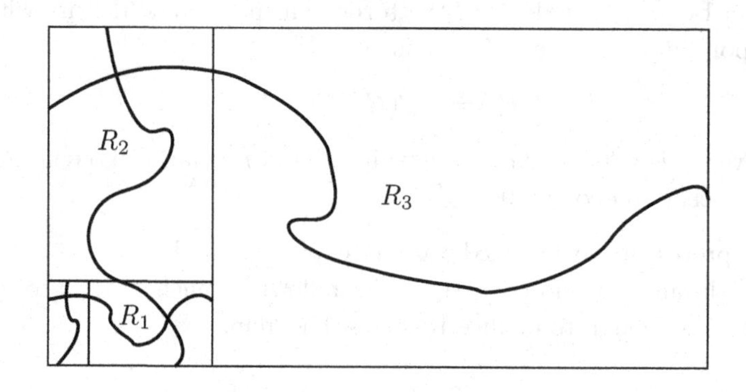

Figure 10. The rectangles R_0 to R_3 (R_0 not labelled) drawn with open paths corresponding to the events E_k.

Let E_k be the event that R_k is crossed the long way by an open path. Note that any two such crossings of R_k and R_{k+1} must meet, so if all the E_k hold, then so does E_∞. If n is large enough then, by Lemma 8,

$$\sum_{k\geq 0} \mathbb{P}_p(E_k^c) \leq \sum_{k\geq 0} (2^k n)^{-\gamma} = \frac{n^{-\gamma}}{1-2^{-\gamma}} < 1,$$

so $\mathbb{P}_p(E_\infty) \geq \mathbb{P}_p(\bigcap_{k\geq 0} E_k) > 0$. □

Together, Theorems 6 and 10 show that $p_{\mathrm{H}}(\mathbb{Z}^2) = 1/2$.

Starting from the weaker Lemma 9, a little more work is needed. Again there are several possible arguments. One is a 'renormalization' argument due to Aizenman, Chayes, Chayes, Fröhlich and Russo [1983]; see also Chayes and Chayes [1986b].

Proof of Theorem 10 – second version. Fix $p > 1/2$, and recall that $h_p(m,n)$ denotes the $\mathbb{P}_p$-probability that an m by n rectangle in $\mathbb{Z}^2$ has an open horizontal crossing. Consider three $2n$ by n rectangles overlapping in two n by n squares as in Figure 11. Noting that the probability that an n by n square has an open vertical crossing is just $h_p(n,n)$, from Harris's Lemma we have

$$h_p(4n,n) \geq h_p(2n,n)^3 h_p(n,n)^2 \geq h_p(2n,n)^5. \tag{12}$$

Placing two disjoint $4n$ by n rectangles side by side to form a $4n$ by

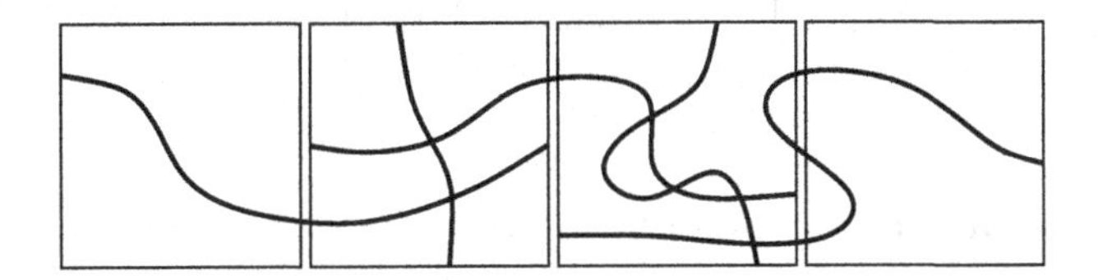

Figure 11. Four n by n squares in $\mathbb{Z}^2$. Each consecutive pair forms a $2n$ by n rectangle. If these three rectangles have open horizontal crossings, and the middle two squares have open vertical crossings, then the union of all four squares has an open horizontal crossing

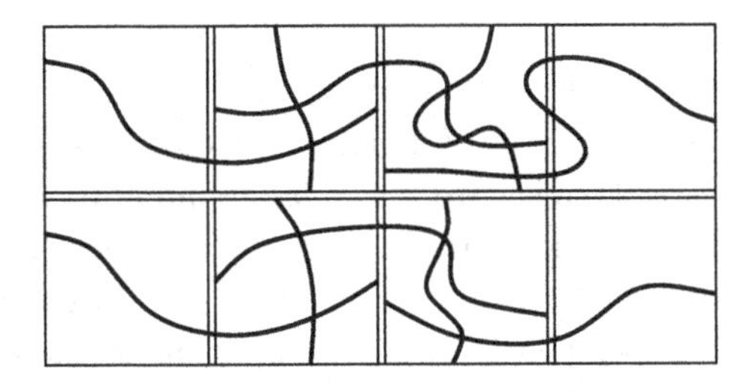

Figure 12. Two disjoint $4n$ by n rectangles forming a $4n$ by $2n$ rectangle.

$2n$ rectangle as in Figure 12, the events that each has an open horizontal crossing are independent. Hence,

$$h_p(4n, 2n) \geq 1 - (1 - h_p(4n, n))^2 \geq 1 - \left(1 - h_p(2n, n)^5\right)^2. \qquad (13)$$

Writing $h_p(2n, n)$ as $1 - \varepsilon$, from (13) we have $h_p(4n, 2n) \geq 1 - 25\varepsilon^2$, which is at least $1 - \varepsilon/2$ if $\varepsilon \leq 1/50$. By Lemma 9 there is an n with $h_p(2n, n) \geq h_p(3n, n) \geq 0.98$; it follows that $h_p(2^{k+1}n, 2^k n) \geq 1 - 2^{-k}/50$ for all $k \geq 0$, and $\mathbb{P}_p(E_\infty) > 0$ follows as in the first proof of Theorem 10. □

The argument above shows that if for a single value of n we have $h_p(2n, n) > 0.951\ldots$ (a root of $x = 1 - (1 - x^5)^2$), then $\theta(p) > 0$.

In fact, arguing as in Chayes and Chayes [1986b], one can do a little better. Dividing a $2n$ by n rectangle into two vertex-disjoint squares, if the rectangle has an open horizontal crossing, so do both squares. As the squares are disjoint, the events that they have open horizontal crossings are independent, so $h_p(2n, n) \leq h_p(n, n)^2$. Therefore the first inequality in (12) implies that $h_p(4n, n) \geq h_p(2n, n)^4$, and the first inequality in

(13) gives $h_p(4n, 2n) \geq 1 - (1 - h_p(2n, n)^4)^2$. Thus the value 0.951... may be replaced by 0.920..., a root of $x = 1 - (1 - x^4)^2$.

3.5 Dependent percolation and exponential decay

In the final section of this chapter we shall prove that for any $p < 1/2$ we have $\mathbb{P}_p(|C_0| \geq n) \leq \exp(-an)$ for some $a = a(p) > 0$. This result is due to Kesten [1980; 1981], although, as we shall see later, it follows easily from Kesten's $p_T^b(\mathbb{Z}^2) = 1/2$ result by an argument of Hammersley [1957b]. Once again, there are several possible proofs; we shall use the concept of dependent percolation – this will be useful in later chapters as well.

Let G be a graph, and let $\widetilde{\mathbb{P}}$ be a site percolation measure on G, i.e., a probability measure on the set of assignments of states (open or closed) to the vertices of G. The measure $\widetilde{\mathbb{P}}$ is *k-independent* if, whenever S and T are sets of vertices of G separated by a graph distance of at least k, the states of the vertices in S are independent of the states of the vertices in T. The term *k-dependent* is also used by some authors. For $k = 1$, the condition is exactly that the states of the vertices are independent, so the first non-trivial case is $k = 2$.

Liggett, Schonmann and Stacey [1997] proved a general result comparing k-independent measures with product measures. We shall need two simple consequences, which are essentially trivial to prove directly. Recall that, when considering site percolation on a graph G, we write C_v for the open cluster containing v, i.e., for the set of sites of G connected to v by paths all of whose vertices are open.

Lemma 11. *Let $k \geq 2$ and $\Delta \geq 2$ be positive integers, and let G be a (finite or infinite) graph with maximum degree at most Δ. There are positive constants $p_1 = p_1(k, \Delta)$ and $a = a(k, \Delta)$ such that if $\widetilde{\mathbb{P}}$ is a k-independent site percolation measure on G in which each site is open with probability at most p_1, then*

$$\widetilde{\mathbb{P}}\big(|C_v| \geq n\big) \leq \exp(-an)$$

for all vertices v of G and all $n \geq 1$.

Furthermore, for $p_1(2)$ we may take any constant such that the quantity $e\Delta(p_1(2))^{1/(\Delta+1)}$ is less than 1.

Proof. If $|C_v| \geq n$, then the subgraph of G induced by the open vertices contains a tree T with n vertices, one of which is v. It is easy to check that the number of such trees in G is at most $(e\Delta)^{n-1}$; see Problem 45 in Bollobás [2006]. Fix such a tree T. If w is any vertex of G, then at most $1 + \Delta + \cdots + \Delta^{k-1} \le \Delta^k$ vertices of G are within graph distance $k - 1$ of w. Hence, there is a subset S of at least n/Δ^k vertices of T such that any $a, b \in S$ are at graph distance at least k; indeed, one can find such a set by a greedy algorithm, choosing vertices one by one. The vertices of S are open independently, so the probability that every vertex of T is open is at most $p_1^{|S|}$. Hence,

$$\widetilde{\mathbb{P}}\big(|C_0| \ge n\big) \le (e\Delta)^{n-1} p_1^{n/\Delta^k} \le \left(e\Delta p_1^{1/\Delta^k}\right)^n.$$

Choosing p_1 small enough that $r = e\Delta p_1^{1/\Delta^k} < 1$, the result follows, with $a = -\log r$.

For the second statement, note that Δ^k appears above simply as a convenient upper bound for $1 + \Delta + \cdots + \Delta^{k-1}$. Hence, when $k = 2$, it may be replaced by $\Delta + 1$. □

In the proof above, we could have assumed without loss of generality that $k = 2$, replacing the graph G by its $(k-1)^{\text{st}}$ *power*, i.e., the graph $G^{(k-1)}$ on $V(G)$ in which two vertices are adjacent if their graph distance in G is at most $k - 1$.

Using Lemma 11, one can easily deduce Kesten's exponential decay result for bond percolation on $\mathbb{Z}^2$ from Lemma 9. The basic idea is to associate a large square S_v to each $v \in \mathbb{Z}^2$, and to assign a state to v depending on the states of the bonds in (and near) S_v, in such a way that the states of sites v separated by at least a certain constant distance k are independent, each v is unlikely to be open, and percolation in the bond model implies an infinite connected set of open sites.

Theorem 12. *In bond percolation on $\mathbb{Z}^2$, for every $p < 1/2$ there is a constant $a = a(p) > 0$ such that $\mathbb{P}_p(|C_0| \ge n) \le \exp(-an)$ for all $n \ge 0$.*

Proof. Fix $p < 1/2$, let $p_1 = p_1(5, 4) > 0$ be a constant for which Lemma 11 holds with $k = 5$, $\Delta = 4$, and set $c = (1 - p_1)^{1/4}$.

Let $\Lambda = \mathbb{Z}^2$ and let $\Lambda^\star$ be its dual, so the bonds of $\Lambda^\star$ are open independently with probability $1-p$. As $1-p > 1/2$, by Lemma 9 there is an m such that $h_{1-p}(3m, m) > c$, so the probability that a $3m$ by m rectangle in $\Lambda^\star$ has an open horizontal crossing (of dual edges) is at least c.

Set $s = m + 1$, and let S be an s by s square in $\mathbb{Z}^2$. We can arrange four $3m$ by m rectangles in $\Lambda^\star$ overlapping in m by m squares, so that their union is an annulus A, as in Figure 8. The interior of A contains a square in Λ with $m+1=s$ sites on a side, which we shall take to be S. By Harris's Lemma, the probability that each of the four rectangles is crossed the long way by a path of open dual edges is at least $c^4 = 1-p_1$. The union of four such crossings contains an open dual cycle surrounding S. Let $B(S)$ be the event that some site in S is connected by an open path to a site at L_∞-distance s from S. Such an open path in Λ cannot cross an open dual cycle in A surrounding S, so we have $\mathbb{P}_p(B(S)) \le p_1$.

Informally, we define a site percolation measure $\widetilde{\mathbb{P}}$ on $\mathbb{Z}^2$ by taking each $v=(x,y)\in\mathbb{Z}^2$ to be open if and only if $B(S)$ holds for the s by s square $S_v = [sx+1, sx+s]\times[sy+1, sy+s]$. More formally, let M denote the independent bond percolation model on $\mathbb{Z}^2$ in which each bond is open with probability p. We define a site percolation model $\widetilde{M}$ on $\mathbb{Z}^2$ as follows. Let $f : 2^{E(\mathbb{Z}^2)} \to 2^{V(\mathbb{Z}^2)}$ be the function from the state space of M to the state space of $\widetilde{M}$ that we have just defined, so $(f(\omega))(v) = 1$ if and only if, in the configuration ω, the event $B(S_v)$ holds. The function f and the measure $\mathbb{P}_p$ induce a probability measure $\widetilde{\mathbb{P}}$ on $2^{V(\mathbb{Z}^2)}$, given by $\widetilde{\mathbb{P}}(A) = \mathbb{P}_p(f^{-1}(A))$. This measure $\widetilde{\mathbb{P}}$ on $2^{V(\mathbb{Z}^2)}$ gives us a site percolation model $\widetilde{M}$ on $2^{V(\mathbb{Z}^2)}$.

Since the event $B(S_v)$ depends only on the states of bonds within L_∞-distance s of S_v, the measure $\widetilde{\mathbb{P}}$ is 5-independent. Furthermore, each $v\in\mathbb{Z}^2$ is open with $\widetilde{\mathbb{P}}$-probability $\mathbb{P}_p(B(S_v)) \le p_1$. Let C_0 be the open cluster of the origin in our original bond percolation M, and let C_0' be the open cluster of the origin in the 5-independent site percolation model $\widetilde{M}$. By Lemma 11 there is an $a>0$ such that

$$\widetilde{\mathbb{P}}(|C_0'| \ge n) \le \exp(-an)$$

for every n.

If $|C_0| > (4s+1)^2$, then every site w of C_0 is joined by an open path to some site at L_∞-distance $2s$ from w. If $w\in S_v$, then it follows that $B(S_v)$ holds. Thus, if $|C_0| > (4s+1)^2$, then $B(S_v)$ holds for every v such that S_v contains sites of C_0. The construction of the model $\widetilde{M}$ gives us a natural coupling of M and $\widetilde{M}$: a site v is open in $\widetilde{M}$ if and only if $S(B_v)$ holds in M. In particular, every v with $S_v \cap C_0 \ne \emptyset$ is open in $\widetilde{M}$, so the set of such v forms an open cluster in $\widetilde{M}$, and is thus a subset of C_0'. Hence, as each S_v contains only s^2 sites, for $n \ge (4s+1)^2$ we have

$$\mathbb{P}_p(|C_0| \ge n) \le \widetilde{\mathbb{P}}\big(|C_0'| \ge n/s^2\big) \le \exp(-an/s^2),$$

completing the proof of Theorem 12. □

Theorem 12 immediately implies that $\chi(p) < \infty$ for $p < 1/2$, so $p_T \geq 1/2$. Together with Theorem 10, this implies the Harris–Kesten Theorem.

Theorem 13. $p_T^b(\mathbb{Z}^2) = p_H^b(\mathbb{Z}^2) = 1/2$. □

Later, we shall see that $p_T = p_H$ holds in a very general context. In fact, Menshikov [1986] proved exponential decay of the radius (and, under an additional assumption, of the volume) of C_0 for $p < p_H$, again in a very general context.

We finish this chapter by noting that dependent percolation gives yet another way of deducing Kesten's Theorem from Lemma 9; again, the key lemma will be useful in other contexts. This time we consider bond percolation. A bond percolation measure on a graph G is *k-independent* if the states of sets S and T of bonds are independent, whenever S and T are at graph distance at least k. This time, the case $k = 1$ is already non-trivial. Indeed, for $k = 1$ the separation condition is exactly that no bond in S shares a site with a bond in T.

Lemma 14. *There is a $p_0 < 1$ such that if $\widetilde{\mathbb{P}}$ is a 1-independent bond percolation measure on $\mathbb{Z}^2$ in which each bond is open with probability at least p_0, then $\widetilde{\mathbb{P}}(|C_0| = \infty) > 0$.*

Proof. Suppose that the open cluster C_0 containing the origin is finite. Then by Lemma 1 of Chapter 1 there is an open dual cycle $S^\star$ surrounding the origin, of length 2ℓ, say. As shown in the proof of Lemma 2 of Chapter 1, there are at most

$$\ell\mu_{2\ell-1} \leq \frac{4\ell}{9}3^{2\ell}$$

dual cycles of length 2ℓ surrounding 0, where $\mu_k \leq 4\times 3^{k-1}$ is the number of paths of length k in $\mathbb{Z}^2$ starting at 0. Let S denote the set of duals of the bonds in $S^\star$, so S is a set of bonds of $\Lambda = \mathbb{Z}^2$. As the graph $\mathbb{Z}^2$ is 4-edge colourable, there is a set of at least $\ell/2$ vertex-disjoint bonds in S. Hence, the $\widetilde{\mathbb{P}}$-probability that $S^\star$ is open, i.e., the $\widetilde{\mathbb{P}}$-probability that all bonds in S are closed, is at most $(1-p_0)^{\ell/2}$.

It follows that the probability that C_0 is finite is at most the expected number of open dual cycles surrounding the origin, which is at most

$$\sum_{\ell\geq 2} \frac{4\ell}{9}3^{2\ell}(1-p_0)^{\ell/2}.$$

If p_0 is close enough to 1 (for example, $p_0 = 0.997$), then this sum is less than 1. □

As stated, Lemma 14 is essentially trivial; it is also immediate from the general comparison result of Liggett, Schonmann and Stacey [1997]. In applications, the value of p_0 is frequently important. Currently, the best known bound is given by the following result of Balister, Bollobás and Walters [2005].

Lemma 15. *If $\widetilde{\mathbb{P}}$ is a 1-independent bond percolation measure on $\mathbb{Z}^2$ in which each bond is open with probability at least 0.8639, then $\widetilde{\mathbb{P}}(|C_0| = \infty) > 0$.* □

Bollobás and Riordan [2006b] pointed out that Lemma 14 may be used to give yet another proof of Kesten's Theorem; for this, the value of p_0 is irrelevant.

Proof of Theorem 10 – third version. Let $p > 1/2$ be fixed, let $p_0 < 1$ be a constant for which Lemma 14 holds, and set $c = p_0^{1/3}$. Given a $3n$ by n rectangle R, let S' and S'' be the two end squares when R is cut into three squares. Note that $H(R)$ certainly implies $H(S')$ so, by Lemma 9,

$$\mathbb{P}_p(V(S'')) = \mathbb{P}_p(V(S')) = \mathbb{P}_p(H(S')) \geq \mathbb{P}_p(H(R)) \geq c$$

if n is large enough, which we shall assume from now on.

Let $G(R)$ be the event $H(R) \cap V(S') \cap V(S'')$; see Figure 13. By

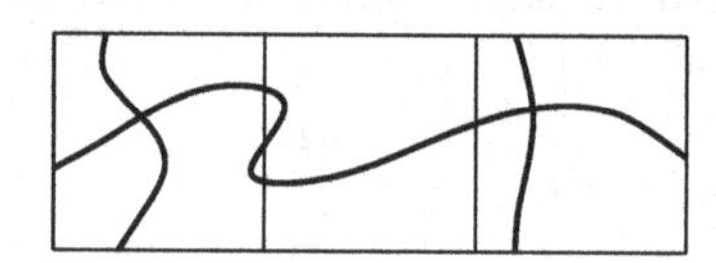

Figure 13. A $3n$ by n rectangle R such that $G(R)$ holds.

Harris's Lemma,

$$\mathbb{P}_p(G(R)) \geq \mathbb{P}_p(H(R))\mathbb{P}_p(V(S'))\mathbb{P}_p(V(S'')) \geq c^3 = p_0.$$

Define $G(R')$ similarly for an n by $3n$ rectangle, so that by symmetry we have $\mathbb{P}_p(G(R')) = \mathbb{P}_p(G(R)) \geq p_0$.

Writing M for the bond percolation model in which each bond of $\mathbb{Z}^2$ is open independently with probability p, let us define a new bond percolation model $\widetilde{M}$ on $\mathbb{Z}^2$ as follows: the edge from (x, y) to $(x+1, y)$ is open in $\widetilde{M}$ if and only if $G(R)$ holds in M for the $3n$ by n rectangle $[2nx+1, 2nx+3n] \times [2ny+1, 2ny+n]$. Similarly, the edge from (x, y) to $(x, y+1)$ is open in $\widetilde{M}$ if and only if $G(R')$ holds in M for the n by $3n$ rectangle $[2nx+1, 2nx+n] \times [2ny+1, 2ny+3n]$.

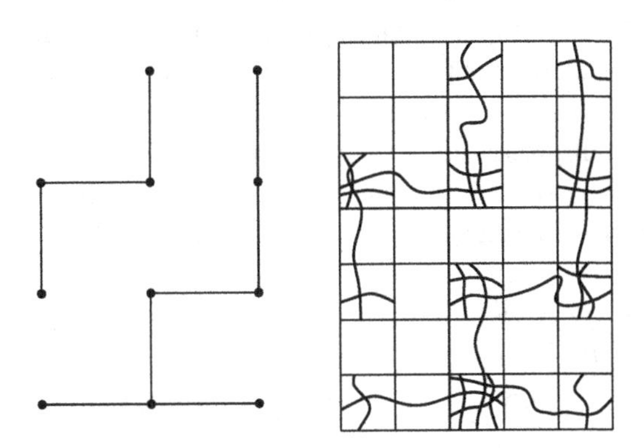

Figure 14. A set of open edges in $\widetilde{M}$ (left), and corresponding rectangles R drawn with $G(R)$ holding in M.

Let $\widetilde{\mathbb{P}}$ be the probability measure on $2^{\mathbb{E}(\mathbb{Z}^2)}$ associated to $\widetilde{M}$. Then $\widetilde{\mathbb{P}}$ is indeed 1-independent, as $G(R)$ depends only on the states of bonds in R, and vertex-disjoint edges of $\mathbb{Z}^2$ correspond to disjoint rectangles.

By Lemma 14, $\widetilde{\mathbb{P}}(|C_0| = \infty) > 0$. However, we have defined $G(R)$ in such a way that an open path in $\widetilde{M}$ guarantees a corresponding (much longer) open path in the original bond percolation M, using the fact that horizontal and vertical crossings of a square must meet; see Figure 14. Hence, $\mathbb{P}_p(E_\infty) \geq \widetilde{\mathbb{P}}(|C_0| = \infty) > 0$, completing the proof of Theorem 10. $\square$

The argument above works with $2n$ by n rectangles; the only reason for using $3n$ by n was to make the figure clearer. Also, in addition to the long crossings, it is enough to require a vertical crossing of the left-hand end square of each rectangle R, and a horizontal crossing of the bottom square of each R'. As $h_p(n, n) \geq h_p(2n, n)^{1/2}$, to prove percolation it thus suffices to find an n with $h_p(2n, n)^{3/2} \geq p_0$, where p_0 is a constant for which Lemma 14 holds. Using the value $p_0 = 0.8639$ from Balister, Bollobás and Walters [2005], $h_p(2n, n) \geq 0.907\ldots$ will do.

3.6 Sub-exponential decay

It is not hard to show that, for $p > 1/2$, with probability 1 there is a unique infinite open cluster; we shall present a very general result of this type in Chapter 5. In this range, all other open clusters are 'small' with probability close to 1; more precisely, the probability that $n \le |C_0| < \infty$ decays rapidly as n increases. In analogy with the situation for the random graph $G(n,p)$, one might expect that this decay mirrors that of open clusters below the critical probability, i.e., that $\mathbb{P}_p(n \le |C_0| < \infty)$ is approximately $\exp(-a'n)$ for some $a' > 0$. This turns out not to be the case. We shall present only the very simplest result in this direction; stronger, more general results have been proved by Aizenman, Delyon and Souillard [1980], among others; see Grimmett [1999].

Theorem 16. *In bond percolation on $\mathbb{Z}^2$, for every $p > 1/2$ there are constants $b = b(p) > 0$ and $c = c(p) > 0$ such that*

$$\exp\bigl(-b\sqrt{n}\bigr) \le \mathbb{P}_p\bigl(n \le |C_0| < \infty\bigr) \le \exp\bigl(-c\sqrt{n}\bigr) \tag{14}$$

for all $n \ge 1$.

Proof. The upper bound is essentially immediate from Theorem 12. Indeed, suppose that $|C_0| = n$, and let $\partial^\infty C_0$ be the external boundary of C_0, as defined in Chapter 1. By Lemma 1 of Chapter 1, $\partial^\infty C_0$ is a cycle in the dual lattice $\mathbb{Z}^2 + (1/2, 1/2)$ containing C_0 in its interior. Thus, the area enclosed by $\partial^\infty C_0$ is at least n, so its length is at least $2\sqrt{n}$. But every (dual) bond in $\partial^\infty C_0$ is open. As $\partial^\infty C_0$ crosses the positive x-axis at some coordinate $1/2 \le x \le n - 1/2$, we have shown that whenever $|C_0| = n$, there is some dual site $v = (x, -1/2)$, $1/2 \le x \le n - 1/2$, such that the open dual cluster containing v contains at least $2\sqrt{n}$ sites. As dual bonds are open independently with probability $1 - p < 1/2$, by Theorem 12 the probability of the latter event is at most $(n-1)\exp\bigl(-2a\sqrt{n}\bigr)$ for some $a > 0$. Thus

$$\mathbb{P}_p\bigl(n \le |C_0| < \infty\bigr) \le \sum_{m \ge n} (m-1)\exp\bigl(-2a\sqrt{m}\bigr),$$

which is at most $\exp\bigl(-b\sqrt{n}\bigr)$ for all n for some $b > 0$.

For the lower bound, suppose that $n = \ell^2$, and let S be the ℓ by ℓ square $[0, \ell-1] \times [0, \ell-1]$. For each of the n sites x in S, let E_x be the event that there is an open path from x to some site in the boundary of S; see Figure 15. Note that E_x holds trivially for all sites in the boundary of S. Whenever C_x is infinite, then E_x holds, so $\mathbb{P}_p(E_x) \ge$

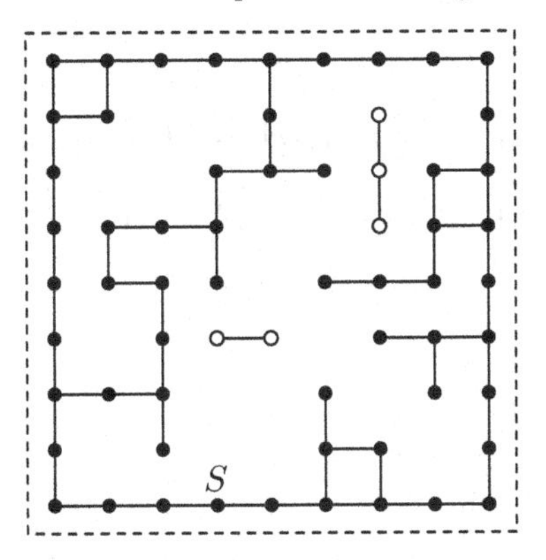

Figure 15. A 9 by 9 square S with bottom-left corner the origin, drawn together with all the open bonds in its interior. The filled circles are the sites x for which E_x holds, i.e., those joined to ∂S by an open path. If the bonds in ∂S are open, and the dual bonds surrounding S (dashed lines) are also open, then the open cluster of the origin consists precisely of the filled circles.

$\theta(p) > 0$. Let N be the number of sites $x \in S$ for which E_x holds. Then $\mathbb{E}_p(N) = \sum_{x \in S} \mathbb{P}_p(E_x) \geq n\theta(p)$. As $N \leq n$ always holds, it follows that

$$\mathbb{P}_p\big(N \geq n\theta(p)/2\big) \geq \theta(p)/2.$$

Let F be the event that each of the $4(\ell - 1)$ boundary bonds of S is open, and that each of the 4ℓ bonds from S to its complement is closed, as in Figure 15. Then

$$\mathbb{P}_p(F) = p^{4(\ell-1)}(1-p)^{4\ell} > (p(1-p))^{4\ell}.$$

Since each event E_x depends only on the states of bonds in the interior of S, the random variable N is independent of the event F, and

$$\begin{aligned}
\mathbb{P}_p\big(n\theta(p)/2 \leq |C_0| < \infty\big) &\geq \mathbb{P}_p\big(F \cap \{N \geq n\theta(p)/2\}\big) \\
&= \mathbb{P}_p(F)\mathbb{P}_p\big(\{N \geq n\theta(p)/2\}\big) \\
&\geq (p(1-p))^{4\ell}\theta(p)/2.
\end{aligned}$$

As $\ell = \sqrt{n}$, this proves the lower bound in (14) for n a square. Since there is a square between n and $3n$ for all $n \geq 1$, the lower bound for all n follows. □

In higher dimensions, i.e., for bond (or site) percolation in $\mathbb{Z}^d$, it is not too hard to guess that $\mathbb{P}_p(n \leq |C_0| < \infty)$ will decay roughly as $\exp\big(-n^{(d-1)/d}\big)$, but this result is not so easy to prove; see Section 8.6 of Grimmett [1999].

4

Exponential decay and critical probabilities – theorems of Menshikov and Aizenman & Barsky

Our aim in this chapter is to show that, for a wide class of percolation models, when $p < p_{\mathrm{H}}$ the cluster size distribution has an exponential tail. Such a result certainly implies that $p_{\mathrm{T}} = p_{\mathrm{H}}$; in fact, it will turn out that exponential decay is relatively easy to prove when $p < p_{\mathrm{T}}$ (at least if 'size' is taken to mean radius, rather than number of sites). Thus the tasks of proving exponential decay and of showing that $p_{\mathrm{T}} = p_{\mathrm{H}}$ are closely related.

Results of the latter type were proved independently by Menshikov [1986] (see also Menshikov, Molchanov and Sidorenko [1986]) and by Aizenman and Barsky [1987] under different assumptions. Here we shall present Menshikov's ingenious argument in detail, and say only a few words about the Aizenman–Barsky approach. As we shall see, Menshikov's proof makes essential use of the Margulis–Russo formula and the van den Berg–Kesten inequality.

4.1 The van den Berg–Kesten inequality and percolation

Let us briefly recall the van den Berg–Kesten inequality, Theorem 5 of Chapter 2, which has a particularly attractive interpretation in the context of percolation. In the context of site (respectively bond) percolation, an increasing event E is one which is preserved by changing the states of one or more sites (bonds) from closed to open, and a *witness* for an increasing event E is just a set W of open sites (bonds) such that the fact that all sites (bonds) in W are open guarantees that E holds. For example, considering bond percolation on $\mathbb{Z}^2$, as in Chapter 3, let R be a rectangle, and let $H(R)$ be the increasing event that R is crossed horizontally by an open path P. Then a witness for $H(R)$ is simply an open path P crossing R horizontally, or a set of open bonds containing such a path.

The box product $E \square F$ of two increasing events is the event that there are disjoint witnesses for E and F. For example, $H(R) \square H(R)$ is the event that there are two edge-disjoint open paths crossing R from left to right. (For site percolation, the paths must be vertex-disjoint.) As we saw in Chapter 2, van den Berg and Kesten [1985] proved that for increasing events in a product probability space, $\mathbb{P}(E \square F) \le \mathbb{P}(E)\mathbb{P}(F)$. This inequality is the driving force behind most of the proofs in this chapter.

Let us illustrate this inequality with a simple application to bond percolation in $\mathbb{Z}^d$. Let $p < p_{\mathrm{T}} = p_{\mathrm{T}}^{\mathrm{b}}(\mathbb{Z}^d)$ be fixed; as usual, when the context is clear, we do not indicate the specific model under consideration in our notation. Since $\chi(p) = \mathbb{E}_p(|C_0|) < \infty$, there is an m such that

$$\mathbb{E}_p\big(|C_0 \cap S_m(0)|\big) \le 1/2, \tag{1}$$

where $S_m(x)$ is the *sphere* of radius m with centre x, i.e., the set of sites at graph distance exactly m from x.

Given a site x and integers $n \ge m \ge 1$, let $\{x \xrightarrow{n}\}$ be the event that there is an open path (i.e., self-avoiding walk) P from x to some site in $S_n(x)$, and let $\{x \xrightarrow{m,n}\}$ be the event that there is an open path P starting at x, visiting some site $y \in S_m(x)$, and ending at a site $z \in S_n(y)$. Then $\{x \xrightarrow{m+n}\} \subset \{x \xrightarrow{m,n}\}$: if P is an open path from x to $S_{m+n}(x)$, we may take y to be the first site of P in $S_m(x)$. As P ends at least a distance n from y, there is a site $z \in P$, coming after y, at distance exactly n from y; see Figure 1.

If $\{x \xrightarrow{m,n}\}$ holds, then for some $y \in S_m(x)$ there are disjoint witnesses for the events $\{x \to y\}$ and $\{y \xrightarrow{n}\}$. Hence,

$$\begin{aligned}
\mathbb{P}_p(0 \xrightarrow{m+n}) \le \mathbb{P}_p(0 \xrightarrow{m,n}) &\le \sum_{y \in S_m(0)} \mathbb{P}_p(\{0 \to y\} \square \{y \xrightarrow{n}\}) \\
&\le \sum_{y \in S_m(0)} \mathbb{P}_p(0 \to y)\mathbb{P}_p(y \xrightarrow{n}) \\
&= \sum_{y \in S_m(0)} \mathbb{P}_p(0 \to y)\mathbb{P}_p(0 \xrightarrow{n}) \\
&= \mathbb{E}_p\big(|C_0 \cap S_m(0)|\big)\mathbb{P}_p(0 \xrightarrow{n}) \le \mathbb{P}_p(0 \xrightarrow{n})/2,
\end{aligned}$$

where the last step is from (1). It follows that $\mathbb{P}_p(0 \xrightarrow{n}) \le 2^{-\lfloor n/m \rfloor}$, so $\mathbb{P}_p(0 \xrightarrow{n})$ decays exponentially as $n \to \infty$.

This result was first proved by Hammersley [1957b] in the much more

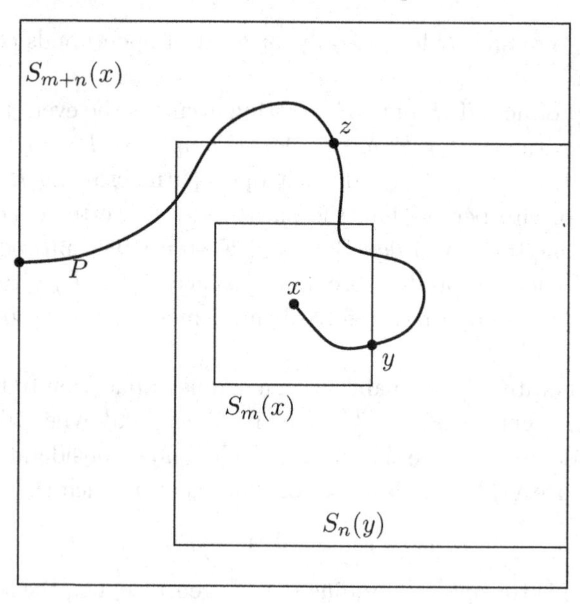

Figure 1. An open path P from x to $S_{m+n}(x)$. The portion of P from x to z shows that $x \xrightarrow{m,n}$ holds.

general setting of Theorem 9 below, well before the van den Berg–Kesten inequality was proved.

4.2 Oriented site percolation

To state the main results of this chapter, we shall consider oriented site percolation, where the sites of an oriented graph $\overrightarrow{\Lambda}$ are taken to be open independently with probability p. The reason for considering oriented site percolation is that results for this model extend immediately to oriented bond percolation, and to unoriented site and bond percolation, by considering suitable transformations of the underlying graph. Before we expand on this observation, let us recall some definitions.

As in Chapter 1, by a *percolation measure* $\mathbb{P}$ on a (possibly oriented) graph Λ we mean a probability measure on the subgraphs of Λ. A random element of this probability space is the *open subgraph* of Λ; for bond percolation, the open subgraph is formed by taking all sites of Λ and only the open bonds; for site percolation, the open subgraph consists of all open sites of Λ and all bonds joining open sites. For a site x of Λ, the open cluster C_x is defined from the open subgraph.

Turning to oriented site percolation, we write $\mathbb{P}_p = \mathbb{P}_{\vec{\Lambda},p} = \mathbb{P}^{\mathrm{s}}_{\vec{\Lambda},p}$ for the probability measure in which each site of our underlying oriented (multi-)graph $\vec{\Lambda}$ is open independently with probability p. As in Chapter 1, we write $\vec{\Lambda}_x^+$ for the *out-subgraph* of $\vec{\Lambda}$ rooted at x, i.e., the subgraph of $\vec{\Lambda}$ containing all sites and bonds that may be reached by (oriented) paths from x. Naturally, we shall consider $\vec{\Lambda}_x^+$ as a rooted oriented graph, with root x. We write $C_x = C_x^+$ for the *open out-cluster* of x, i.e., the set of all sites y reachable from x by open paths in $\vec{\Lambda}$. Note that a path in $\vec{\Lambda}$ is always oriented. As usual in site percolation, a path is open if all of its sites are open. We write $\theta_x(p) = \theta_x^+(\vec{\Lambda};p)$ for the $\mathbb{P}_p$-probability that C_x is infinite, and $\chi_x(p) = \chi_x^+(\vec{\Lambda};p)$ for the expectation of $|C_x|$.

For each site x we have two critical probabilities, $p_{\mathrm{H}}^{\mathrm{s}}(\vec{\Lambda};x) = \sup\{p : \theta_x(p) = 0\}$ and $p_{\mathrm{T}}^{\mathrm{s}}(\vec{\Lambda};x) = \sup\{p : \chi_x(p) < \infty\}$. As $\theta_x(p) > 0$ implies $\chi_x(p) = \infty$, trivially $p_{\mathrm{T}}^{\mathrm{s}}(\vec{\Lambda};x) \le p_{\mathrm{H}}^{\mathrm{s}}(\vec{\Lambda};x)$. In general, the critical probabilities $p_{\mathrm{H}}^{\mathrm{s}}(\vec{\Lambda};x)$ and $p_{\mathrm{T}}^{\mathrm{s}}(\vec{\Lambda};x)$ do depend on the site x. However, as noted in Chapter 1, if $\vec{\Lambda}$ is strongly connected, i.e., if there is an (oriented) path from x to y for any two sites x and y, then the critical probabilities are independent of the sites. We shall see later that a weaker condition is enough. When the critical probabilities are independent of the site x, we write $p_{\mathrm{H}}^{\mathrm{s}}(\vec{\Lambda})$ and $p_{\mathrm{T}}^{\mathrm{s}}(\vec{\Lambda})$ for their common values.

In order to spell out in detail the connections between oriented and unoriented site and bond percolation, we shall need one more definition. Let us say that two percolation measures $\mathbb{P}_1$, $\mathbb{P}_2$ on (possibly oriented) graphs Λ_1, Λ_2 are *equivalent* if there is an absolute constant $A > 0$ such that, whenever $\{i,j\} = \{1,2\}$ and x is a site of Λ_i, there is a site y of Λ_j and a set S of at most A sites of Λ_j, such that

$$\mathbb{P}_j\big(|C_y| \ge An\big)/A \le \mathbb{P}_i(|C_x| \ge n) \le A\mathbb{P}_j\Big(\Big|\bigcup_{z\in S} C_z\Big| \ge n/A\Big) \tag{2}$$

for all $n \ge 1$. This definition is roughly analogous to that of equivalence for metric spaces. The reason for considering a set S of sites in the upper bound is that, in bond percolation, we consider the open cluster containing a given *site*, which is the union of the open clusters containing all bonds incident with that site, or, in the oriented case, directed from that site.

Note that if $\mathbb{P}_1$ and $\mathbb{P}_2$ are equivalent, and $\chi_x < \infty$ for all sites x of Λ_1, then $\chi_y < \infty$ for all sites y of Λ_2. (Here, of course, χ_x is defined using the expectation corresponding to $\mathbb{P}_1$, and χ_y that corresponding to $\mathbb{P}_2$.) Roughly speaking, it follows that equivalence preserves the critical probabilities p_{H} and p_{T}.

For an unoriented graph Λ, let $\varphi_{\mathrm{s}}(\Lambda)$ be the oriented graph on the same vertex set, where each bond xy of Λ is replaced by two oriented bonds, $\overrightarrow{xy}$ and $\overrightarrow{yx}$. Site percolation on Λ and oriented site percolation on $\varphi_{\mathrm{s}}(\Lambda)$ are equivalent. Indeed, (2) holds with $A = 1$: the distributions of C_x in the two measures $\mathbb{P}^{\mathrm{s}}_{\Lambda,p}$ and $\mathbb{P}^{\mathrm{s}}_{\overrightarrow{\Lambda},p}$, $\overrightarrow{\Lambda} = \varphi_{\mathrm{s}}(\Lambda)$, are identical.

If $\overrightarrow{\Lambda}$ is an oriented (multi-)graph, let $\varphi_{\mathrm{b}}(\overrightarrow{\Lambda})$ be the *oriented line-graph* of $\overrightarrow{\Lambda}$, which has a site for each (oriented) bond of $\overrightarrow{\Lambda}$, and an oriented edge from $\overrightarrow{e}$ to $\overrightarrow{f}$ whenever $\overrightarrow{e}$ and $\overrightarrow{f}$ are bonds of $\overrightarrow{\Lambda}$ with the head of $\overrightarrow{e}$ equal to the tail of $\overrightarrow{f}$. The subscript b in the notation φ_{b} reminds us that it is oriented *bond* percolation on $\overrightarrow{\Lambda}$ that we shall model by oriented site percolation on $\varphi_{\mathrm{b}}(\overrightarrow{\Lambda})$.

For an unoriented graph Λ, let $L(\Lambda)$ be the usual line-graph of Λ, with a site for each bond e of Λ, and a bond joining e and f if e and f share an end-vertex in Λ, and set $\varphi_{\mathrm{b}}(\Lambda) = \varphi_{\mathrm{s}}(L(\Lambda))$.

It is not hard to show that, provided the degrees of the underlying graphs are bounded, oriented bond percolation on $\overrightarrow{\Lambda}$ is equivalent to oriented site percolation on $\varphi_{\mathrm{b}}(\overrightarrow{\Lambda})$, and that bond percolation on Λ is equivalent to site percolation on $L(\Lambda)$, and hence to oriented site percolation on $\varphi_{\mathrm{b}}(\Lambda) = \varphi_{\mathrm{s}}(L(\Lambda))$. We give a formal proof of the first of these statements.

Theorem 1. *Let $\overrightarrow{\Lambda}$ be an oriented (multi-)graph in which every site has out-degree at least 1 and at most $\Delta < \infty$. For each $p \in (0,1)$, the percolation measures $\mathbb{P}^{\mathrm{b}}_{\overrightarrow{\Lambda},p}$ and $\mathbb{P}^{\mathrm{s}}_{\overrightarrow{M},p}$ are equivalent, where $\overrightarrow{M} = \varphi_{\mathrm{b}}(\overrightarrow{\Lambda})$.*

Proof. For a bond $\overrightarrow{e} \in E(\overrightarrow{\Lambda})$, let us write $\varphi(\overrightarrow{e})$ for the corresponding site of $\overrightarrow{M}$. We couple the measures $\mathbb{P}_1 = \mathbb{P}^{\mathrm{b}}_{\overrightarrow{\Lambda},p}$ and $\mathbb{P}_2 = \mathbb{P}^{\mathrm{s}}_{\overrightarrow{M},p}$ by taking each site of $\overrightarrow{M}$ to be open if and only if the corresponding bond of $\overrightarrow{\Lambda}$ is open.

For a site $x \in \overrightarrow{\Lambda}$, let $E(C_x)$ denote the edge-set of the open cluster C_x of the bond percolation on $\overrightarrow{\Lambda}$. In other words, $E(C_x)$ is the set of bonds $\overrightarrow{e}$ of $\overrightarrow{\Lambda}$ such that there is a path P in $\overrightarrow{\Lambda}$ with initial site x and final *bond* $\overrightarrow{e}$, all of whose bonds are open. Since $\overrightarrow{\Lambda}$ has maximum out-degree at most Δ, we have

$$|C_x| - 1 \leq |E(C_x)| \leq \Delta |C_x|$$

for any open cluster C_x in $\overrightarrow{\Lambda}$.

We shall show that (2) holds with $A = \max\{\Delta, 2, 1/p\}$. If x is a site of $\overrightarrow{\Lambda}$, let $E^+(x)$ denote the set of bonds of $\overrightarrow{\Lambda}$ having x as initial site. As an open path P in $\overrightarrow{\Lambda}$ corresponds to a path P' of open sites in $\overrightarrow{M}$, we have

$$E(C_x) = \bigcup_{\overrightarrow{e} \in E^+(x)} C_{\varphi(\overrightarrow{e})}.$$

Hence,

$$\mathbb{P}_1\big(|C_x| \geq n\big) \leq \mathbb{P}_1\big(|E(C_x)| \geq n-1\big) = \mathbb{P}_2\left(\Big|\bigcup_{\overrightarrow{e} \in E^+(x)} C_{\varphi(\overrightarrow{e})}\Big| \geq n-1\right).$$

As $|E^+(x)| \leq \Delta$, this gives the upper bound in (2) with $i = 1$, $j = 2$, for any $n \geq 2$. The condition for $n = 1$ is trivial, as $\mathbb{P}_1\big(|C_x| \geq 1\big) = 1$, while $\mathbb{P}_2\big(|C_{\varphi(\overrightarrow{e})}| \geq 1\big) = p$ for any bond $\overrightarrow{e}$ of $\overrightarrow{\Lambda}$, and we have chosen $A \geq 1/p$.

For the lower bound, pick any $\overrightarrow{e} \in E^+(x)$. Then

$$\mathbb{P}_1\big(|C_x| \geq n\big) \geq \mathbb{P}_1\big(|E(C_x)| \geq \Delta n\big) \geq \mathbb{P}_2\big(|C_{\varphi(\overrightarrow{e})}| \geq \Delta n\big).$$

It remains to prove (2) with $i = 2$, $j = 1$. Let y be any site of $\overrightarrow{M}$, so $y = \varphi(\overrightarrow{ab})$ for some bond $\overrightarrow{ab}$ of $\overrightarrow{\Lambda}$. Then $C_y \subset E(C_a)$, and, whenever y is open, $E(C_b) \subset C_y$. It follows that

$$p\mathbb{P}_1\big(|C_b| \geq n+1\big) \leq \mathbb{P}_2\big(|C_y| \geq n\big) \leq \mathbb{P}_1\big(|C_a| \geq n/\Delta\big),$$

completing the proof. □

The proof of the unoriented analogue of Theorem 1 is similar.

Theorem 1 implies that $p_{\mathrm{H}}^{\mathrm{b}}(\overrightarrow{\Lambda}) = p_{\mathrm{H}}^{\mathrm{s}}(\varphi_{\mathrm{b}}(\overrightarrow{\Lambda}))$, whenever one of these critical probabilities is well defined. Similarly, $p_{\mathrm{T}}^{\mathrm{b}}(\overrightarrow{\Lambda}) = p_{\mathrm{T}}^{\mathrm{s}}(\varphi_{\mathrm{b}}(\overrightarrow{\Lambda}))$. Also, for an unoriented graph Λ, we have $p_{\mathrm{H}}^{\mathrm{b}}(\Lambda) = p_{\mathrm{H}}^{\mathrm{s}}(\varphi_{\mathrm{b}}(\Lambda))$, and so on. Thus, when proving results of the type '$p_{\mathrm{T}} = p_{\mathrm{H}}$', it suffices to consider oriented site percolation.

There is a reason for the maximum degree restriction in Theorem 1. If each bond of a complete graph K_n on n sites is open independently with probability p, and C_x is the open cluster containing a given site $x \in K_n$, then $\mathbb{E}(|C_x|) = \Theta(n)$ as $n \to \infty$ with p fixed, while $\mathbb{E}(|E(C_x)|) = \Theta(n^2)$. Let Λ be the graph formed by attaching a sequence of complete graphs with sizes $n_1, n_2, \ldots$ to an infinite path, as in Figure 2. Taking $n_i = 2^i$,

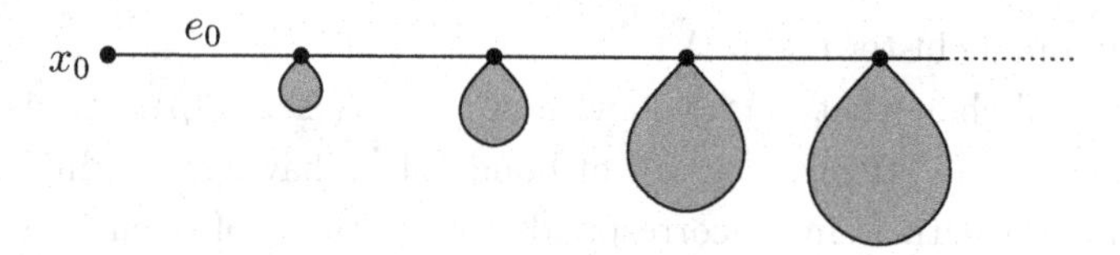

Figure 2. A series of complete graphs attached to an infinite path.

say, we have $\chi^{\mathrm{b}}_{x_0}(\Lambda;p) = \Theta\left(\sum_i p^i 2^i\right)$, while $\chi^{\mathrm{s}}_{e_0}(L(\Lambda);p) = \Theta\left(\sum_i p^i 4^i\right)$, so

$$1/4 = p^{\mathrm{s}}_{\mathrm{T}}(\varphi_{\mathrm{b}}(\Lambda)) = p^{\mathrm{s}}_{\mathrm{T}}(L(\Lambda)) < p^{\mathrm{b}}_{\mathrm{T}}(\Lambda) = 1/2.$$

The construction may be adapted to the oriented case by orienting every bond 'away from x_0', orienting each complete graph transitively.

Returning to the general study of oriented site percolation on a graph $\vec{\Lambda}$, let us say that two sites x and y are *out-like* if the out-subgraphs $\vec{\Lambda}^+_x$ and $\vec{\Lambda}^+_y$ are isomorphic as rooted oriented graphs. (Broadbent and Hammersley [1957] used this term in a slightly different way.) The distribution of $C_x = C^+_x$ depends only on $\vec{\Lambda}^+_x$, so for out-like sites x and y we have

$$\theta_x(p) = \theta_y(p) \text{ and } \chi_x(p) = \chi_y(p).$$

By the *out-class* $[x]$ of a site x, we mean the equivalence class under the out-like relation that contains x. In other words, $[x]$ is the set of sites y such that x and y are out-like. We write $C_{\vec{\Lambda}}$ for the *out-class graph* of $\vec{\Lambda}$, whose vertices are the out-classes, in which there is an oriented edge from $[x]$ to $[y]$ if and only if there are sites $x' \in [x]$ and $y' \in [y]$ with $\overrightarrow{x'y'}$ a bond of $\vec{\Lambda}$. We allow $[x] = [y]$, so the out-class graph may contain loops. Since the out-classes appearing among the sites in $\vec{\Lambda}^+_x$ depend only on the isomorphism class of $\vec{\Lambda}^+_x$ and hence only on $[x]$, there is an edge from $[x]$ to $[y]$ whenever $\overrightarrow{xy'} \in E(\vec{\Lambda})$ for some $y' \in [y]$.

For example, if $\vec{\Lambda}$ is $\vec{\mathbb{Z}}^d$, then there is a single out-class, and the out-class graph has one vertex with a loop. If $\vec{\Lambda}$ is a rooted tree in which sites at even levels have 2 children and sites at odd levels have 3, as in Figure 3, then there are two out-classes, one corresponding to all sites at even levels and one to all sites at odd levels, and in $C_{\vec{\Lambda}}$ there is an oriented edge from each out-class to the other.

Much of the time, we shall be interested in oriented graphs with $C_{\vec{\Lambda}}$ finite. In the unoriented case, the analogous condition is somewhat

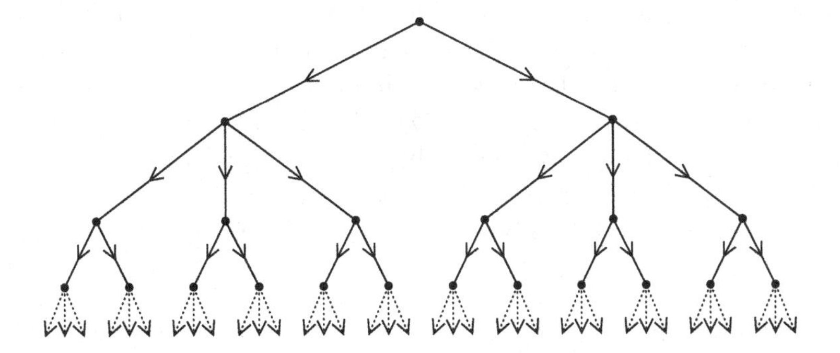

Figure 3. A rooted oriented tree in which out-degrees 2 and 3 alternate along any directed path.

simpler: recall from Chapter 1 that two sites x and y of a graph Λ are *equivalent* if there is an automorphism of Λ mapping x into y. The graph Λ is of *finite type* if there are finitely many equivalence classes of sites under this relation. In particular, vertex-transitive graphs are of finite type, with only one equivalence class. In the most interesting examples, Λ will be vertex-transitive. For example, the Archimedean lattices studied in the next chapter, shown in Figure 18 of that chapter, have this property. Occasionally, however, we shall study more general finite-type graphs, for example, the duals of the Archimedean lattices other than the square, hexagonal and triangular lattices. For oriented percolation, $\overrightarrow{\Lambda}$ will almost always be such that any two sites are out-like, so $|C_{\overrightarrow{\Lambda}}| = 1$. However, the greater generality allowed by assuming only that $C_{\overrightarrow{\Lambda}}$ is finite does not complicate the proofs, so we shall always work with this assumption, rather than requiring that $|C_{\overrightarrow{\Lambda}}| = 1$.

The next lemma gives a condition on an oriented graph $\overrightarrow{\Lambda}$ that will play the same role as the condition than an unoriented graph Λ be connected. Recall that an oriented graph is *strongly connected* if, for any two vertices x and y, there is an oriented path from x to y.

Lemma 2. *Let $\overrightarrow{\Lambda}$ be an infinite, locally finite oriented multi-graph with $C_{\overrightarrow{\Lambda}}$ strongly connected. Then there are real numbers $p_{\mathrm{H}}^{\mathrm{s}}(\overrightarrow{\Lambda})$ and $p_{\mathrm{T}}^{\mathrm{s}}(\overrightarrow{\Lambda})$ such that*

$$p_{\mathrm{H}}^{\mathrm{s}}(\overrightarrow{\Lambda};x) = p_{\mathrm{H}}^{\mathrm{s}}(\overrightarrow{\Lambda}) \quad \textit{and} \quad p_{\mathrm{T}}^{\mathrm{s}}(\overrightarrow{\Lambda};x) = p_{\mathrm{T}}^{\mathrm{s}}(\overrightarrow{\Lambda})$$

for all sites x of $\overrightarrow{\Lambda}$.

Proof. Let x and y be any two sites of $\vec{\Lambda}$. As there is an oriented path from $[x]$ to $[y]$ in $C_{\vec{\Lambda}}$, there is an oriented path P in $\vec{\Lambda}$ from x to a site $y' \in [y]$. If P has length ℓ, then $\theta_x(p) \geq p^\ell \theta_{y'}(p) = p^\ell \theta_y(p)$, and $\chi_x(p) \geq p^\ell \chi_{y'}(p) = p^\ell \chi_y(p)$. As x and y are arbitrary, we find that $\theta_x(p) > 0$ for some site x if and only if $\theta_x(p) > 0$ for all sites x, and that $\chi_x(p) < \infty$ for some site x if and only if $\chi_x(p) < \infty$ for all sites x. Hence, the critical probabilities $p_{\mathrm{H}}^{\mathrm{s}}(\vec{\Lambda}; x)$ and $p_{\mathrm{T}}^{\mathrm{s}}(\vec{\Lambda}; x)$ do not depend on the site x, as claimed. □

Lemma 2 should be contrasted with the observation in Chapter 1 that in the unoriented case the critical probability p_{H} or p_{T} does not depend on the initial site for *any* connected underlying graph Λ. Note that there are many interesting cases in which $C_{\vec{\Lambda}}$ is strongly connected while $\vec{\Lambda}$ itself is not, for example, $\vec{\mathbb{Z}}^d$, and the tree in Figure 3.

For sites x and y of an oriented graph $\vec{\Lambda}$, we say that y is at distance n from x, and write $d(x, y) = n$, if there is an oriented path from x to y, and the shortest such path has length n. We write $S_n^+(x)$ for the *out-sphere* of radius n centred at x, i.e., for the set of sites y at distance n from x. Note that $y \in S_n^+(x)$ does not imply $x \in S_n^+(y)$; indeed, there may be no oriented path from y to x at all, in which case $d(y, x)$ is undefined. When $\vec{\Lambda}$ is obtained from an unoriented graph Λ by replacing each bond with two bonds, one oriented in each direction, then $d(x, y)$ is just the usual graph distance on Λ, and $S_n^+(x)$ is just $S_n(\Lambda; x)$, the set of sites of Λ at graph distance n from x in Λ. In fact, the results and proofs in this chapter may be read for unoriented percolation simply by ignoring all references to the orientation of edges.

Although it so happens that such complications will not be important, let us note that the behaviour of the distance function $d(x, y)$ on oriented graphs may be somewhat counterintuitive. For example, let $\vec{\Lambda}$ be the oriented graph shown in Figure 4. Although this is perhaps not obvious from the figure, all sites in $\vec{\Lambda}$ are out-like. Taking $x = (0, 0)$, $y = (0, b)$ and $z = (2^b n, b)$, we have $d(x, y) = b$, $d(y, z) = 2^b n$, but $d(x, z) = b + n$. Thus, a site z may be a very long way from a site y close to x, but still z is not far from x.

Let us write $r(C_x)$ for the *radius* of the open out-cluster C_x, i.e.,

$$r(C_x) = \sup\{n : C_x \cap S_n^+(x) \neq \emptyset\}.$$

Our main aim in this chapter is to study the tail of the distribution of $r(C_x)$, i.e., to study the probability of the event $r(C_x) \geq n$, which is exactly the event $\{x \xrightarrow{n}\}$ that there is an oriented path from a given site

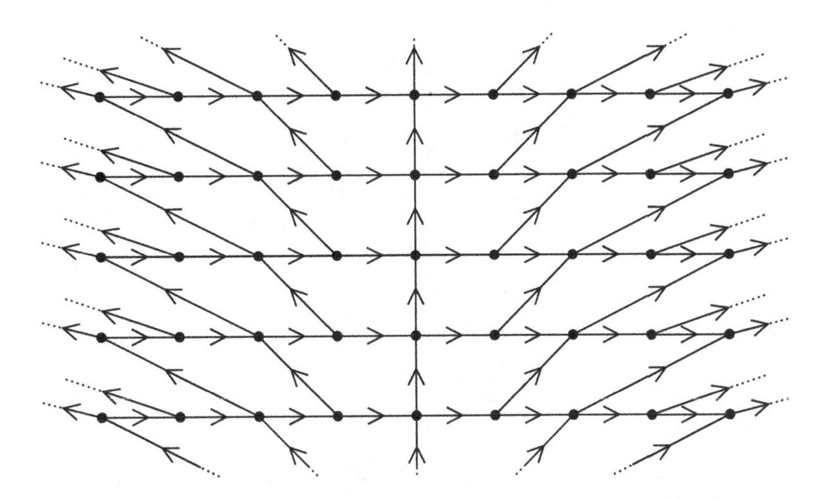

Figure 4. The oriented graph $\overrightarrow{\Lambda}$ on $\mathbb{Z}^2$ in which two bonds leave each site (a, b), one going to $(a+1, b)$, and the other to $(2a, b+1)$. This example was given by Paul Balister.

x to some site $y \in S_n^+(x)$. Note that, as $\overrightarrow{\Lambda}$ is locally finite, $r(C_x)$ is finite if and only if C_x is finite. Note also that the length of the shortest open path from x to $S_n^+(x)$ may be much longer than n.

In many of the arguments, it will be convenient to work with a closely related event that does not depend on the state of x. Let $R_n(x)$ be the event that there is an oriented path P from x to some site $y \in S_n^+(x)$, with every site of P *other than* x open. This event is illustrated in Figure 5.

Equivalently, $R_n(x)$ is the event that there is an open path P' from some (out-)neighbour of x to a site in $S_n^+(x)$. (It makes no difference whether or not we constrain P' not to use the site x: if there is such a path P' visiting x, and y is the site after x along P', then the portion P'' of P' starting at y is a path from an out-neighbour of x to $S_n^+(x)$ not visiting x.) In the light of the latter definition, we will sometimes write $\{x^+ \xrightarrow{n}\}$ for $R_n(x)$. We shall study the quantity

$$\rho_n(x, p) = \mathbb{P}_p^{\mathrm{s}}(R_n(x)).$$

As the event $R_n(x)$ does not depend on the state of x, and $\{x \xrightarrow{n}\}$ holds if and only if x is open and $R_n(x)$ holds, we have

$$\mathbb{P}_p^{\mathrm{s}}\big(r(C_x) \geq n\big) = \mathbb{P}_p^{\mathrm{s}}\big(x \xrightarrow{n}\big) = p\mathbb{P}_p^{\mathrm{s}}\big(R_n(x)\big) = p\rho_n(x, p).$$

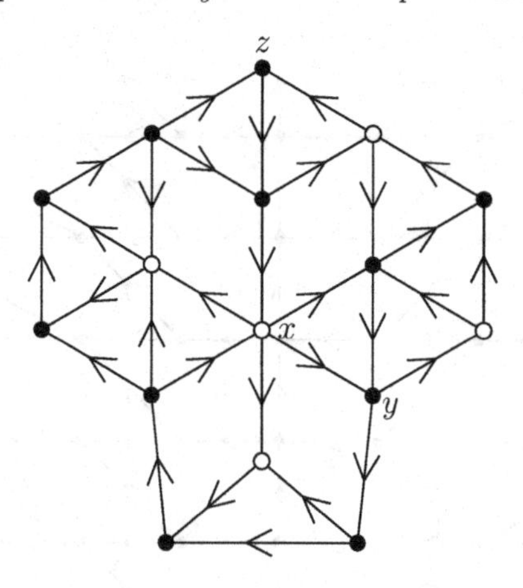

Figure 5. An illustration of the event $R_4(x)$. Solid circles represent open sites. As $d(x,z) = 4$, the (unique) open path from the out-neighbour y of x to z is a witness for $R_4(x)$.

The reason for considering $R_n(x)$ rather than $\{x \xrightarrow{n}\}$ is that we shall look for disjoint witnesses for pairs of events such as $R_n(x)$, $R_k(x)$, which would be impossible for $\{x \xrightarrow{n}\}$ and $\{x \xrightarrow{k}\}$.

If $p < p_{\mathrm{H}}^{\mathrm{s}}(\overrightarrow{\Lambda}; x)$, then $r(C_x) < \infty$ with probability 1, so $\mathbb{P}_p^{\mathrm{s}}\big(r(C_x) \geq n\big) \to 0$ as $n \to \infty$. Suppose that $C_{\overrightarrow{\Lambda}}$ is strongly connected. Then, by Lemma 2, there is a single critical probability $p_{\mathrm{H}}^{\mathrm{s}}(\overrightarrow{\Lambda})$ independent of the initial site x. For a fixed $p < p_{\mathrm{H}}^{\mathrm{s}}(\overrightarrow{\Lambda})$, we thus have $\lim_{n\to\infty} \rho_n(x,p) = 0$ for every site x. In the arguments below, we shall need bounds on

$$\rho_n(p) = \sup_{x \in \overrightarrow{\Lambda}} \rho_n(x,p).$$

To obtain such a bound, we shall assume that $C_{\overrightarrow{\Lambda}}$ is finite.

Lemma 3. *Let $\overrightarrow{\Lambda}$ be an infinite, locally finite oriented multi-graph with $C_{\overrightarrow{\Lambda}}$ finite and strongly connected, and let $p \in (0,1)$. Then there is a constant $a > 0$ such that*

$$\mathbb{P}_p\big(|C_x| \geq n\big) \geq a\mathbb{P}_p\big(|C_y| \geq n\big) \tag{3}$$

for all sites x and y and integers $n \geq 1$, where $\mathbb{P}_p$ denotes $\mathbb{P}^{\mathrm{s}}_{\overrightarrow{\Lambda},p}$. If $p < p_{\mathrm{H}}^{\mathrm{s}}(\overrightarrow{\Lambda})$, then

$$\rho_n(p) \to 0 \tag{4}$$

as $n \to \infty$.

Proof. Since $C_{\vec{\Lambda}}$ is finite and strongly connected, there is a constant ℓ such that for any two out-classes $[x]$ and $[y]$, there is a path of length at most ℓ from $[x]$ to $[y]$ in $C_{\vec{\Lambda}}$. Hence, for any sites x and y, there is a path P of length $\ell' \leq \ell$ from x to a site $y' \in [y]$. Let E be the event that all sites of P other than (perhaps) y' are open. Then

$$\begin{aligned}\mathbb{P}_p(|C_x| \geq n) \geq \mathbb{P}_p(E \cap \{|C_{y'}| \geq n\}) &= p^{\ell'}\mathbb{P}_p(|C_{y'}| \geq n \mid E) \\ &\geq p^{\ell}\mathbb{P}_p(|C_{y'}| \geq n) = p^{\ell}\mathbb{P}_p(|C_y| \geq n),\end{aligned}$$

and (3) follows with $a = p^{\ell}$.

For (4), note first that $p_{\mathrm{H}}^{\mathrm{s}}(\vec{\Lambda})$ is well defined by Lemma 2. For $p < p_{\mathrm{H}}^{\mathrm{s}}(\vec{\Lambda})$, we have $\rho_n(x,p) \to 0$ for each x. But $\rho_n(x,p)$ depends only on the out-class $[x]$ of x. Thus $\rho_n(p)$ is the pointwise maximum of finitely many functions each of which tends to zero, so $\rho_n(p) \to 0$. □

Let us remark that (3) implies $\theta_x \geq a\theta_y$ and $\chi_x \geq a\chi_y$, with $a = p^{\ell}$ the constant given by the proof of Lemma 3. It would be tempting to think that (3) holds with $|C_x|$ replaced by $r(C_x)$. In the case of unoriented percolation, if $d(x,y) = \ell' \leq \ell$ and $r(C_y) \geq n + \ell'$, then there is an open path P from y to a site z with $d(y,z) \geq n + \ell'$, and hence $d(x,z) \geq n$. Considering, as usual, one particular path from x to y, it follows that

$$\mathbb{P}(r(C_x) \geq n) \geq p^{\ell'}\mathbb{P}(r(C_y) \geq n + \ell') \geq p^{\ell}\mathbb{P}(r(C_y) \geq n + \ell).$$

Since this holds for all x and y, we see that the behaviour of $\mathbb{P}(r(C_x) \geq n)$ as $n \to \infty$ is similar for all sites x. In the oriented case, however, we may have $d(x,z)$ much smaller than $d(y,z) - d(x,y)$, as in the example in Figure 4, and there does not seem to be an obvious reason why $\mathbb{P}(r(C_x) \geq n)$ should decay at roughly the same rate for different sites x. Menshikov, Molchanov and Sidorenko [1986] state in their Lemma 6.1 that this does hold for the oriented case, but their proof is for the unoriented case. Fortunately, it turns out that their Lemma 6.1 is not needed in the subsequent arguments, since it may be replaced by (4). (In [1986], Menshikov outlined his proof for the unoriented case, where the corresponding problem does not arise.)

Although oriented percolation is sometimes harder to work with than unoriented, it so happens that the proofs in this chapter are equally simple for the oriented and unoriented models. We state the results for the oriented case, as this implies the unoriented case. In fact, the reader interested only in the latter case may simply ignore all references to orientation, and so read the proofs as if they had been given for unoriented percolation.

4.3 Almost exponential decay of the radius – Menshikov's Theorem

Our main aim in this section is to present Menshikov's fundamental result that, under mild assumptions, below the critical probability $p_{\rm H}$ the open cluster containing the origin is very unlikely to be large. A consequence of this result is that the critical probabilities $p_{\rm T}$ and $p_{\rm H}$ are equal. Independently of Menshikov, Aizenman and Barsky proved similar results in a very different way: we shall outline their approach at the end of this section.

Recall that $R_n(x)$ is the event that there is an open path from an out-neighbour of x to some site in $S_n^+(x)$, so $\{x \xrightarrow{n}\}$ holds if and only if x is open and $R_n(x)$ holds. We shall start with a study of $\rho_n(x,p) = \mathbb{P}_p(R_n(x))$, and how this changes with p. As we shall use the Margulis–Russo formula, the first step is to understand what it means for a site to be pivotal for the event $R_n(x)$. As before, the context is site percolation on an oriented graph $\overrightarrow{\Lambda}$, so $\mathbb{P}_p = \mathbb{P}^{\rm s}_{\overrightarrow{\Lambda},p}$.

Suppose that the increasing event $R_n(x)$ holds. For convenience, suppose also that x is open, although this will not be relevant. As $R_n(x)$ is an increasing event, and we are assuming that $R_n(x)$ holds, a site y is pivotal for $R_n(x)$ if and only if $R_n(x)$ would not hold if the state of y were changed from open to closed, i.e., if all open paths from x to $S_n^+(x)$ pass through y. Let P be any open path from x to $S_n^+(x)$. Then all pivots (pivotal sites) appear on P. Let $b_1, b_2, \ldots, b_r$, $r \geq 0$, be the pivots in the order in which they appear along P. (Our notation b_i is chosen as we think of the pivots as *bridges* that any open path from x to $S_n^+(x)$ must cross.) Note that if P' is any other open path from x to $S_n^+(x)$, then not only must P' pass through each of the sites $b_1, \ldots, b_r$, but these sites appear along P' in the same order $b_1, b_2, \ldots$. Otherwise, there is an $s \geq 0$ such that the order of the pivots along P' starts with $b_1, b_2, \ldots, b_s, b_t$, where $t > s+1$. But then the union of the initial segment of P' up to b_t and the final segment of P starting at b_t contains an open path from x to $S_n^+(x)$ avoiding b_{s+1}, which is impossible; see Figure 6.

In fact, the more detailed picture is as follows: the set of sites on open paths from x to $S_n^+(x)$ forms a graph in which $b_1, \ldots, b_r$ are cut-vertices; see Figure 7. Taking $b_0 = x$, for $1 \leq i \leq r$, let T_i denote the set of sites on open paths from b_{i-1} to b_i, excluding b_{i-1} and b_i. Also, let T_{r+1} denote the set of sites other than b_r on open paths from b_r to $S_n^+(x)$. Then, for $1 \leq i \leq r$, the set $T_i \cup \{b_{i-1}, b_i\}$ contains two paths

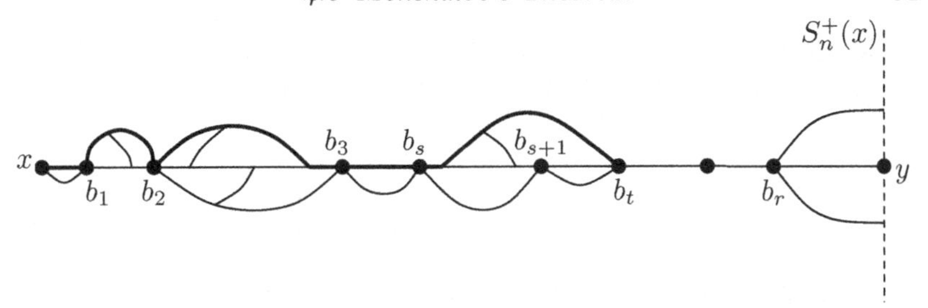

Figure 6. An impossible configuration: the pivots b_i appear in different orders along two open paths P, P' from x to $S_n^+(x)$. The path P is the straight line segment from x to y; the section of the path P' from x to b_t is drawn with thick lines. The path P' cannot 'jump' from b_s to b_t, $t > s+1$, since $P \cup P'$ would then contain a path from x to y avoiding b_{s+1}.

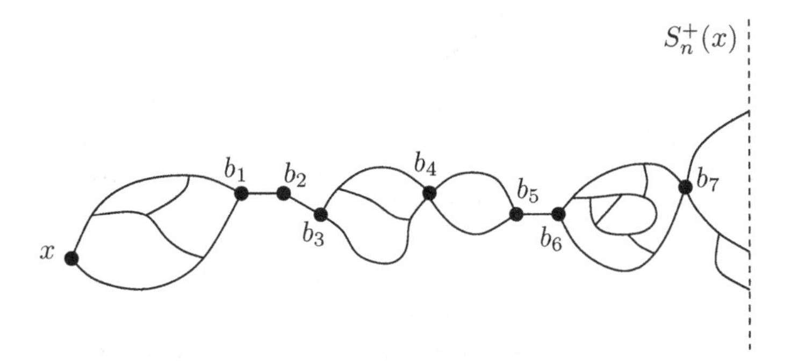

Figure 7. The set of sites on open paths from x to $S_n^+(x)$, shown with a possible arrangement of the pivots b_i for the event $R_n(x)$. Note that $\overrightarrow{b_1b_2}$, $\overrightarrow{b_2b_3}$ and $\overrightarrow{b_5b_6}$ are edges of $\overrightarrow{\Lambda}$.

from b_{i-1} to b_i that are vertex-disjoint apart from their endpoints; in the case where $\overrightarrow{b_{i-1}b_i} \in E(\overrightarrow{\Lambda})$, this holds trivially: both paths consist of this single bond. Also, T_{r+1} contains two paths from b_r to $S_n^+(x)$ that are vertex-disjoint except at b_r. For $i < j$, there is no (oriented) bond from a site in T_i to a site in T_j.

Let $b_0 = x$. Whenever $R_n(x)$ holds, let D_i, $i = 1, 2, \ldots$ denote the distance from b_{i-1} to b_i, where we set $D_i = \infty$ if there are fewer than i pivots. Let $1 \leq k \leq n$. Since D_1 is the distance from x to the first pivot, if $D_1 > k$ then either there is no pivot, i.e., there are (at least) two disjoint open paths from neighbours of x to $S_n^+(x)$, or the first pivot b_1 is at distance at least $k+1$ from x. In the latter case, there are two disjoint open paths from neighbours of x to b_1, and an open path from b_1 to $S_n^+(x)$ disjoint from T_1 and hence from these paths; see Figure 8.

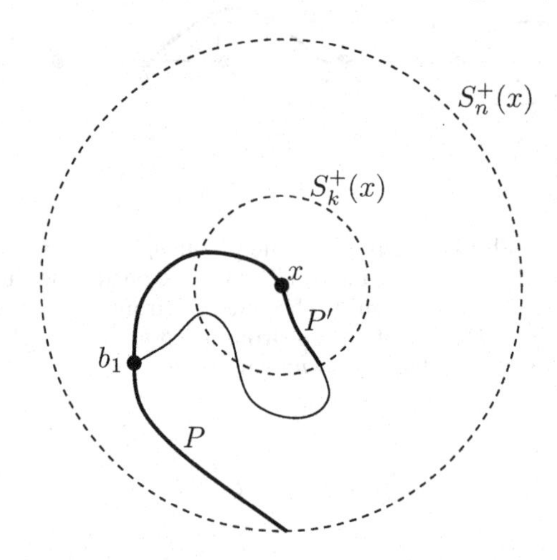

Figure 8. If $D_1 > k$, then, if there are any pivots for the event $R_n(x)$, the first, b_1, is at distance more than k from x. Thus there are open paths P from x to $S_n^+(x)$ and P' from x to $S_k^+(x)$ that are disjoint except at x.

In either case, there are paths P and P' from x to $S_n^+(x)$ and to $S_k^+(x)$ respectively, with P and P' disjoint except at x and all sites of P and P' other than x open. Thus, the event $R_n(x) \cap \{D_1 > k\}$ is contained in the event $R_n(x) \Box R_k(x)$. Hence, by the van den Berg–Kesten inequality, Theorem 5 of Chapter 2, we have

$$\mathbb{P}_p\big(R_n(x), D_1 > k\big) \leq \mathbb{P}_p\big(R_n(x)\big)\mathbb{P}_p\big(R_k(x)\big),$$

i.e.,

$$\mathbb{P}_p\big(D_1 > k \mid R_n(x)\big) \leq \rho_k(x, p).$$

The main tool in Menshikov's proof is a generalization of this observation.

When $R_n(x)$ holds and there are at least t pivots, we write $\mathbf{I}_t$ for the tth *interior cluster*, meaning the set of sites z such that there is a path P from x to z with $b_t \notin P$ and all sites of P other than x open; see Figure 9. Thus, if x is open, then $\mathbf{I}_t$ is the set of sites reachable from x by open paths not passing through b_t. We use bold font for the random variable $\mathbf{I}_t$ because it will be particularly important to distinguish $\mathbf{I}_t$ from its possible values I_t. (With other random variables, such as b_t, we do not bother.) When $b_t = y$ and $\mathbf{I}_t = I_t$, then every site in $I_t \cup \{y\}$ is

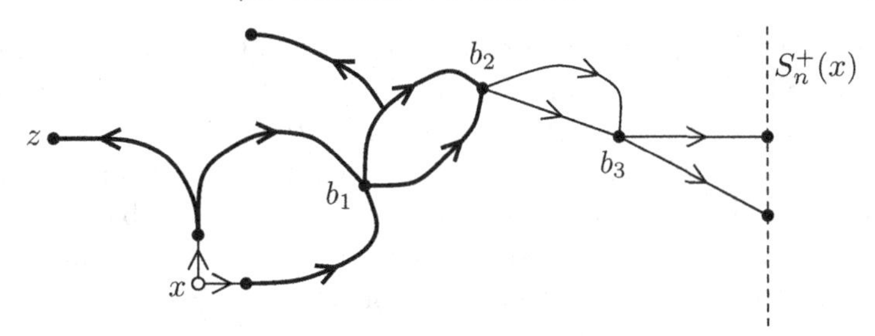

Figure 9. The 2nd interior cluster $\mathbf{I}_2$ consists of the thick lines including all of their endpoints except b_2. Note that $z \in \mathbf{I}_2$, even though z is not on an open path from a neighbour of x to $S_n^+(x)$.

open, and all sites in ∂I_t other than y are closed, where ∂I_t is the set of out-neighbours of $I_t \cup \{x\}$.

With this preparation, we are ready to present the key lemma for Menshikov's main theorem. As usual, $\mathbb{P}_p^{\mathrm{s}}$ denotes the probability measure in which the sites of $\overrightarrow{\Lambda}$ are open independently with probability p.

Lemma 4. *Let x be a site of an oriented graph $\overrightarrow{\Lambda}$, and let n, r and d_i, $1 \le i < r$, be positive integers. For $1 \le k \le n - \sum_{i=1}^{r-1} d_i$, we have*

$$\mathbb{P}_p^{\mathrm{s}}\big(D_r > k \mid E\big) \le \sup\{\rho_k(y,p) : y \in \overrightarrow{\Lambda}\}, \tag{5}$$

where E is the event

$$E = R_n(x) \cap \{D_1 = d_1, D_2 = d_2, \ldots, D_{r-1} = d_{r-1}\}$$

and, as before, $\rho_k(y,p) = \mathbb{P}_p^{\mathrm{s}}\big(R_k(y)\big)$.

Proof. Throughout this proof, we work within the finite subgraph of $\overrightarrow{\Lambda}$ consisting of sites at distance at most n from x. We write $\mathbb{P}_p$ for $\mathbb{P}_p^{\mathrm{s}}$. We have already dealt with the case $r = 1$ above, so we assume that $r > 1$.

We shall partition the event E according to the location y of the $(r-1)^{\mathrm{st}}$ pivot b_{r-1}, and according to the interior cluster $\mathbf{I} = \mathbf{I}_{r-1}$. For any possible value I of $\mathbf{I}$ consistent with $E \cap \{b_{r-1} = y\}$, let

$$E_{y,I} = R_n(x) \cap \{b_{r-1} = y\} \cap \{\mathbf{I} = I\}.$$

If $\mathbf{I} = I$, then $D_i = d_i$ for $1 \le i \le r-1$, so the events $E_{y,I}$ form a partition of E.

If $E_{y,I}$ holds then, as noted above, so does the event $F_{y,I}$ that all sites in $I \cup \{y\}$ are open, and all sites in $\partial I \setminus \{y\}$ are closed. In fact, $E_{y,I}$ holds if and only if $F_{y,I}$ holds and there is an open path P from a neighbour of y to $S_n^+(x)$ disjoint from $I \cup \partial I$. (Note that $y \in \partial I$.) Let us write $G_{y,I}$ for the event that there is such a path P.

Let $X = (I \cup \partial I)^c$, and let $\mathbb{P}_p^X$ denote the product probability measure in which each site of X is open independently with probability p. We may regard the event $G_{y,I}$ as an event in this probability space. Note that, in the measure $\mathbb{P}_p$, if we condition on $F_{y,I}$, then every site of X is open independently with probability p. (The event $F_{y,I}$ is defined 'without looking at' sites in X.) Hence,

$$\mathbb{P}_p\big(E_{y,I} \mid F_{y,I}\big) = \mathbb{P}_p^X(G_{y,I}),$$

and, as $E_{y,I} \subset F_{y,I}$,

$$\mathbb{P}_p(E_{y,I}) = \mathbb{P}_p(F_{y,I})\mathbb{P}_p^X(G_{y,I}).$$

Let $H_{y,I}$ denote the event that X contains an open path from a neighbour of y to $S_k^+(y)$. Suppose that $E_{y,I}$ holds and that $D_r > k$. Then $F_{y,I}$ holds, and, as in the case $r = 1$ above, $X \cup \{y\}$ contains open paths P, P', disjoint except at y, with P joining y to $S_n^+(x)$, and P' joining y to $S_k^+(y)$; see Figure 10. Thus, X contains disjoint witnesses for the

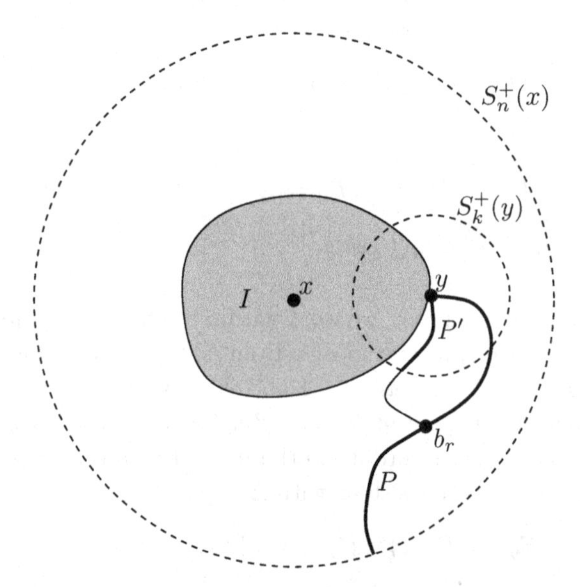

Figure 10. The shaded region represents the interior cluster $\mathbf{I} = \mathbf{I}_{r-1}$. If $E_{y,I}$ holds, then $\mathbf{I} = I$ and y is the $(r-1)^{\text{st}}$ pivot b_{r-1} for the event $R_n(x)$. If, in addition, $D_r > k$ with $d(x,y) + k \le n$, then there are disjoint paths P, P' from y to $S_n^+(x)$ and to $S_k^+(y)$, with P, $P' \subset X = (I \cup \partial I)^c$.

events $G_{y,I}$ and $H_{y,I}$, so $G_{y,I} \Box H_{y,I}$ holds. As $E_{y,I} \cap \{D_r > k\}$ is a subset of $F_{y,I}$, we have shown that

$$\mathbb{P}_p\big(E_{y,I} \cap \{D_r > k\}\big) \leq \mathbb{P}_p(F_{y,I})\mathbb{P}_p^X\big(G_{y,I} \Box H_{y,I}\big).$$

Applying the van den Berg–Kesten inequality , Theorem 5 of Chapter 2, to the product probability measure $\mathbb{P}_p^X$, we have

$$\mathbb{P}_p^X\big(G_{y,I} \Box H_{y,I}\big) \leq \mathbb{P}_p^X(G_{y,I})\mathbb{P}_p^X(H_{y,I}).$$

Combining the three relations above,

$$\mathbb{P}_p\big(E_{y,I} \cap \{D_r > k\}\big) \leq \mathbb{P}_p(E_{y,I})\mathbb{P}_p^X(H_{y,I}).$$

Now for any possible y and I, we have $\mathbb{P}_p^X(H_{y,I}) = \mathbb{P}_p(H_{y,I}) \leq \mathbb{P}_p\big(R_k(y)\big) = \rho_k(y,p)$. Thus

$$\mathbb{P}_p\big(E_{y,I} \cap \{D_r > k\}\big) \leq \mathbb{P}_p(E_{y,I})\rho_k(y,p),$$

i.e.,

$$\mathbb{P}_p\big(D_r > k \mid E_{y,I}\big) \leq \rho_k(y,p).$$

As the events $E_{y,I}$ partition E, the bound (5) follows. □

Lemma 4 is the key ingredient in the proof of Menshikov's main theorem. Although this lemma gives detailed information about the distribution of the distances to successive pivots, all we shall need is the simple consequence that, on average, there are many pivots if n is large. To state this lemma, let $N(E)$ denote the number of sites that are pivotal for an event E.

Lemma 5. *Let x be a site of an oriented graph $\vec{\Lambda}$, and let n and k be positive integers. Then*

$$\mathbb{E}_p\Big(N\big(R_n(x)\big) \mid R_n(x)\Big) \geq \lfloor n/k \rfloor \big(1 - \sup_y \rho_k(y,p)\big)^{\lfloor n/k \rfloor}.$$

Proof. Let $D_1, D_2, \ldots$ be as in Lemma 4. As the event

$$\{D_1, \ldots, D_{r-1} \leq k\} \cap R_n(x)$$

is a disjoint union of events of the form

$$\{D_1 = d_1, \ldots, D_{r-1} = d_{r-1}\} \cap R_n(x),$$

Lemma 4 implies that

$$\mathbb{P}_p\Big(D_r \leq k \mid D_1, \ldots, D_{r-1} \leq k, R_n(x)\Big) \geq 1 - \sup_y \rho_k(y,p), \qquad (6)$$

whenever $rk \le n$. In particular,

$$\mathbb{P}_p\Big(D_1 \le k \mid R_n(x)\Big) \ge 1 - \sup_y \rho_k(y,p).$$

Taking $r = 2$ in (6) we obtain

$$\begin{aligned}\mathbb{P}_p\Big(D_1, D_2 \le k \mid R_n(x)\Big) & \\ &= \mathbb{P}_p\Big(D_2 \le k \mid D_1 \le k, R_n(x)\Big)\mathbb{P}_p\Big(D_1 \le k \mid R_n(x)\Big) \\ &\ge \big(1 - \sup_y \rho_k(y,p)\big)^2.\end{aligned}$$

Continuing in this way, we see that

$$\mathbb{P}_p\Big(D_1, \dots, D_r \le k \mid R_n(x)\Big) \ge \big(1 - \sup_y \rho_k(y,p)\big)^r$$

for any $r \le \lfloor n/k \rfloor$.

If $D_1, \dots, D_r \le k$, then there are at least r pivotal sites for the event $R_n(x)$, so $N\big(R_n(x)\big) \ge r$. Thus,

$$\begin{aligned}\mathbb{E}_p\Big(N\big(R_n(x)\big) \mid R_n(x)\Big) &\ge \lfloor n/k \rfloor \mathbb{P}_p\Big(D_1, \dots, D_{\lfloor n/k \rfloor} \le k \mid R_n(x)\Big) \\ &\ge \lfloor n/k \rfloor \big(1 - \sup_y \rho_k(y,p)\big)^{\lfloor n/k \rfloor},\end{aligned}$$

as claimed. □

We are now ready to present the two fundamental results of Menshikov [1986] (see also Menshikov, Molchanov and Sidorenko [1986]). The first shows that, under very mild conditions, we have almost exponential decay of the radius of the open cluster below p_{H}. As we shall see, an immediate consequence of this result is that, for a large class of graphs, the critical probabilities p_{T} and p_{H} coincide. The results below are stated for site percolation on oriented graphs. Using the equivalences between various percolation models described in Section 2, corresponding results for bond percolation and for unoriented graphs follow easily.

Here, then, is Menshikov's main theorem. In this result, $p_{\mathrm{H}}^{\mathrm{s}}(\overrightarrow{\Lambda})$ denotes the common value of the critical probabilities $p_{\mathrm{H}}^{\mathrm{s}}(\overrightarrow{\Lambda}; x)$, whose existence is guaranteed by Lemma 2.

Theorem 6. *Let $\overrightarrow{\Lambda}$ be an infinite, locally finite oriented multi-graph with $C_{\overrightarrow{\Lambda}}$ finite and strongly connected, and let $p < p_{\mathrm{H}}^{\mathrm{s}}(\overrightarrow{\Lambda})$. Then there is an $\alpha > 0$ such that*

$$\mathbb{P}_p^{\mathrm{s}}(x \xrightarrow{n}) \le \exp\big(-\alpha n/(\log n)^2\big) \tag{7}$$

for all sites x and integers $n \geq 2$.

The proof we present follows that given by Menshikov, Molchanov and Sidorenko [1986] very closely. The basic idea is to fix $p < p_0 < p_{\mathrm{H}}^{\mathrm{s}}(\overrightarrow{\Lambda})$, and to use Lemma 5 to show that at probability p_0, if n is large then the expected number of pivots for the event $R_n(x)$ is large: in doing this, we use $\theta(p_0) = 0$ to show that $\sup_y \rho_k(y, p_0)$ is small for k large. Then, from the Margulis–Russo formula, it will follow that $\rho_n(x, p') = \mathbb{P}_{p'}(R_n(x))$ decreases rapidly as p' decreases. Finally, we feed the new bounds on the function ρ back into Lemma 5. Amazingly, even though we have no a priori information on the rate at which $\sup_y \rho_k(y, p_0)$ tends to 0 as $k \to \infty$, applying Lemma 5 again and again with carefully chosen parameters, (7) can be deduced.

Proof. Since throughout our argument we work with site percolation, we shall write $\mathbb{P}_p$ for $\mathbb{P}_p^{\mathrm{s}}$ and p_{H} for $p_{\mathrm{H}}^{\mathrm{s}}$.

As before, for $p \in (0,1)$, let $\rho_n(p) = \sup_{x \in \overrightarrow{\Lambda}} \rho_n(x, p)$. Note that, by Lemma 3, for any $p < p_{\mathrm{H}}(\overrightarrow{\Lambda})$ we have

$$\rho_n(p) \to 0 \tag{8}$$

as $n \to \infty$.

By Lemma 5, we have

$$\mathbb{E}_p\Big(N\big(R_n(x)\big) \mid R_n(x)\Big) \geq \lfloor n/k \rfloor (1 - \rho_k(p))^{\lfloor n/k \rfloor}$$

for any site x and positive integers n and k. By the Margulis–Russo formula (Lemma 9 of Chapter 2), for any increasing event E we have

$$\frac{\mathrm{d}}{\mathrm{d}p}\mathbb{P}_p(E) = \mathbb{E}_p\big(N(E)\big) \geq \mathbb{E}_p\big(N(E)\mathbf{1}_E\big) = \mathbb{E}_p\big(N(E) \mid E\big)\mathbb{P}_p(E),$$

so

$$\frac{\mathrm{d}}{\mathrm{d}p} \log \mathbb{P}_p(E) = \frac{1}{\mathbb{P}_p(E)} \frac{\mathrm{d}}{\mathrm{d}p}\mathbb{P}_p(E) \geq \mathbb{E}_p\big(N(E) \mid E\big).$$

In particular, taking $E = R_n(x)$,

$$\frac{\mathrm{d}}{\mathrm{d}p} \log \rho_n(x, p) \geq \lfloor n/k \rfloor (1 - \rho_k(p))^{\lfloor n/k \rfloor}.$$

Fix $p_- < p_+$. As $\rho_k(p)$ is an increasing function of p, for any $p \in [p_-, p_+]$ the right-hand side above is at least the value at p_+, so

$$\frac{\rho_n(x, p_-)}{\rho_n(x, p_+)} \leq \exp\Big(-(p_+ - p_-)\lfloor n/k \rfloor (1 - \rho_k(p_+))^{\lfloor n/k \rfloor}\Big).$$

As the bound on the ratio is independent of x, it follows that

$$\rho_n(p_-) \le \rho_n(p_+) \exp\left(-(p_+ - p_-)\lfloor n/k\rfloor(1-\rho_k(p_+))^{\lfloor n/k\rfloor}\right). \qquad (9)$$

Menshikov's proof of (7) involves no further combinatorics: 'all' we need to do is apply (9) repeatedly. However, it is far from easy to find the right way to do this.

Continuing with the proof of (7), let $p < p_{\mathrm{H}}(\overrightarrow{\Lambda})$ be fixed from now on. Pick p_0 satisfying $p < p_0 < p_{\mathrm{H}}(\overrightarrow{\Lambda})$. By (8) we have $\rho_n(p_0) \to 0$ as $n \to \infty$. Writing ρ_0 for $\rho_{n_0}(p_0)$, it follows that if n_0 is sufficiently large, then $\rho_0 \le 1/100$ and $\rho_0 \log(1/\rho_0) \le (p_0 - p)/6$. Let us fix such an n_0 from now on.

Writing ρ_i for $\rho_{n_i}(p_i)$, we inductively define two sequences $n_0 \le n_1 \le n_2 \le \cdots$ and $p_0 \ge p_1 \ge p_2 \ge \cdots$ by

$$n_{i+1} = n_i \lceil 1/\rho_i \rceil \quad \text{and} \quad p_{i+1} = p_i - \rho_i \log(1/\rho_i),$$

as long as $p_{i+1} > 0$; in fact, as we shall see later, $p_i > p$ for all p. Let us apply (9) with $k = n_i$, $n = n_{i+1}$, $p_+ = p_i$ and $p_- = p_{i+1}$. Thus $\lfloor n/k \rfloor = \lfloor n_{i+1}/n_i \rfloor = \lceil 1/\rho_i \rceil$, and $\rho_k(p_+) = \rho_{n_i}(p_i) = \rho_i$, so (9) gives

$$\rho_{n_{i+1}}(p_{i+1}) \le \rho_{n_{i+1}}(p_i) \exp\left(-(p_i - p_{i+1})\lceil 1/\rho_i \rceil (1-\rho_i)^{\lceil 1/\rho_i \rceil}\right).$$

The sequence ρ_i is decreasing, so $\rho_i \le \rho_0 \le 1/100$ for every i, from which it follows that $(1-\rho_i)^{\lceil 1/\rho_i \rceil} \ge 1/3$. Thus we have

$$\begin{aligned}\rho_{n_{i+1}}(p_{i+1}) &\le \rho_{n_{i+1}}(p_i) \exp(-\rho_i \log(1/\rho_i)\lceil 1/\rho_i \rceil/3) \\ &\le \rho_{n_{i+1}}(p_i) \exp(-\log(1/\rho_i)/3) = \rho_{n_{i+1}}(p_i)\rho_i^{1/3}.\end{aligned}$$

As $n_{i+1} \ge n_i$, the event $R_{n_{i+1}}(x)$ is a subset of $R_{n_i}(x)$ for any x, so

$$\rho_{i+1} = \rho_{n_{i+1}}(p_{i+1}) \le \rho_{n_i}(p_i)\rho_i^{1/3} = \rho_i^{4/3}. \qquad (10)$$

From (10) we have $\rho_i \le \rho_0^{(4/3)^i}$, so, very crudely, $\rho_i \le \rho_0/e^i$ for every i. As $x \log(1/x)$ is increasing on $(0, 1/e)$,

$$\begin{aligned}p_0 - p_i &= \sum_{j<i} \rho_j \log(1/\rho_j) \le \sum_{j<i} e^{-j}\rho_0 \log(e^j/\rho_0) \\ &= \sum_{j<i} e^{-j}\rho_0(j + \log(1/\rho_0)) \le 2\rho_0\log(1/\rho_0) + \rho_0 < (p_0 - p)/2,\end{aligned}$$

where the second last inequality uses $\sum_{j\ge 0} je^{-j} \le 1$ and the final inequality holds by our choice of n_0. It follows that the construction of the sequences n_i, p_i continues indefinitely, and $p_i > (p_0 + p)/2$ for every i.

Let $p' = (p_0 + p)/2 > p$. At this point we have constructed an increasing sequence n_i, and a decreasing sequence ρ_i, such that

$$\rho_{n_i}(p') < \rho_{n_i}(p_i) = \rho_i$$

for $i = 0, 1, 2, \ldots$. To understand the significance of this bound, we should compare n_i and ρ_i.

Let $s_i = \lceil 1/\rho_i \rceil$, so $n_i = n_0 \prod_{j<i} s_i$, and $s_0 = \lceil 1/\rho_0 \rceil \geq 100$. Since $s_i \geq s_0 \geq 100$, the rounding in the definition of s_i makes little difference. In particular, (10) implies that $s_{i+1} \geq s_i^{5/4}$, say. Hence, for $1 \leq j \leq i$, we have $s_{i-j} \leq s_i^{(4/5)^j}$, which gives

$$\prod_{j<i} s_j = \prod_{1 \leq j \leq i} s_{i-j} \leq s_i^{\sum_{j \geq 1}(4/5)^j} = s_i^4,$$

so $n_{i+1} = s_i n_i \leq s_i^5 n_0$.

Let $n \geq n_0$ be arbitrary. Then there is an i with

$$n_i \leq n < n_{i+1} = s_i n_i \leq s_i^5 n_0.$$

But

$$\rho_n(p') \leq \rho_{n_i}(p') \leq \rho_{n_i}(p_i) = \rho_i \leq 2/s_i \leq 2(n/n_0)^{-1/5}.$$

As n_0 is fixed, it follows that there is a constant c such that

$$\rho_n(p') \leq cn^{-1/5} \tag{11}$$

for all $n \geq 1$. This weak polynomial bound is a far cry from (7), but in fact the proof is almost complete. All that remains is to use (9), this time in a straightforward way.

Fix p'' with $p < p'' < p'$. For $n \geq 1$, let $k_1 = k_1(n) = \lfloor n^{5/6} \rfloor$. Then $\lfloor n/k_1 \rfloor = \Theta(n^{1/6})$, and, from (11), $\rho_{k_1}(p') = O(n^{-1/6})$. It follows that $(1 - \rho_{k_1}(p'))^{\lfloor n/k_1 \rfloor}$ is bounded away from zero. Hence, applying (9) with $k = k_1(n)$, $p_+ = p'$ and $p_- = p''$,

$$\rho_n(p'') \leq \exp\bigl(-(p' - p'')\Omega(n^{1/6})\bigr) = \exp\bigl(-\Omega(n^{1/6})\bigr) \tag{12}$$

as $n \to \infty$.

Fix p''' with $p < p''' < p''$, and let $k_2 = k_2(n) = \lfloor (\log n)^7 \rfloor$ (for $n \geq 2$). From (12), we have

$$\rho_{k_2}(p'') = \exp\bigl(-\Omega\bigl((\log n)^{7/6}\bigr)\bigr) = o(n^{-1}),$$

so $(1 - \rho_{k_2}(p''))^{\lfloor n/k_2 \rfloor} \to 1$ as $n \to \infty$. Applying (9) with $k = k_2(n)$ it follows that

$$\rho_n(p''') = \exp\bigl(-\Omega(n/(\log n)^7)\bigr).$$

One final iteration now gives the result: taking $k_3 = \log n(\log\log n)^8$ it follows that $\rho_{k_3}(p''') = o(n^{-1})$, and applying (9) once more we find that

$$\rho_n(p) = \exp\bigl(-\Omega\bigl(n/(\log n(\log\log n)^8)\bigr)\bigr), \tag{13}$$

and (7) follows. □

The method above gives a slightly stronger bound than the inequality (7) claimed in Theorem 6. Indeed, the unappealing final estimate (13) above is stronger than (7). Iterating further, one can push the bound almost to $\exp(-n/\log n)$, but this is as far as the method seems to go.

As noted earlier, under a very mild additional assumption, Theorem 6 immediately implies another theorem of Menshikov, that $p_{\mathrm{T}} \geq p_{\mathrm{H}}$, and so $p_{\mathrm{T}} = p_{\mathrm{H}}$.

Theorem 7. *Let $\overrightarrow{\Lambda}$ be an infinite, locally finite oriented multi-graph with $C_{\overrightarrow{\Lambda}}$ finite and strongly connected. If there is a constant C such that*

$$|S_n^+(x)| \leq \exp\bigl(Cn/(\log n)^3\bigr) \tag{14}$$

for every site x and integer n, then $p_{\mathrm{T}}^{\mathrm{s}}(\overrightarrow{\Lambda}) = p_{\mathrm{H}}^{\mathrm{s}}(\overrightarrow{\Lambda})$.

Proof. By Lemma 2, the critical probabilities $p_{\mathrm{H}}^{\mathrm{s}}(\overrightarrow{\Lambda};x)$ and $p_{\mathrm{T}}^{\mathrm{s}}(\overrightarrow{\Lambda};x)$ are independent of the site x. Let $p < p_{\mathrm{H}}^{\mathrm{s}}(\overrightarrow{\Lambda})$, and let x be any site of $\overrightarrow{\Lambda}$. Then

$$\chi_x(\overrightarrow{\Lambda};p) = \mathbb{E}_p^{\mathrm{s}}(|C_x|) = \sum_n \sum_{y\in S_n^+(x)} \mathbb{P}_p^{\mathrm{s}}(y \in C_x) \leq \sum_n |S_n^+(x)|\rho_n(x,p).$$

The final sum converges by Theorem 6, so $p \leq p_{\mathrm{T}}^{\mathrm{s}}(\overrightarrow{\Lambda};x) = p_{\mathrm{T}}^{\mathrm{s}}(\overrightarrow{\Lambda})$. As $p < p_{\mathrm{H}}^{\mathrm{s}}(\overrightarrow{\Lambda})$ was arbitrary, it follows that $p_{\mathrm{T}}^{\mathrm{s}}(\overrightarrow{\Lambda}) \geq p_{\mathrm{H}}^{\mathrm{s}}(\overrightarrow{\Lambda})$, so $p_{\mathrm{T}}^{\mathrm{s}}(\overrightarrow{\Lambda}) = p_{\mathrm{H}}^{\mathrm{s}}(\overrightarrow{\Lambda})$. □

Using the equivalence between percolation models discussed in Section 2, Theorems 6 and 7 immediately imply corresponding statements for bond percolation on $\overrightarrow{\Lambda}$, and for unoriented site and bond percolation. Many of the most interesting percolation models satisfy the assumptions of Theorem 7 (after the appropriate translation to oriented site percolation). Indeed, in many cases (for example site or bond percolation on $\mathbb{Z}^d$), the class-graph has only a single vertex, and the sizes of the neighbourhoods $S_n^+(x)$ in $\overrightarrow{\Lambda}$ grow polynomially in n.

It is tempting to think that for any finite-type graph, either the 'growth function' $\gamma(n) = \sup_{x\in\Lambda} |S_n(x)|$ will be bounded by a polynomial, or $\gamma(n) \geq \exp(an)$ for some $a > 0$. Indeed, Milnor [1968] conjectured that this assertion holds in the special case of Cayley graphs of finitely generated groups. Surprisingly, even this is not true: Grigorchuk [1983; 1984] gave a counterexample. Furthermore, solving in the negative a problem of Gromov [1981], Wilson [2004] proved that a group of exponential growth need not be of uniformly exponential growth. (See also Muchnik and Pak [2001], Pyber [2004], and Eskin, Mozes and Oh [2005].)

The condition that $C_{\overrightarrow{\Lambda}}$ be finite is essential for Theorem 7. We illustrate this for unoriented bond percolation. Let G_i be a 2^{i^2} by 2^{i^2} 'grid graph', i.e., a square subgraph of $\mathbb{Z}^2$ with 2^{2i^2} vertices, and let Λ be the graph obtained by stringing together the graphs G_i as in Figure 11.

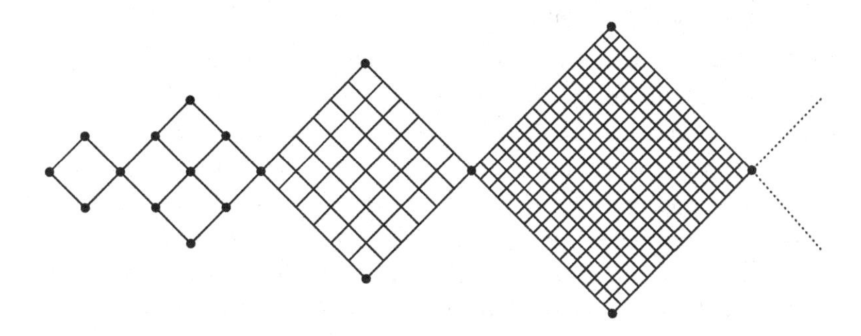

Figure 11. A series of grid-graphs strung together by their opposite corners; for the example, the sizes of the grids grow super-exponentially.

Note that Λ may be embedded into $\mathbb{Z}^2$ in a way that preserves graph distance, so $|S_n(\Lambda; x)| \leq |S_n(\mathbb{Z}^2; 0)| = 4n$ for $n \geq 1$, and the growth condition (14) is satisfied. If $p > 1/2$, then it follows easily from the results in Chapter 3 that the probability that there is an open path from one corner of G_i to the opposite corner is bounded below by a constant. (This may be shown by using exponential decay for $p < 1/2$ and considering the dual lattice.) Also, the expected number of sites of G_i that may be reached by open paths from a given corner is $\Theta(|G_i|)$. Since $|G_i|$ grows super-exponentially, it follows that $p_{\mathrm{T}}^{\mathrm{b}}(\Lambda) \leq 1/2$, so, as $\Lambda \subset \mathbb{Z}^2$, we have $p_{\mathrm{T}}^{\mathrm{b}}(\Lambda) = 1/2$. On the other hand, $p_{\mathrm{H}}^{\mathrm{b}}(\Lambda) = 1$, since, for $p < 1$, the probability that each of the infinitely many cut vertices is incident with at least two open edges is 0. A slight variant of this construction works for oriented site percolation: take the line graph, and then replace each bond by two oriented bonds.

In the unoriented case, we do require the underlying graph Λ to be of finite type. However, there is little reason to consider the graph structure of C_Λ, defined analogously to $C_{\vec{\Lambda}}$: the graphs Λ we study are always connected, so C_Λ is also connected. Note that if $\vec{\Lambda} = \varphi_s(\Lambda)$ is obtained from Λ by replacing each bond by two oriented bonds, as in Section 2, then two sites x and y are out-like in $\vec{\Lambda}$ if and only if they are equivalent in Λ. Thus, if Λ is a connected finite-type graph, then $C_{\vec{\Lambda}}$ is finite and strongly connected. For locally finite graphs, the transformation mapping an oriented or unoriented graph to its line graph also preserves the finiteness of the class-graph. Finally, these transformations also preserve the growth rate of the neighbourhoods, so Theorem 7 does indeed imply corresponding results for bond percolation, and for unoriented percolation.

In the statement and proofs of Theorems 6 and 7, we took every site to have the same probability p of being open. These results extend easily to a somewhat more general setting, in which different sites x have different probabilities p_x of being open, with the states of the sites independent. Recall that the definition of the class-graph $C_{\vec{\Lambda}}$, or the equivalence of sites in the unoriented case, involves isomorphisms between subgraphs of $\vec{\Lambda}$, or automorphisms of Λ. Naturally, we now require any such isomorphism to preserve the 'weights', i.e., the probabilities that the sites are open. Thus, out-like sites still behave in the same way for percolation. As we require $C_{\vec{\Lambda}}$ to be finite, there is one probability p_i for each out-like class, so we obtain a percolation model parametrized by a vector $\mathbf{p} = (p_1, \ldots, p_k)$. For example, one could take $\vec{\Lambda} = \vec{\mathbb{Z}}^2$, with alternate sites having probabilities p_1 and p_2 of being open, resulting in two out-like classes.

Theorem 6 carries over to this 'finite-type weighted' context in a natural way: using self-explanatory notation, the result is that if for some $p_1, \ldots, p_k$ we have $\theta(\vec{\Lambda}; p_1, \ldots, p_k) = 0$ then, whenever $p_i' < p_i$ for each i, an assertion equivalent to (7) holds in the probability measure $\mathbb{P}_{\mathbf{p}'}$. Here $\mathbb{P}_{\mathbf{p}'}$ is the measure in which sites of class i are open with probability p_i', with the states of all sites independent. There are two ways to see this. One is to note that the proof given above carries over *mutatis mutandis.* In particular, the Margulis–Russo formula (Lemma 9 of Chapter 2) states that for an increasing event E, the sum of the partial derivatives of $\mathbb{P}_{\mathbf{p}}(E)$ is exactly the expected number $\mathbb{E}(N(E))$ of sites that are pivotal for E. If we decrease each p_i at the same rate, then the calculations in the proofs above are exactly the same as for the uniform case.

Alternatively, one can realize an arbitrarily good approximation to the weighted model as an unweighted model, by choosing p close to 1 and replacing each site open with probability p_x by an oriented path of length ℓ chosen so that p^ℓ is within a factor p of p_x. Using this idea it is easy to deduce the weighted version of Menshikov's Theorem from the unweighted one.

Note that we require $p'_i < p_i$ for *all* i: in general, it is not enough to require $p'_i \le p_i$ for all i and $p'_i < p_i$ for some i. This may be seen by considering any graph in which sites of one type are 'useless', for example, the graph $\vec{\Lambda}$ obtained from $\vec{\mathbb{Z}}^2$ by adding a directed cycle to each site, where the sites of $\vec{\mathbb{Z}}^2$ are open with probability p_1, and the new sites with probability p_2.

Many of the other results we shall present also have natural weighted versions. Most of the time, we shall present only the unweighted version, and shall not discuss the simple modifications or deductions needed for the weighted versions.

Aizenman and Barsky [1987] gave a result closely related to Theorem 7. Their proof uses very different methods to those of Menshikov, although it also relies on the van den Berg–Kesten inequality and the Margulis–Russo formula. A key idea of the proof is the introduction of a 2-variable generalization of the percolation probability $\theta(p)$. Considering bond percolation on a graph Λ in which all sites are equivalent, this may be written as

$$M(\beta, h) = 1 - \sum_{n=1}^{\infty} e^{-nh} \mathbb{P}^{\mathrm{b}}_{\Lambda,p}\big(|C_x| = n\big),$$

where $p = 1 - e^{-\beta}$. The reason for the reparametrization is that Aizenman and Barsky consider a more general model of percolation on $\mathbb{Z}^d$, where 'long-range' bonds are allowed: each edge $\mathbf{xy}$ of the complete graph on $\mathbb{Z}^d$ is open with probability $1 - \exp(-\beta J(\mathbf{x} - \mathbf{y}))$, where J is a non-negative symmetric function on $\mathbb{Z}^d$. Taking $h = 0$, the sum above gives exactly $\mathbb{P}^{\mathrm{b}}_{\Lambda,p}\big(|C_x| < \infty\big) = \theta(p)$. Writing $|J|$ for $\sum_{\mathbf{x}\in\mathbb{Z}^d} J(\mathbf{x})$, Aizenman and Barsky prove two differential inequalities for $M = M(\beta, h)$, namely

$$\frac{\partial M}{\partial \beta} \le |J| M \frac{\partial M}{\partial h},$$

and

$$M \le h\frac{\partial M}{\partial h} + M^2 + \beta M \frac{\partial M}{\partial \beta}.$$

The latter inequality generalizes an earlier result of Chayes and Chayes [1986a] that

$$\theta(p) \le \theta(p)^2 + \theta(p)\frac{\partial \theta(p)}{\partial p}$$

for bond percolation on $\mathbb{Z}^d$. (For site percolation, the term $\theta(p)^2$ is replaced by $p^{-1}\theta(p)^2$.) Using these differential inequalities, Aizenman and Barsky show that $M(\beta_{\mathrm{T}}, h) \ge ch^{1/2}$ for some $c \ge 0$, where β_{T} corresponds to p_{T}. Finally, they deduce that $p_{\mathrm{T}} = p_{\mathrm{H}}$. Note that the quantity $M(\beta, h)$ has a simple combinatorial description: if we extend the percolation model by adding a single new vertex G (the 'ghost' vertex), and join each site of $\mathbb{Z}^d$ to G independently with probability $1-e^{-h}$, then $M(\beta, h)$ is the probability that there is no open path from 0 to G.

4.4 Exponential decay of the radius

Our aim in this section is to show that, under mild conditions, for $p < p_{\mathrm{T}}$ the distribution of the radius of the open cluster containing a given site has an exponential tail. Theorem 6 does not quite give this, although it comes close. However, exponential decay follows from Theorem 7 by a result of Hammersley [1957b]; a special case of this result was described at the start of the chapter. The statement and proof that follow appear considerably more complicated than those for the special case of bond percolation on $\mathbb{Z}^d$ considered at the start of the chapter. There are two reasons: first, there is a minor technical complication that arises in the case of site percolation. Second, we shall separate out the heart of the result as a lemma, and we wish to state a reasonably strong form of this lemma for future reference. If our aim were only to prove Hammersley's result, Theorem 9 below, then a weaker form of the lemma would suffice. In fact, the much greater generality considered here introduces no essential complications.

Given a site x of a directed graph Λ, let

$$B_r^+(x) = \bigcup_{k=0}^{r} S_k^+(x) = \{y \in \overrightarrow{\Lambda} : d(x,y) \le r\}$$

be the *out-ball* of radius r centred at x. Let $N_r^+(x)$ denote the number of sites in $S_r^+(x)$ that may be reached by an open path (i.e., self-avoiding walk) P in $B_r^+(x)$ starting at an out-neighbour y of x.

Lemma 8. *Let $\overrightarrow{\Lambda}$ be an oriented multi-graph, $r \geq 1$ an integer, and $\gamma < 1$ a real number. If $\mathbb{E}_p^{\mathrm{s}}(N_r^+(x)) \leq \gamma$ for every site x of $\overrightarrow{\Lambda}$, then*

$$\mathbb{P}_p^{\mathrm{s}}(x \xrightarrow{n}) \leq p\gamma^{\lfloor n/r \rfloor}$$

for every site x of $\overrightarrow{\Lambda}$ and every $n \geq 1$.

Proof. As usual, we write $\mathbb{P}_p$ for $\mathbb{P}_p^{\mathrm{s}}$. Let $r, n \geq 1$. Recall that $R_m(x) = \{x^+ \xrightarrow{m}\}$ denotes the event that there is an open path P from an out-neighbour of x to a site $y \in S_m^+(x)$, and that $\rho_m(x,p) = \mathbb{P}_p(R_m(x))$. By splitting the path P at the point it first reaches $S_r^+(x)$, we see that $R_{r+n}(x)$ is the union of the events E_y, $y \in S_r^+(x)$, where E_y is the event that there is an open path P in $B_{r+n}^+(x)$ from an out-neighbour of x to $S_{r+n}^+(x)$ that first meets $S_r^+(x)$ at y. (This is not, in general, a disjoint union, since there may be many such paths P.) Splitting such a path P at y, the initial segment P' from a neighbour of x to y lies in $B_r^+(x)$, and is thus a witness for the event E'_y that there is a path in $B_r^+(x)$ from an out-neighbour of x to y; see Figure 12. The remainder, P'', of P is a path starting at an out-neighbour of y and ending in $S_{r+n}^+(x)$. Thus P'' must contain a site of $S_n^+(y)$, so P'' (or an initial segment of this path) is a witness for $R_n(y)$. As P' and P'' are disjoint, we have

$$E_y \subset E'_y \,\square\, R_n(y).$$

Writing $\rho_n(p)$ for $\sup_y \rho_n(y,p)$, as before, by the van den Berg–Kesten inequality it follows that

$$\begin{aligned}\rho_{r+n}(x,p) &\leq \sum_{y \in S_r^+(x)} \mathbb{P}_p(E_y) \leq \sum_{y \in S_r^+(x)} \mathbb{P}_p(E'_y)\mathbb{P}_p(R_n(y)) \\ &\leq \sum_{y \in S_r^+(x)} \mathbb{P}_p(E'_y)\rho_n(p) = \mathbb{E}_p(N_r^+(x))\rho_n(p) \leq \gamma\rho_n(p).\end{aligned}$$

As this holds for every x, we have $\rho_{r+n}(p) \leq \gamma\rho_n(p)$, so $\rho_n(p) \leq \gamma^{\lfloor n/r \rfloor}$ for every n. As $\mathbb{P}_p(x \xrightarrow{n}) = p\rho_n(x)$, the result follows. □

Note that we work with paths starting at a neighbour of a given site to avoid 're-using' a site when we concatenate paths. This complication does not arise for bond percolation, where the corresponding result is that, if the expected number of sites of $S_r^+(x)$ that may be reached

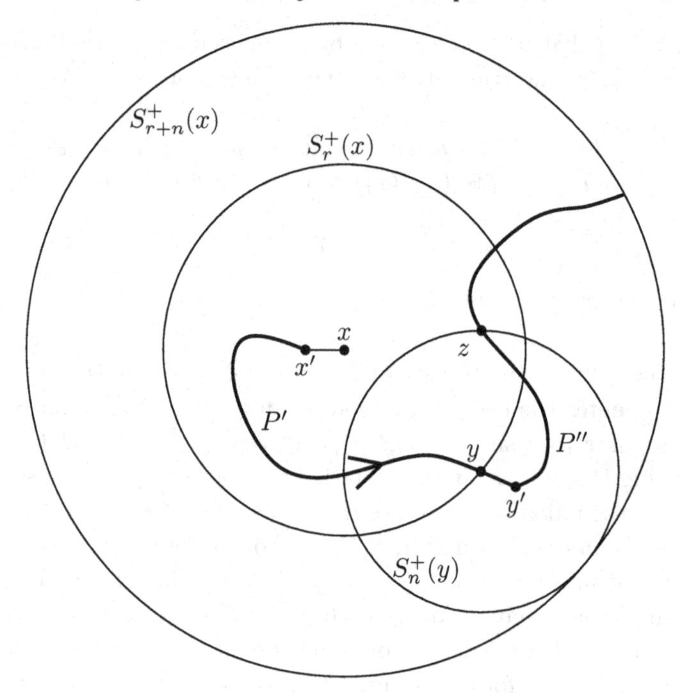

Figure 12. The thick lines show an open path P witnessing the event $R_{r+n}(x)$. The sub-paths P' from x' to y and P'' from y' to z are disjoint witnesses for the events E'_y and $R_n(y) = \{y^+ \xrightarrow{n}\}$ respectively.

from x by open paths within $B^+_r(x)$ is at most γ for every x, then $\mathbb{P}_p(x \xrightarrow{n}) \le \gamma^{\lfloor n/r \rfloor}$ holds for every site x and integer n. From Lemma 8, it is very easy to deduce Hammersley's result.

Theorem 9. *Let $\overrightarrow{\Lambda}$ be an oriented multi-graph with $C_{\overrightarrow{\Lambda}}$ finite and strongly connected, and let $p < p^{\mathrm{s}}_{\mathrm{T}}(\overrightarrow{\Lambda})$. Then there is an $\alpha > 0$ such that*

$$\mathbb{P}^{\mathrm{s}}_p(x \xrightarrow{n}) \le \exp(-\alpha n) \tag{15}$$

for every site x and integer $n \ge 1$.

Proof. Let x be a site of $\overrightarrow{\Lambda}$. By Lemma 2 we have $p^{\mathrm{s}}_{\mathrm{T}}(\overrightarrow{\Lambda}; x) = p^{\mathrm{s}}_{\mathrm{T}}(\overrightarrow{\Lambda})$, so $\chi^{\mathrm{s}}_x(p) < \infty$. Now

$$\chi^{\mathrm{s}}_x(p) = \sum_r \sum_{y \in S^+_r(x)} \mathbb{P}^{\mathrm{s}}_p(x \to y) = p \sum_r \sum_{y \in S^+_r(x)} \mathbb{P}^{\mathrm{s}}_p(x^+ \to y),$$

where $\{x^+ \to y\}$ is the event that there is an open path from an out-neighbour of x to y. (Thus, $R_r(x) = \bigcup_{y \in S_r^+(x)} \{x^+ \to y\}$. Also, as before, if $\{x^+ \to y\}$ holds then there is an open path from $S_1^+(x)$ to y not visiting x, so the events 'x is open' and $\{x^+ \to y\}$ are independent.) Let

$$\gamma_r(x) = \sum_{y \in S_r^+(x)} \mathbb{P}_p^{\mathrm{s}}(x^+ \to y).$$

Then $\sum_r \gamma_r(x)$ is convergent for each x, so $\gamma_r(x) \to 0$. As $\gamma_r(x)$ depends only on the out-class $[x]$ of x, and there are only finitely many out-classes, there is an r such that $\gamma_r(x) \leq 1/2$ for every site x. Since $\mathbb{E}_p^{\mathrm{s}}(N_r^+(x)) \leq \gamma_r(x)$, the result then follows from Lemma 8. □

Let us remark that, under the assumptions of Theorem 9, if the expected number of open paths starting at each site x is finite, then (15) can be deduced without using the van den Berg–Kesten inequality. However, it is perfectly possible for this expectation to be infinite even when $\chi_x^{\mathrm{s}}(p)$ is finite, as shown by the graph in Figure 13. Indeed, for this

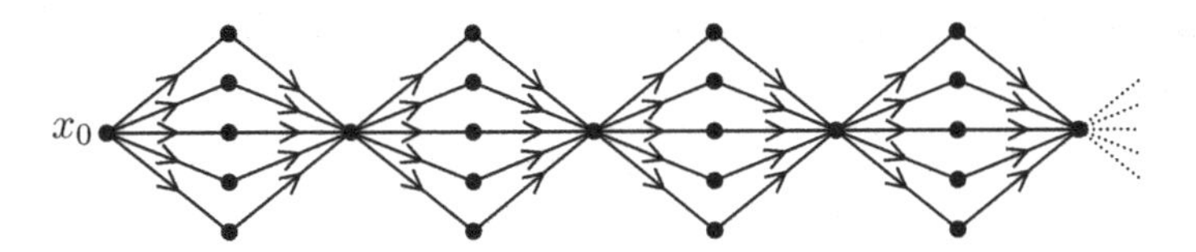

Figure 13. A graph with $p_{\mathrm{T}}^{\mathrm{s}} = 1$ in which there are exponentially many paths of length ℓ from a given site.

graph $p_{\mathrm{H}}^{\mathrm{s}} = p_{\mathrm{T}}^{\mathrm{s}} = 1$, but taking each site to be open independently with probability p, the expected number of open paths of length ℓ starting at x_0 is $5^{\lceil \ell/2 \rceil} p^{\ell+1}$, which tends to infinity as $\ell \to \infty$ for any $p > 1/\sqrt{5}$. Performing the same construction starting with a binary tree rather than a path gives a similar example with $p_{\mathrm{H}}^{\mathrm{s}} = p_{\mathrm{T}}^{\mathrm{s}} < 1$.

4.5 Exponential decay of the volume – the Aizenman–Newman Theorem

Our aim in this section is to strengthen Theorem 9: under suitable conditions, not only does the radius of the open cluster C_x containing a given site x decay exponentially, but so does its 'volume', $|C_x|$. Aizenman and Newman [1984] gave a very general result of this form, which we shall come to later. However, a special case of their theorem may be proved in a very different way, using 2-independence: we present this first.

We shall formulate the next result for unoriented site percolation on a graph Λ. Recall that, in this context, two sites are *equivalent* if there is an automorphism of Λ mapping one into the other. (This corresponds to being out-like if we replace each edge by two oriented edges.) A graph is of *finite type* if there are finitely many equivalence classes of sites under this relation. As usual, we write $S_n(x)$ for the set of sites at graph distance n from x. We write $B_n(x)$ for $\bigcup_{i=0}^n S_i(x)$, i.e., for the *ball* of radius n centred at x.

The conditions in the result below are generous enough to ensure that it applies to the most interesting cases, including site and bond percolation on $\mathbb{Z}^d$, and on Archimedean lattices. The proof is based on Theorem 6 or Theorem 9, and the concept of 2-independent site percolation. Although it is rather easy, and strongly reminiscent of the proof of Theorem 12 of Chapter 3, here we have to work harder to get a covering corresponding to the cover of $\mathbb{Z}^2$ with large squares.

Theorem 10. *Let Λ be a connected, infinite, locally-finite, finite-type unoriented graph. Suppose that*

$$\sup_{x\in\Lambda} |B_r(x)| \le r^{(\log r)/100} \tag{16}$$

for all sufficiently large r. Then, for every $p < p_{\mathrm{H}}^{\mathrm{s}}(\Lambda)$, there is an $a = a(\Lambda, p) > 0$ such that

$$\mathbb{P}_p^{\mathrm{s}}(|C_x| \ge n) \le \exp(-an) \tag{17}$$

for all sites x and all $n \ge 0$.

Proof. The overall plan of the proof is as follows: we shall cover $V(\Lambda)$ by a set of balls $B_{2r}(w)$, $w \in W$, that are reasonably well 'spread out': forming a graph $D(W)$ on W by joining w, w' if they are at distance at most $6r$, say, we shall show that the maximum degree of $D(W)$ is not too large. We then construct a 2-independent site percolation measure on $D(W)$ as follows: a site $w \in W$ will be *active* if some $x \in B_{2r}(w)$ is joined by an open path to a 'distant' site, i.e., a site at distance at least r from x; these events are independent if $d(w, w') > 6r$. Also, any site in a large enough open cluster is joined to a distant site, so a large open cluster in Λ implies a large active cluster in $D(W)$. From Theorem 6 (Menshikov's Theorem) and inequality (16), each w is very unlikely to be active if r is chosen large enough; it will then follow from Lemma 11 of Chapter 3 that open clusters in this 2-independent measure are small, giving the result.

To carry out this plan, we shall first show that balls of radius $7r$ in Λ are not too much larger than those of radius r. Let $\Delta < \infty$ be the maximum degree of Λ. As Λ is connected, for each pair $\{[x],[y]\}$ of equivalence classes of sites there is a path P from some $x' \in [x]$ to some $y' \in [y]$. Thus there is an integer L such that for any x and y, there are $x' \in [x]$ and $y' \in [y]$ with $d(x', y') \leq L$. It follows that, crudely,

$$|B_r(x)| = |B_r(x')| \leq |B_{r+L}(y')| = |B_{r+L}(y)| \leq (1+\Delta+\cdots+\Delta^L)|B_r(y)|$$

for every $r \geq 1$. Let $b_r^+ = \max\{|B_r(x)| : x \in \Lambda\}$ and $b_r^- = \min\{|B_r(x)| : x \in \Lambda\}$. Then we have $1 \leq b_r^+/b_r^- \leq C$ for every r, where C depends only on Λ.

We claim that

$$b_{7r}^+/b_r^- \leq \sqrt{r} \tag{18}$$

holds for infinitely many r. Otherwise, there is an r_0 such that for all $r \geq r_0$ we have

$$b_{7r}^+/b_r^+ \geq b_{7r}^+/(Cb_r^-) \geq \sqrt{r}/C.$$

Thus, for r large enough, adding $\log 7$ to $\ell = \log r$ increases $\log b_r^+$ by at least $(1/3)\log r = \ell/3$, which implies that $\log b_r^+ \geq \ell^2/100$ for large r, contradicting (16), and proving the claim. From now on we consider only r for which (18) holds.

Let $W \subset V(\Lambda)$ be a maximal (infinite) set of sites subject to the balls $\{B_r(w) : w \in W\}$ being disjoint. (Such a set W exists in any graph by Zorn's Lemma. In fact, since Λ is connected and locally finite, Λ is countable, so W may be constructed step by step.) Note that the balls $\{B_{2r}(w) : w \in W\}$ cover $V(\Lambda)$: if some site y did not belong to $\bigcup_{w\in W} B_{2r}(w)$, then y could have been added to W. We define an auxiliary graph $D(W)$ with vertex set W as follows: two sites w, $u \in W$ are adjacent in $D(W)$ if $d(w,u) \leq 6r$, where $d(x,y)$ denotes the graph distance in Λ. If Γ_w denotes the set of neighbours of w in $D(W)$, then the balls $\{B_r(u) : u \in \Gamma_w\}$ are disjoint subsets of $B_{7r}(w)$; see Figure 14. Hence,

$$|\Gamma_w| b_r^- \leq \Big| \bigcup_{u\in\Gamma_w} B_r(u) \Big| \leq |B_{7r}(w)| \leq b_{7r}^+.$$

Therefore, using (18), we have $|\Gamma_w| \leq b_{7r}^+/b_r^- \leq \sqrt{r}$, i.e., the maximum degree Δ_1 of $D(W)$ is at most $\sqrt{r}$.

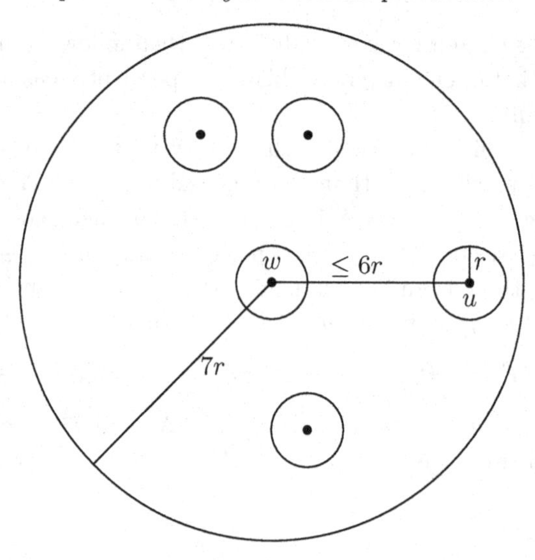

Figure 14. Balls of radius r centred at w and at various sites $u \in \Gamma_w$. The balls are disjoint as $\{w\} \cup \Gamma_w \subset W$.

Let us say that a site $w \in W$ is *active* if there is at least one site $y \in B_{2r}(w)$ for which $\{y \xrightarrow{r}\}$ holds. Otherwise, w is *passive*. Note that the *disposition* (active or passive) of $w \in W$ depends only on the states (open or closed) of the sites of Λ within distance $3r$ of w. Let U and U' be subsets of W such that no edge of $D(W)$ joins U to U'. Then no site of Λ is within distance $3r$ of both U and U', so the dispositions of the sites in U are independent of the dispositions of the sites in U'. Let $\widetilde{\mathbb{P}}$ be the site percolation measure on $D(W)$ defined by taking the active sites to be open. What we have just shown is that $\widetilde{\mathbb{P}}$ is a 2-independent site percolation measure.

The probability that a given site in W is active is at most $p_1 = b_{2r}^{+} \sup_y \mathbb{P}_p^{\mathrm{s}}(y \xrightarrow{r})$. If r is large enough then, from (16) and Theorem 6, we have

$$p_1 \le (2r)^{\log(2r)/100} \exp\bigl(-r/(\log r)^2\bigr) \le \exp(-r^{2/3}).$$

To apply Lemma 11 of Chapter 3, let us define

$$f(r) = e\Delta_1 p_1^{1/(\Delta_1+1)}.$$

Since $\Delta_1 = \Delta(D(W)) \le \sqrt{r}$ whenever (18) holds, we have

$$f(r) \le \exp\bigl(-r^{1/6+o(1)}\bigr).$$

In particular, $f(r) < 1$ if r is large enough, and, as (18) holds for infinitely many r, there is some r for which (18) holds and $f(r) < 1$. From now on, we fix such an r.

Applying Lemma 11 of Chapter 3 to the 2-independent site percolation measure $\widetilde{\mathbb{P}}$ on the graph $D(W)$, we see that there is a constant $c > 0$ such that, for any $z \in W$ and any $m \geq 0$,

$$\widetilde{\mathbb{P}}(|C_z(D(W))| \geq m) \leq \exp(-cm),$$

where $C_z(D(W))$ is the open (active) cluster of $D(W)$ containing z.

To complete the proof, note that if x is a site of Λ and $|C_x| > b^+_{r-1}$, then for every site $y \in C_x$, the event $\{y \xrightarrow{r}\}$ holds. (There is no room for all of C_x inside $B_{r-1}(y)$.) Let $W' = \{w \in W : B_{2r}(w) \cap C_x \neq \emptyset\}$. Then every site $w \in W'$ is active. Also, as the balls $B_{2r}(w)$, $w \in W$, cover $V(\Lambda)$, the balls $B_{2r}(w)$, $w \in W'$, cover C_x.

We claim that W' induces a connected subgraph of $D(W)$. Indeed, if w, $w' \in W'$, then there are sites y, $y' \in C_x$ with $y \in B_{2r}(w)$ and $y' \in B_{2r}(w')$. As C_x is connected, there is a path $y = y_1 y_2 \cdots y_t = y'$ with every y_i in C_x. We may choose $w_1, \ldots, w_t \in W'$ with $y_i \in B_{2r}(w_i)$, and $w_1 = w$, $w_t = w'$. For $1 \leq i \leq t-1$ we have $d(w_i, w_{i+1}) \leq d(y_i, y_{i+1}) + 4r = 4r + 1 \leq 6r$, so either $w_i = w_{i+1}$, or $w_i w_{i+1}$ is an edge of $D(W)$. Hence, w and w' are joined by a path in the subgraph of $D(W)$ induced by W'.

Fixing $x \in \Lambda$, let $z \in W$ be any site such that $x \in B_{2r}(z)$. We have shown that W' is a set of active sites that is connected in $D(W)$, so W' is a subset of $C_z(D(W))$. Also, as C_x is covered by $|W'|$ balls of radius $2r$, we have $|W'| \geq |C_x|/b^+_{2r}$. Hence, for $n > b^+_{r-1}$,

$$\begin{aligned}\mathbb{P}^{\mathrm{s}}_p(|C_x| \geq n) &\leq \widetilde{\mathbb{P}}(|W'| \geq n/b^+_{2r}) \\ &\leq \widetilde{\mathbb{P}}(|C_z(D(W))| \geq n/b^+_{2r}) \leq \exp(-cn/b^+_{2r}).\end{aligned}$$

As r is constant, the proof is complete. □

In the proof of Theorem 10, we used Menshikov's Theorem, Theorem 6, to obtain a good bound on $\mathbb{P}^{\mathrm{s}}_p(x \xrightarrow{n})$ for $p < p^{\mathrm{s}}_{\mathrm{H}}$. Instead, we could have used Hammersley's much simpler result, Theorem 9. Modifying the proof in this way would give the same conclusion, (17), but only for $p < p^{\mathrm{s}}_{\mathrm{T}}$. (Of course, under the assumptions of Theorem 10, $p^{\mathrm{s}}_{\mathrm{T}} = p^{\mathrm{s}}_{\mathrm{H}}$ by Menshikov's Theorem.)

Next we shall present the general result of Aizenman and Newman [1984] mentioned at the beginning of the section.

Theorem 11. *Let $\overrightarrow{\Lambda}$ be infinite, locally-finite oriented multi-graph with $C_{\overrightarrow{\Lambda}}$ finite and strongly connected, and let $p < p_{\mathrm{T}}^{\mathrm{s}}(\overrightarrow{\Lambda})$. Then there is an $\alpha > 0$ such that*

$$\mathbb{P}_p^{\mathrm{s}}\big(|C_x| \geq n\big) \leq \exp(-\alpha n) \tag{19}$$

for all sites x and integers $n \geq 1$.

Proof. As $p < p_{\mathrm{T}}^{\mathrm{s}}(\overrightarrow{\Lambda})$ is fixed, we suppress the dependence on p in our notation.

Suppose first that $C_{\overrightarrow{\Lambda}}$ has only a single vertex, i.e., that all sites are out-like. This is the main part of the proof; the extension to finitely many out-classes is a minor variation.

As before, let $\{x^+ \to y\}$ denote the event that there is an open path from an out-neighbour of x to y. We interpret $\{x^+ \to x\}$ as the event which always holds. Thus $x \to y$ holds if and only if $x^+ \to y$ and x is open. Note that

$$\chi' = \sum_{y \in \overrightarrow{\Lambda}} \mathbb{P}(x^+ \to y) = \sum_{y \in \overrightarrow{\Lambda}} \mathbb{P}(x \to y)/p = \chi_x(p)/p < \infty.$$

We shall estimate the moments of $|C_x|$ in terms of χ', using the van den Berg–Kesten inequality. We start with the second moment.

Suppose that y_1, $y_2 \in C_x$. We do not assume that x, y_1 and y_2 are distinct, although the picture is clearer if we do. As $y_1 \in C_x$, there is an open (oriented, as always) path P_1 from x to y_1. Let P_2 be a shortest open path from a site u on P_1 to y_2: such a path exists as $x \to y_2$. Then P_1 and P_2 are vertex-disjoint except at u (otherwise, P_2 could be shortened). Let us split P_1 into two paths, a path P_1' from x to u, and a path P_1'' from u to y_1; see Figure 15. Omitting the initial vertices from the paths P_1', P_1'' and P_2 gives disjoint witnesses for the events $\{x^+ \to u\}$, $\{u^+ \to y_1\}$ and $\{u^+ \to y_2\}$, respectively. Thus, if y_1, $y_2 \in C_x$, then the event

$$\{x^+ \to u\} \,\square\, \{u^+ \to y_1\} \,\square\, \{u^+ \to y_2\}$$

holds for some $u \in \overrightarrow{\Lambda}$. (Recall that $\square$ is associative on increasing events.) Using the van den Berg–Kesten inequality, it follows that

$$\mathbb{P}(y_1, y_2 \in C_x) \leq \sum_{u \in \overrightarrow{\Lambda}} \mathbb{P}(x^+ \to u)\mathbb{P}(u^+ \to y_1)\mathbb{P}(u^+ \to y_2).$$

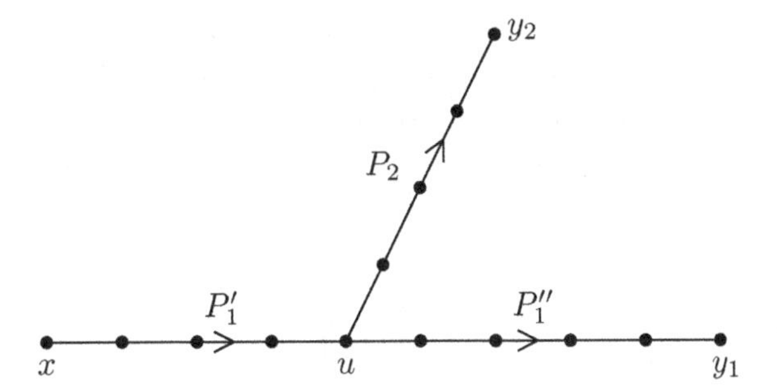

Figure 15. Open paths $P_1 = P'_1 \cup P''_1$ and P_2 showing that $y_1, y_2 \in C_x$. After deleting the initial sites from P'_1, P''_1 and P_2, these paths are disjoint.

Summing over $y_1, y_2 \in \overrightarrow{\Lambda}$, it follows that

$$\begin{aligned} \mathbb{E}(|C_x|^2) &= \sum_{y_1}\sum_{y_2} \mathbb{P}(y_1, y_2 \in C_x) \\ &\le \sum_{y_1,y_2,u} \mathbb{P}(x^+ \to u)\mathbb{P}(u^+ \to y_1)\mathbb{P}(u^+ \to y_2), \end{aligned}$$

where, from now on, all summation variables run over all sites of $\overrightarrow{\Lambda}$.

As all vertices are out-like, we have $\sum_w \mathbb{P}(z^+ \to w) = \chi'$ for any $z \in \overrightarrow{\Lambda}$. Evaluating the triple sum above by first summing over y_2, then over y_1, and then over u, it follows that this sum is *exactly* $(\chi')^3$, so $\mathbb{E}(|C_x|^2) \le (\chi')^3$.

For higher moments the argument is very similar, but slightly harder to write down: the only additional complication is that if $y_1, y_2 \in C_x$ and P_1, P_2 and $u_1 = u$ are defined as above, then if y_3 is also in C_x, the shortest path P_3 from a site $u_2 \in P_1 \cup P_2$ to y_3 may start from a site u_2 on P'_1, on P''_1, or on P_2. It will be convenient to write y_0 for x, and to denote the 'branch vertices' u_i by $y_{-1}, y_{-2}, \ldots$.

We shall say that an oriented tree T on the set

$$[-k+1, k] = \{0, 1, 2, \ldots, k, -1, -2, \ldots, -(k-1)\}.$$

is a *k-template* if it satisfies the following recursive definition: for $k = 1$, the only 1-template is the tree consisting of the oriented edge $\overrightarrow{01}$. For $k \ge 2$, the tree T is a k-template if it may be obtained from some $(k-1)$-template T' by first inserting the vertex $-(k-1)$ to subdivide an edge of T, and then adding an oriented edge from $-(k-1)$ to k. Note that the number of k templates is exactly $N_k = 1 \times 3 \times 5 \times \cdots \times (2k-3) = (2k-3)!!$, as there are $e(T') = 2k-3$ choices for the edge of T' to subdivide.

Fixing the site $x = y_0$ throughout, a *realization* R of a template T is a sequence $y_1, \dots, y_k, y_{-1}, \dots, y_{-k+1}$ of (not necessarily distinct) sites of $\vec{\Lambda}$ such that there are disjoint open paths $P_{\vec{ij}}$, $\vec{ij} \in E(T)$, with $P_{\vec{ij}}$ a (minimal) witness for $y_i^+ \to y_j$; see Figure 16. We call $y_1, \dots, y_k$ the

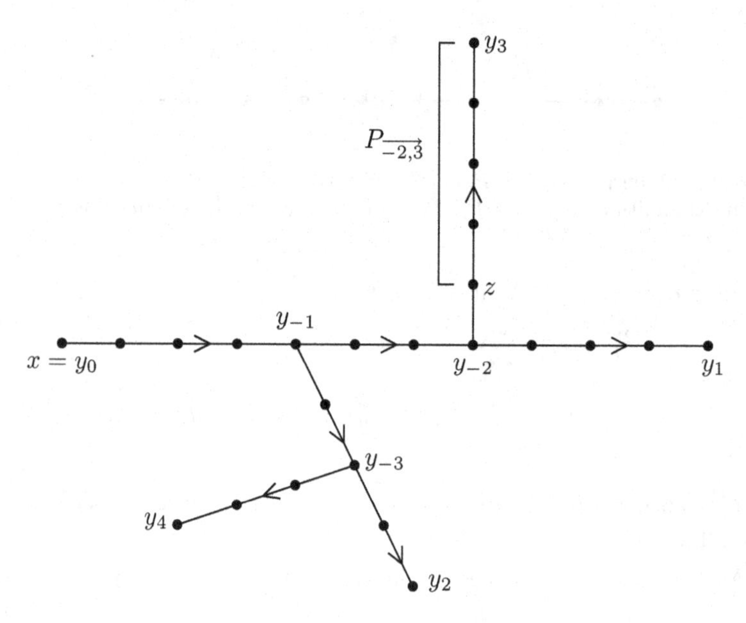

Figure 16. A realization of a 4-template. The directed open path $P_{\overrightarrow{-2,3}}$ from z to y_3 is a witness for $\{y_{-2}^+ \to y_3\}$.

leaves of the realization. Note that we may have $y_i = y_j$ for $i \neq j$, in which case $P_{\vec{ij}}$ is the 'empty path' with no sites or bonds.

Recalling that $y_0 = x$, we claim that, whenever $y_1, \dots, y_k \in C_x$, there is a realization R of some template T such that R has leaves $y_1, \dots, y_k$. The proof is by induction on k. For $k = 1$ the claim is immediate: as $y_1 \in C_x$ and $x = y_0$, there is a witness for $\{y_0^+ \to y_1\}$, which is all that is required. For the induction step, given a realization R' with leaves $y_1, \dots, y_{k-1}$, let P be a shortest open path from a site y_{-k+1} appearing in R' (i.e., from x or from a site in some $P_{\vec{ij}}$ in R') to y_k. Splitting the witness (path) on which y_{-k+1} appears as in the case $k = 2$ above, and taking P (without its initial vertex) as the witness $P_{\overrightarrow{(-k+1)k}}$ for $\{y_{-k+1}^+ \to y_k\}$, we find a realization R of some template T.

By the van den Berg–Kesten inequality, the probability that T has a realization R corresponding to a particular sequence y_i, is

$$\mathbb{P}\left(\Box_{\overrightarrow{ij}\in T}\{y_i^+ \to y_j\}\right) \le \prod_{\overrightarrow{ij}\in T} \mathbb{P}(y_i^+ \to y_j).$$

Hence, the probability that $y_1, \ldots, y_k \in C_x$ is at most

$$\sum_T \sum_{Y_k^-} \prod_{\overrightarrow{ij}\in T} \mathbb{P}(y_i^+ \to y_j),$$

where $\sum_T$ denotes summation over k-templates, and $\sum_{Y_k^-}$ summation over sites $y_{-1}, y_{-2}, \ldots, y_{-k+1)} \in \overrightarrow{\Lambda}$. Thus, summing over $y_1, \ldots, y_k$,

$$\mathbb{E}\big(|C_x|^k\big) \le \sum_T \pi(T), \tag{20}$$

where

$$\pi(T) = \sum_{Y_k^\pm} \prod_{\overrightarrow{ij}\in T} \mathbb{P}(y_i^+ \to y_j),$$

with the sum $\sum_{Y_k^\pm}$ running over sites $y_1, \ldots, y_k, y_{-1}, y_{-2}, \ldots, y_{-k+1}$.

The definition of $\pi(T)$ may be extended to define $\pi(U)$ for any oriented labelled tree U with 0 as a vertex: we sum over a variable y_i for each vertex $i \ne 0$ of U. It is easy to see that if every edge is oriented away from 0, as is the case if U is a template, then

$$\pi(U) = (\chi')^{e(U)}. \tag{21}$$

Indeed, there is a leaf $j \ne 0$ of U, so $\overrightarrow{ij} \in E(U)$ for some i. As

$$\sum_{y_j} \mathbb{P}(y_i^+ \to y_j) = \chi' \tag{22}$$

for any y_i, we have $\pi(U) = \chi'\pi(U')$ where $U' = U - \overrightarrow{ij}$, and (21) follows by induction.

From (20) and (21), we have

$$\mathbb{E}\big(|C_x|^k\big) \le \sum_T (\chi')^{e(T)} = N_k(\chi')^{2k-1} = (2k-3)!!(\chi')^{2k-1}$$

for every k. Hence, for any $t > 0$ we have

$$\mathbb{E}\big(\exp(t|C_x|)\big) = \sum_k \mathbb{E}\big(t^k|C_x|^k/k!\big) \le \sum_k \frac{(2k-3)!!}{k!}\big(t(\chi')^2\big)^k/\chi'.$$

As $(2k-3)!!/k! \leq 2^k$, this expectation is finite for $0 < t < (\chi')^{-2}/2$. As

$$\mathbb{E}\big(\exp(t|C_x|)\big) = \sum_n \exp(tn)\mathbb{P}\big(|C_x| = n\big),$$

it follows that $\mathbb{P}\big(|C_x| = n\big) \leq \exp(-tn)$ for all sufficiently large n, and the result follows.

So far, we assumed that all vertices are out-like. As promised, extending the result from this case to the general case is straightforward. Indeed, the proof is the same, except that we replace χ' by

$$\chi'' = \sup_w \sum_y \mathbb{P}(w^+ \to y) = \sup_w \chi_w(p)/p.$$

In place of (22), we then have

$$\sum_{y_j} \mathbb{P}(y_i^+ \to y_j) \leq \chi'',$$

so $\pi(U) \leq (\chi'')^{e(U)}$. As shown in Lemma 2, when $C_{\vec{\Lambda}}$ is strongly connected, then $\chi_w(p)$ is finite for one site w if and only if it is finite for all sites, i.e., if $p < p_{\mathrm{T}}^{\mathrm{s}}$. Furthermore, as $\chi_w(p)$ depends only on the out-class of w, and $C_{\vec{\Lambda}}$ is finite, the supremum above may be taken over a finite set, so χ'' is finite. The rest of the proof is unchanged. □

Let us recap very briefly the main theorems of this section; all these results concern finite-type graphs with strongly connected class-graphs. We have proved results of Hammersley and of Aizenman and Newman that, under mild conditions, for $p < p_{\mathrm{T}} \leq p_{\mathrm{H}}$, both the radius and the volume of the open cluster containing a given site decay exponentially. We have also proved Menshikov's Theorem that, under a very weak condition on the growth of the neighbourhoods, the critical probabilities p_{T} and p_{H} coincide. Thus, for a very wide class of ground graphs, $p_{\mathrm{T}} = p_{\mathrm{H}}$, and if p is less than this common value p_{c}, then the size of the open cluster of the origin decays exponentially.

A very interesting problem that we have not touched upon is the speed of this decay. More precisely, what are the best constants α in (15) and in (19), and how do these constants depend on p as p approaches p_{T} from below. The arguments above give some bounds on the constants in terms of other quantities, but their behaviour as $p \to p_{\mathrm{T}}$ is a very difficult question, about which rather little is known. We shall return to this briefly in Chapter 7.

5

Uniqueness of the infinite open cluster and critical probabilities

Our first aim in this chapter is to present a result of Aizenman, Kesten and Newman [1987] that, under mild conditions, above the critical probability there is a unique infinite open cluster; Burton and Keane [1989] have given a very simple and elegant proof of this result. Together with Menshikov's Theorem, this uniqueness result gives an alternative proof of the Harris–Kesten Theorem; this proof is easily adapted to determine the critical probabilities of certain other lattices. The key consequence of uniqueness is that, under a symmetry assumption, the critical probabilities for bond percolation on a planar lattice and on its dual must sum to 1. We shall prove this, along with a corresponding result for site percolation, assuming only order two symmetry. Finally, we discuss the star-delta transformation, which may be used to find the critical probabilities for certain lattices that are not self-dual.

5.1 Uniqueness of the infinite open cluster – the Aizenman–Kesten–Newman Theorem

Throughout this chapter the underlying graph Λ will be unoriented, and of finite-type. In other words, there are finitely many equivalence classes of sites under the relation in which two sites x and y are equivalent if there is an automorphism of Λ mapping x to y, in which case we write $x \sim y$. As usual, Λ will be infinite, locally finite, and connected.

We start this chapter with the result of Aizenman, Kesten and Newman [1987] that, in this setting, with probability 1, there is at most one infinite open cluster. For the special case of bond percolation on $\mathbb{Z}^2$, this was proved by Harris [1960]; Fisher [1961] noted that Harris's argument carries over to site percolation on $\mathbb{Z}^2$. A little later, Gandolfi, Grimmett and Russo [1988] simplified the original proof of Aizenman,

Kesten and Newman. A very different and extremely simple proof was then given by Burton and Keane [1989]; we shall present their elegant argument below. Related results for dependent measures were proved by Gandolfi, Keane and Russo [1988] and Gandolfi [1989].

This section is the only place where we shall use properties of infinite product spaces in a non-trivial way (although, of course, we could re-write even these arguments in terms of limits of probabilities in finite spaces if we wanted to).

Recall that the probability measure $\mathbb{P}_p = \mathbb{P}^{\mathrm{s}}_{\Lambda,p}$ is a product measure on the infinite product space $\Omega = \{0,1\}^{V(\Lambda)}$. Thus, the σ-field Σ of measurable subsets of Ω is generated by the cylindrical sets

$$C(F,\sigma) = \{\omega \in \Omega : \omega_f = \sigma_f \text{ for } f \in F\},$$

where F is a finite subset of $V(\Lambda)$ and $\sigma \in \{0,1\}^F$.

As usual, an *event* is just a measurable subset of Ω. Thus, any property of the open subgraph $\mathcal{O}$ that depends on the states of only finitely many sites is an event, and countable unions and intersections of events are events. In fact, any property of $\mathcal{O}$ that one would ever want to consider in percolation is an event. For example, recall that $\{x \xrightarrow{n}\}$ denotes the property that there is an open path from the site x to a site at graph distance n from x. For a given site x and integer n, this property depends on the states of finitely many sites, and so is measurable. It follows that $\{x \to \infty\}$ is an event. Similarly, 'there is an infinite open cluster' and 'there are exactly two infinite open clusters' are events.

An event E is *automorphism invariant* if, for every automorphism φ of the underlying graph Λ, the induced automorphism $\varphi^* : \Omega \to \Omega$ maps E into itself. In particular,

$$I_k = \{ \text{ there are exactly } k \text{ infinite open clusters } \}$$

is an automorphism-invariant event for $k = 0, 1, \ldots, \infty$. The only property of automorphism-invariant events we shall use is that any automorphism-invariant event has probability 0 or 1. We state this as a lemma for site percolation. The corresponding result for bond percolation follows by considering the line graph, noting that if Λ is of finite type, then so is $L(\Lambda)$.

Lemma 1. *Let Λ be a locally finite, finite-type infinite graph, and let $E \subset \Omega = \{0,1\}^{V(\Lambda)}$ be an automorphism-invariant event. Then $\mathbb{P}^{\mathrm{s}}_{\Lambda,p}(E) \in \{0,1\}$.*

Proof. We shall write $\mathbb{P}$ for $\mathbb{P}^{\mathrm{s}}_{\Lambda,p}$. Let x_0 be a site of Λ. Note that, as Λ is infinite, locally finite, and of finite type, there are infinitely many sites x equivalent to x_0.

Let $\varepsilon > 0$ be given. Since E is measurable, there is a finite set F of sites of Λ and a cylindrical event E_F depending only on the states of the sites in F, such that

$$\mathbb{P}(E \triangle E_F) \leq \varepsilon. \tag{1}$$

Let $M = \max\{d(x_0, y) : y \in F\}$. Since Λ is locally finite, the ball $B_{2M}(x_0) = \{z : d(x_0, z) \leq 2M\}$ is finite. Thus there is a site x with $x \sim x_0$ and $d(x, x_0) > 2M$. Let φ be an automorphism of Λ mapping x_0 to x. For $y \in F$ we have

$$\begin{aligned} d(x_0, \varphi(y)) &\geq d(x_0, \varphi(x_0)) - d(\varphi(x_0), \varphi(y)) \\ &= d(x_0, x) - d(x_0, y) > 2M - M = M, \end{aligned}$$

so $\varphi(y) \notin F$. Thus the sets of sites F and $\varphi(F)$ are disjoint. It follows that the event $\varphi(E_F)$, defined in the natural way, is independent of E_F, so

$$\mathbb{P}(E_F \cap \varphi(E_F)) = \mathbb{P}(E_F)\mathbb{P}(\varphi(E_F)) = \mathbb{P}(E_F)^2.$$

Thus,

$$\begin{aligned} |\mathbb{P}(E) - \mathbb{P}(E_F)^2| &= |\mathbb{P}(E \cap E) - \mathbb{P}(E_F \cap \varphi(E_F))| \\ &\leq \mathbb{P}\big((E \cap E) \triangle (E_F \cap \varphi(E_F))\big). \end{aligned}$$

For any sets A, B, C, D we have $(A \cap B) \triangle (C \cap D) \subset (A \triangle C) \cup (B \triangle D)$. Thus, as E is automorphism-invariant,

$$\begin{aligned} |\mathbb{P}(E) - \mathbb{P}(E_F)^2| &\leq \mathbb{P}(E \triangle E_F) + \mathbb{P}\big(E \triangle \varphi(E_F)\big) \\ &= \mathbb{P}(E \triangle E_F) + \mathbb{P}\big(\varphi(E) \triangle \varphi(E_F)\big) \\ &= \mathbb{P}(E \triangle E_F) + \mathbb{P}(E \triangle E_F) \\ &= 2\mathbb{P}(E \triangle E_F) \leq 2\varepsilon, \end{aligned}$$

where the final inequality is from (1).

Since $|\mathbb{P}(E_F) - \mathbb{P}(E)| \leq |\mathbb{P}\big(E \triangle E_F\big)| \leq \varepsilon$, we have

$$\begin{aligned} |\mathbb{P}(E) - \mathbb{P}(E)^2| &\leq |\mathbb{P}(E) - \mathbb{P}(E_F)^2| + |\mathbb{P}(E_F)^2 - \mathbb{P}(E)^2| \\ &\leq 2\varepsilon + 2\varepsilon = 4\varepsilon. \end{aligned}$$

Since $\varepsilon > 0$ is arbitrary, it follows that $\mathbb{P}(E) = \mathbb{P}(E)^2$, so $\mathbb{P}(E)$ is 0 or 1.

□

In the proof above, we did not need E to be invariant under all automorphisms of Λ, just under a set of automorphisms large enough that any finite set can be separated from itself by such an automorphism. In the context of lattices in $\mathbb{R}^d$, for example, invariance under a single automorphism φ of Λ corresponding to a translation of $\mathbb{R}^d$ through any vector $(a_1, a_2, \ldots, a_d) \neq (0, 0, \ldots, 0)$ is enough. In particular, Lemma 1 is often stated for *translation-invariant* events. In the terminology of ergodic theory, we have shown that the induced automorphism $\varphi^* : \Omega \to \Omega$ is ergodic.

We now turn to the particular event I_k that there are exactly k open clusters, where $0 \leq k \leq \infty$. We start with a lemma of Newman and Schulman [1981], showing that, with probability 1, there are 0, 1 or infinitely many open clusters.

Lemma 2. *Let Λ be a connected, infinite, locally finite, finite-type graph, and let $p \in (0,1)$. Then*

$$\mathbb{P}^{\mathrm{s}}_{\Lambda,p}\left(\bigcup_{2\leq k<\infty} I_k\right) = 0.$$

Hence, either $\mathbb{P}^{\mathrm{s}}_{\Lambda,p}(I_0) = 1$, $\mathbb{P}^{\mathrm{s}}_{\Lambda,p}(I_1) = 1$, or $\mathbb{P}^{\mathrm{s}}_{\Lambda,p}(I_\infty) = 1$.

Proof. By Lemma 1, it suffices to prove the first statement. As before, we write $\mathbb{P}$ for $\mathbb{P}^{\mathrm{s}}_{\Lambda,p}$.

Suppose for a contradiction that $\mathbb{P}(I_k) > 0$ for some $2 \leq k < \infty$. Let x_0 be any fixed site of Λ, and let $T_{n,k}$ be the event that I_k holds, and each infinite cluster contains a site in $B_n(x_0)$. As the balls $B_n(x_0)$ cover Λ, we have $I_k = \bigcup_n T_{n,k}$, so $\mathbb{P}(T_{n,k}) \nearrow \mathbb{P}(I_k)$, and there is an n such that $\mathbb{P}(T_{n,k}) > 0$.

Changing the state of every site in $B_n(x_0)$ to open, we see that $\mathbb{P}(I_1) > 0$. Indeed, spelling everything out in great detail, the event $T_{n,k}$ is the disjoint union of the events

$$T_{n,k,\mathbf{s}} = T_{n,k} \cap \{S = \mathbf{s}\},$$

where $\mathbf{s} = (s_x)_{x\in B_n(x_0)} \in \{0,1\}^{V(B_n(x_0))}$, and $S = (S_x)_{x\in B_n(x_0)}$ is the vector giving the states of all sites in $B_n(x_0)$. Thus there is an $\mathbf{s}$ for which $\mathbb{P}(T_{n,k,\mathbf{s}}) > 0$. Now if $\omega \in T_{n,k,\mathbf{s}}$ and ω' is the configuration obtained from ω by changing the state of each of the closed sites in $B_n(x_0)$ from closed to open, then $\omega' \in I_1$. Thus

$$\mathbb{P}(I_1) \geq \mathbb{P}(\{\omega' : \omega \in T_{n,k,\mathbf{s}}\}) = (p/(1-p))^c \mathbb{P}(T_{n,k,\mathbf{s}}) > 0,$$

where c is the number of sites in $B_n(x_0)$ that are closed when $S = \mathbf{s}$.

We have shown that if $\mathbb{P}(I_k) > 0$ for some $2 \leq k < \infty$, then $\mathbb{P}(I_1) > 0$. But then $\mathbb{P}(I_k) = \mathbb{P}(I_1) = 1$ by Lemma 1. As $I_k \cap I_1 = \emptyset$, this is impossible. □

To prove the main result of this section, we shall need a simple deterministic lemma concerning finite graphs.

Lemma 3. *Let G be a finite graph with k components, and let L and $C = \{c_1, \ldots, c_s\}$ be disjoint sets of vertices of G, with at least one c_i in each component of G. Let $m_1, \ldots, m_s$ be integers, each at least 2. Suppose that for each i, deleting the vertex c_i disconnects the component containing c_i into smaller components, m_i of which contain vertices of L. Then*

$$|L| \geq 2k + \sum_{i=1}^{s} (m_i - 2).$$

Proof. By considering each component of G separately, we may assume that $k = 1$, i.e., that G is connected. Removing an edge from G can only increase the number of components of $G - c_i$ containing elements of L, so we may assume that G is minimal subject to G being connected and containing $C \cup L$. Thus G is a tree all of whose leaves are in $C \cup L$. As $m_i \geq 2$, no vertex $c_i \in C$ can be a leaf of G, so all leaves are in L (there may also be internal vertices in L).

Since $G - c_i$ has m_i components, the vertex c_i has degree m_i. But for any tree with at least one edge, the number of leaves is exactly $2 + \sum_v (d(v) - 2)$, where the sum runs over internal vertices. As $d(v) \geq 2$ for each internal vertex, the sum is at least $\sum_{i=1}^{s} (m_i - 2)$, and the result follows. □

Let us say that an infinite graph Λ is *amenable* if $|S_n(x)|/|B_n(x)| \to 0$ as $n \to \infty$ for each site x, i.e., if large balls contain many more sites than their boundary spheres. There are several notions of amenability for graphs; in the present context this variant seems to be the most useful. In fact, the concept of amenability originated in group theory, where it is defined somewhat differently. Note that, if Λ is amenable and of finite type, then the limit above is automatically uniform in x.

The following result is due to Aizenman, Kesten and Newman [1987]; the proof we shall present is that of Burton and Keane [1989].

Theorem 4. *Let Λ be a connected, locally finite, finite-type, amenable* *infinite graph, and let $p \in (0,1)$. Then either $\mathbb{P}^{s}_{\Lambda,p}(I_0) = 1$ or $\mathbb{P}^{s}_{\Lambda,p}(I_1) = 1$, where I_k is the event that there are exactly k infinite open clusters in the site percolation on Λ.*

Proof. As usual, let us write $\mathbb{P}$ for $\mathbb{P}^{s}_{\Lambda,p}$. In the light of Lemma 2, this result is equivalent to the assertion that $\mathbb{P}(I_\infty) = 0$. In fact, we shall show that the probability that there are at least three (possibly infinitely many) infinite open clusters is zero. Suppose for a contradiction that this is not the case. Let x_0 be any site of Λ, and let X_0 be the set of sites equivalent to x_0. As the balls $B_r(x_0)$, $r \geq 1$, cover Λ, there is an r such that, with positive probability, $B_r(x_0)$ contains sites from (at least) three infinite open clusters. For the rest of the argument, we fix such an r.

Let $T_r(x)$ be the event that every site in $B_r(x)$ is open, and there is an infinite open cluster $\mathcal{O}$ such that when the states of all the sites in $B_r(x)$ are changed from open to closed, $\mathcal{O}$ is disconnected into at least three infinite open clusters; see Figure 1. If ω is a configuration in which

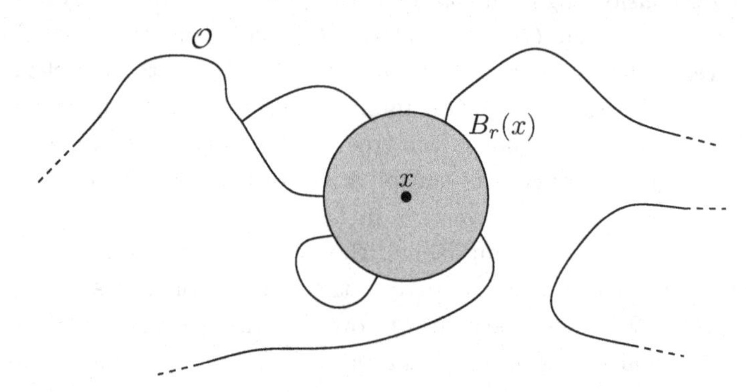

Figure 1. A site x for which $T_r(x)$ holds. Every site in $B_r(x)$ is open; the rest of the open subgraph is shown by solid lines. If the sites in $B_r(x)$ are deleted, or their states changed to closed, then the infinite open cluster $\mathcal{O}$ falls into four pieces, three of which are infinite.

$B_r(x_0)$ meets at least three infinite open clusters, and ω' is obtained from ω by changing the states of all the sites in $B_r(x_0)$ to open, then $\omega' \in T_r(x_0)$. Hence, $\mathbb{P}\big(T_r(x_0)\big) > 0$. Thus, for all sites $x \in X_0$ we have

$$\mathbb{P}\big(T_r(x)\big) = a, \tag{2}$$

for some constant $a > 0$.

Our next aim is to show that, if n is much larger than r, then we can find many disjoint balls $B_r(x)$, $x \in X_0$, inside the ball $B_n(x_0)$. In fact, to make the picture clearer, we shall find balls $B_r(x)$ that are far from each other. To do this, let $W \subset X_0 \cap B_{n-r}(x_0)$ be maximal subject to the balls $B_{2r}(w)$, $w \in W$, being disjoint. If $w' \in X_0 \cap B_{n-r}(x_0)$, then $d(w', w) \leq 4r$ for some $w \in W$; otherwise, w' could have been added to W. As Λ is connected and of finite type, there is a constant ℓ such that every site is within distance ℓ of a site in X_0. Thus, every $y \in B_{n-r-\ell}(x_0)$ is within distance ℓ of some $w' \in X_0 \cap B_{n-r}(x_0)$, and hence within distance $4r+\ell$ of some $w \in W$. In other words, for $n \geq r+\ell$, the balls $B_{4r+\ell}(w)$, $w \in W$, cover $B_{n-r-\ell}(x_0)$. Thus,

$$|W| \geq |B_{n-r-\ell}(x_0)|/|B_{4r+\ell}(x_0)|.$$

Very crudely,

$$|B_{n+1}(x_0)| \leq |B_{n-r-\ell}(x_0)|\bigl(1 + \Delta + \Delta^2 + \cdots + \Delta^{r+\ell+1}\bigr),$$

where $\Delta < \infty$ is the maximum degree of Λ. Thus, since r is fixed, there is a $c > 0$ such that

$$|W| \geq c|B_{n+1}(x_0)|$$

for all $n \geq r + \ell$. As Λ is amenable, it follows that

$$|W| \geq a^{-1}|S_{n+1}(x_0)|$$

if n is large enough, where a is the constant in (2). Let us fix such an n.

Let us call a ball $B_r(w)$ a *cut-ball* if $w \in W \subset B_{n-r}(x_0)$ and $T_r(w)$ holds. Note that if $B_r(w)$ is a cut-ball, then $B_r(w) \subset B_n(x_0)$, and every site in $B_r(w)$ is open. Since $w \sim x_0$ for every $w \in W$, by linearity of expectation the expected number of cut-balls is

$$\sum_{w \in W} \mathbb{P}(T_r(w)) = a|W| \geq |S_{n+1}(x_0)|.$$

Hence, as $\mathbb{P}(Z \geq \mathbb{E}(Z)) > 0$ for any random variable Z, there is a configuration ω such that, in this configuration, we have

$$s \geq |S_{n+1}(x_0)|, \tag{3}$$

where s is the number of cut-balls. As we shall soon see, this contradicts Lemma 3. For the rest of the argument, we consider one particular configuration ω for which (3) holds: in the rest of the argument there is no randomness.

Let $\mathcal{O}$ denote the union of all infinite open clusters of the configuration ω meeting $B_n(x_0)$, considered as a subgraph of Λ. In the configuration

ω' obtained from ω by changing the states of all sites in cut-balls to closed, the (perhaps already disconnected) cluster $\mathcal{O}$ is disconnected into several open clusters, some infinite and some finite. Let the infinite ones be $L_1, L_2, \ldots, L_t$, and the finite ones $F_1, F_2, \ldots, F_u$. Each L_i contains a site in $S_{n+1}(x_0)$, so

$$t \leq |S_{n+1}(x_0)|. \tag{4}$$

Let the cut-balls be $C_1, \ldots, C_s$. We define a graph H from $\mathcal{O}$ by contracting each cut-ball C_i to a single vertex c_i, each F_i to a single vertex f_i, and each L_i to a single vertex ℓ_i; see Figure 2. In the graph H, there is an edge from ℓ_i to c_j, for example, if and only if some site of L_i is adjacent to some site of C_j.

Infinite components of $\mathcal{O}$ correspond to components of H containing at least one vertex in $L = \{\ell_1, \ldots, \ell_t\}$. Thus, the condition that C_i is a cut-ball says exactly that deleting c_i from H disconnects a component into at least three components containing vertices of L. Thus we may apply Lemma 3 with $m_i \geq 3$, $i = 1, 2, \ldots, s$, to conclude that

$$t = |L| \geq 2 + \sum_{i=1}^{s} (3 - 2) = s + 2.$$

This contradicts (3) and (4), completing the proof. □

Simply put, Theorem 4 tells us that, above the critical probability $p_{\mathrm{H}}^{\mathrm{s}}$, almost surely there is a unique infinite open cluster. To conclude this section we remark that, by a simple argument of van den Berg and Keane [1984], Theorem 4 implies that $\theta(\Lambda; p)$ is a continuous function of p, except possibly at $p = p_{\mathrm{H}}^{\mathrm{s}}(\Lambda)$.

5.2 The Harris–Kesten Theorem revisited

Combined with Menshikov's Theorem, Theorem 4 leads to yet another proof of the Harris–Kesten result that $p_{\mathrm{H}}^{\mathrm{b}}(\mathbb{Z}^2) = p_{\mathrm{T}}^{\mathrm{b}}(\mathbb{Z}^2) = 1/2$. This proof will adapt easily to give the exact values of the critical probabilities for certain other planar lattices. We start by reproving the 'easy' inequality, that $p_{\mathrm{H}}^{\mathrm{b}}(\mathbb{Z}^2) \geq 1/2$. More precisely, we shall deduce Harris's Theorem, restated below, from Theorem 4. The argument we give is due to Zhang; see Grimmett [1999, p. 289].

Theorem 5. *For bond percolation on $\mathbb{Z}^2$ we have $\theta(1/2) = 0$.*

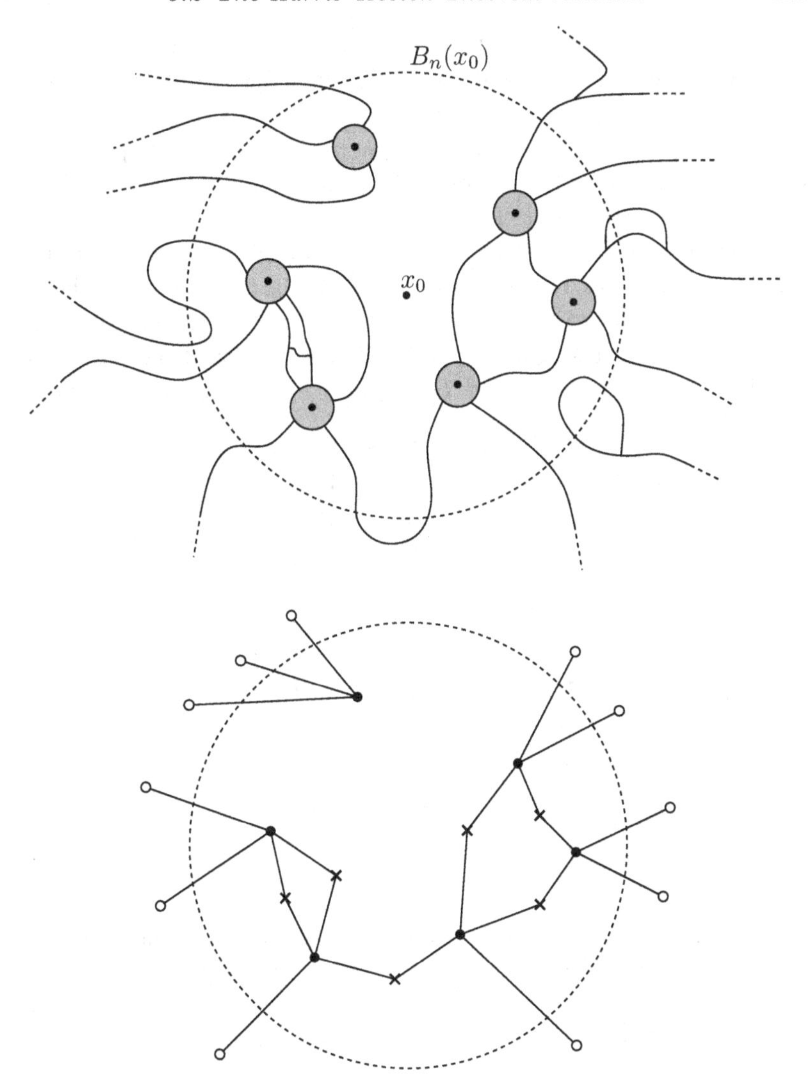

Figure 2. The upper figure shows the union $\mathcal{O}$ of all infinite open clusters meeting $B_n(x_0)$. The shaded balls, in which all sites are open, are the cut-balls. The lower figure shows the corresponding graph H. The filled circles are the vertices c_i corresponding to the cut-balls, the hollow circles the vertices ℓ_i corresponding to the infinite clusters L_i, and the crosses the vertices f_i corresponding to the finite clusters F_i.

Proof. Suppose not. Then, applying Theorem 4 to site percolation on the line graph of $\mathbb{Z}^2$, we see that $\mathbb{P}_{1/2}(I_1) = 1$, where I_1 is the event that there is exactly one infinite open cluster, and $\mathbb{P}_{1/2} = \mathbb{P}^{\mathrm{b}}_{\mathbb{Z}^2,1/2}$. It follows that there is an n_0 such that, if $n \geq n_0$, then the probability that an infinite open cluster meets a given n by n square is at least $1 - 10^{-4}$.

Let $n = n_0 + 1$, and let S be an n by n square in $\mathbb{Z}^2$. Suppose that some site x in S is in an infinite open cluster. Then there is an infinite open path P starting at x. Let y be the last site on P that is in S, and let P' be the sub-path of P starting at y; then P' is an open path from S to infinity, using only bonds outside S. Let L_1 be the event that there is an infinite open path P' as above *leaving S upwards*, i.e., with the initial site y on the upper side of S, the initial bond vertical, and all bonds outside S, as in Figure 3. Let L_2, L_3 and L_4 be defined analogously,

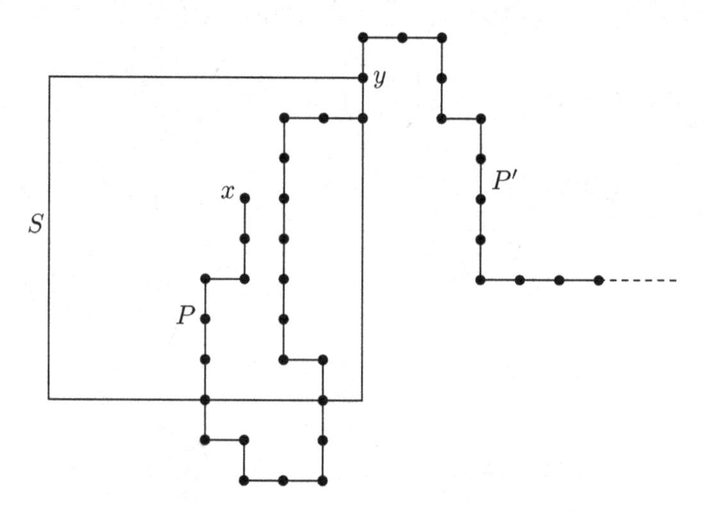

Figure 3. An infinite open path P starting at a site x in a square S. The sub-path P' from the last site y of P in S leaves S upwards.

rotating S though 90 degrees each time. Thus $\mathbb{P}_{1/2}(L_i) = \mathbb{P}_{1/2}(L_1)$ for all i. We have

$$\mathbb{P}_{1/2}\big(L_1 \cup L_2 \cup L_3 \cup L_4\big) \geq 1 - 10^{-4}. \tag{5}$$

As the L_i are increasing events, it follows from Harris's Lemma by the 'nth-root trick' (equation (8) of Chapter 2) that $\mathbb{P}(L_i) \geq 1 - 1/10$ for each i: otherwise, $\mathbb{P}(L_i^c) > 1/10$ for each i, and, as the L_i^c are decreasing and hence positively correlated, $\mathbb{P}(\bigcap_i L_i^c) > 10^{-4}$, contradicting (5).

Recall that the planar dual of the square lattice $\mathbb{Z}^2$ is the lattice $\mathbb{Z}^2 + (1/2, 1/2)$, and that we take a dual bond $e^\star$ crossing a bond e of $\mathbb{Z}^2$ to be open if and only if e is closed. Let S' be an $n-1$ by $n-1$

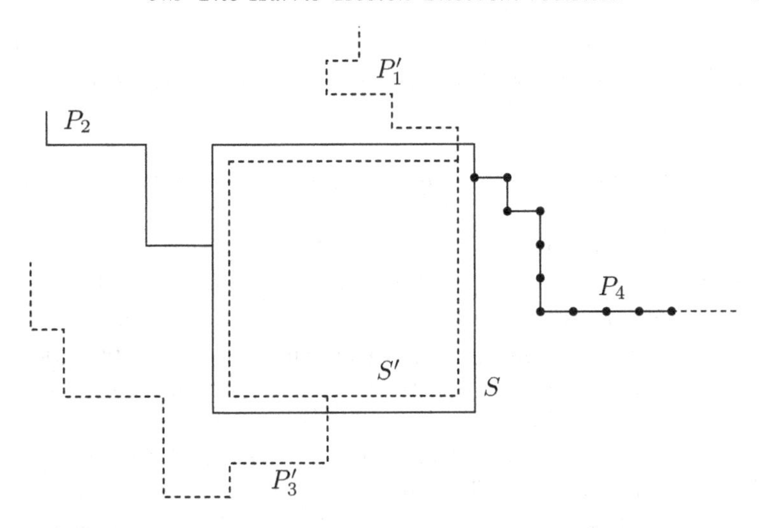

Figure 4. Infinite open paths P_2 and P_4 in the lattice $\mathbb{Z}^2$, leaving a square S to the right and to the left. The infinite open paths P_1', P_3' in the dual lattice leave S' upwards and downwards. If P_1' and P_3' may be connected by open dual bonds in S', then there are at least two infinite open clusters, one containing P_2, and one containing P_4.

square in the dual lattice inside S, as in Figure 4. As the bonds of the dual lattice are also open independently with probability 1/2, and as $n - 1 = n_0 \geq n_0$, the argument above shows that $\mathbb{P}_{1/2}(L_i') \geq 9/10$, where L_i' is the event that an infinite open dual path leaves the ith side of S_i. Thus the event $E = L_1' \cap L_2 \cap L_3' \cap L_4$ illustrated in Figure 4 has probability at least $1 - 4 \times (1 - 9/10) = 6/10 > 0$. The event E depends only on the states of bonds outside S'. Thus, with positive probability, E holds and every dual bond in S' is open. But then the paths P_1' and P_3' may be joined to form a doubly infinite path P' that separates the plane into two pieces. As P' consists of open dual edges, and an open edge of $\mathbb{Z}^2$ cannot cross an open dual edge, the open paths P_2 and P_4 lie on opposite sides of P', and thus in separate open clusters. Hence, there are at least two infinite open clusters.

In short, starting from the assumption $\mathbb{P}_{1/2}(I_1) = 1$, we have shown that with positive probability there are at least two infinite open clusters: a contradiction. It follows that $\mathbb{P}_{1/2}(I_1) = 0$, and hence that $\theta(1/2) = 0$. $\square$

Using Menshikov's Theorem, it is very easy to complete yet another proof of the Harris–Kesten Theorem, Theorem 13 of Chapter 3, restated below.

Theorem 6. *For bond percolation on the square lattice, $p_{\mathrm{H}} = p_{\mathrm{T}} = 1/2$.*

Proof. As noted in Chapter 4, Theorems 7 and 9 of that chapter, stated for oriented site percolation, apply also to unoriented bond percolation, and in particular, to bond percolation on $\mathbb{Z}^2$. (Formally, one can apply the theorems to the graph obtained from the line graph of $\mathbb{Z}^2$ by replacing each bond by two oppositely oriented bonds.) In particular, Menshikov's result, Theorem 7 of Chapter 4, gives $p_{\mathrm{H}}^{\mathrm{b}}(\mathbb{Z}^2) = p_{\mathrm{T}}^{\mathrm{b}}(\mathbb{Z}^2)$. Since $\theta(1/2) = 0$ implies $p_{\mathrm{H}}^{\mathrm{b}}(\mathbb{Z}^2) \geq 1/2$, it thus suffices to prove that $p_{\mathrm{T}}^{\mathrm{b}}(\mathbb{Z}^2) \leq 1/2$.

This is immediate from Theorem 9 of Chapter 4 and Lemma 1 of Chapter 3. Indeed, suppose that $p_{\mathrm{T}}^{\mathrm{b}}(\mathbb{Z}^2) > 1/2$. Then, by the first of these results, as $1/2$ is less than the critical probability, there is an $\alpha > 0$ such that $\mathbb{P}_{1/2}(x \xrightarrow{n}) \leq \exp(-\alpha n)$ for all sites x and integers n, where $\{x \xrightarrow{n}\}$ denotes the event that x is joined by an open path to some site at graph distance n from x. Taking n large enough, we have

$$\mathbb{P}_{1/2}(0 \xrightarrow{n-1}) \leq 1/(100n).$$

Let S be an n by n square in $\mathbb{Z}^2$. As before, let $H(S)$ be the event that there is an open horizontal crossing of S. If $H(S)$ holds, then one of the n sites on the left of S is joined by an open path in S to a site on the right of S, at distance at least $n-1$. Hence,

$$\mathbb{P}_{1/2}\big(H(S)\big) \leq n\mathbb{P}_{1/2}(0 \xrightarrow{n-1}) \leq 1/100.$$

But this contradicts the basic fact that $\mathbb{P}_{1/2}\big(H(S)\big) \geq 1/2$, which we know from Corollary 3 of Lemma 1 of Chapter 3. □

As the Harris–Kesten Theorem is so fundamental, let us briefly summarize the different approaches to its proof that we have presented here. All the proofs start from the basic fact that either a rectangle has an open horizontal crossing, or its dual has an open vertical crossing. Then, to prove that $p_{\mathrm{H}}^{\mathrm{b}}(\mathbb{Z}^2) \geq 1/2$, one may prove a Russo–Seymour–Welsh (RSW) type theorem, as in Chapter 3, relating crossings of rectangles to crossings of squares. Alternatively, one can deduce the result from the Aizenman–Kesten–Newman Theorem, Theorem 4. To prove that $p_{\mathrm{H}}^{\mathrm{b}}(\mathbb{Z}^2) \leq 1/2$, having proved an RSW type theorem, one can apply one of various sharp-threshold results as in Chapter 3. Alternatively, one can deduce that $p_{\mathrm{H}}^{\mathrm{b}}(\mathbb{Z}^2) = p_{\mathrm{T}}^{\mathrm{b}}(\mathbb{Z}^2) \geq 1/2$ directly from Menshikov's Theorem. (For the last part, we do not need exponential decay of the radius as used above: the almost exponential decay given directly by Theorem 6 of Chapter 4 is more than enough.) The approach used in Chapter 3 is perhaps more down-to-earth, and simpler in any one given case. The advantage of the Menshikov–Aizenman–Kesten–Newman approach illustrated in this chapter is that the tools are very general, so generalizing the result to other settings is easier. Of course, one still needs a suitable starting point, given by some kind of self-duality.

Let us note that, in this latter approach, proving that percolation does occur under suitable conditions, which was historically much the harder part of the Harris–Kesten result, is very easy: the deduction from Menshikov's Theorem is very simple, and can be applied in a great variety of settings. In contrast, showing that percolation does not occur, which was historically the easier part of the result, is more difficult: the deduction from the Aizenman–Kesten–Newman Theorem is not quite so simple, and requires more assumptions. We shall see this phenomenon again when we consider percolation on other lattices.

5.3 Site percolation on the triangular and square lattices

We next consider the (equilateral) triangular lattice $T \subset \mathbb{R}^2$. For definiteness, let us rotate and scale T so that $(0,0)$ and $(1,0)$ are sites of T, and all bonds have length 1. Portions of T are illustrated in Figures 5 and 7 below. Our aim is to show that $p_{\mathrm{H}}^{\mathrm{s}}(T) = p_{\mathrm{T}}^{\mathrm{s}}(T) = 1/2$. As a starting point, we need a suitable self-duality property.

In bond percolation on $\mathbb{Z}^2$, the outer boundary of a finite open cluster can be viewed as an open cycle in the dual lattice $\mathbb{Z}^2 + (1/2, 1/2)$. For site percolation on T, it is easy to see that any finite open cluster is bounded by a closed cycle in the *same* lattice T. Also, an open path in T cannot start inside and end outside a closed cycle in T: indeed, the latter statement holds for site percolation on any plane graph, as a cycle in a plane graph separates the plane into two components. These observations give a sufficient starting point to enable us to prove that $p_{\mathrm{H}}^{\mathrm{s}}(T) = p_{\mathrm{T}}^{\mathrm{s}}(T) = 1/2$, using the results of Menshikov and of Aizenman, Kesten and Newman. In fact, as in Chapter 3, it is easy to prove a striking 'large-scale' consequence of the self-duality. As usual, we write $\mathbb{P}_p$ for the percolation measure under consideration, in this case, for $\mathbb{P}_{T,p}^{\mathrm{s}}$.

Lemma 7. *Let R_n be the rhombus in T with n sites on a side shown* *in Figure 5, and let $H(R_n)$ be the event that there is an open path in T consisting of sites in R_n, starting at a site on the left-hand side of R_n, and ending at a site on the right-hand side. Then $\mathbb{P}_{1/2}(H(R_n)) = 1/2$ for every $n \geq 1$.*

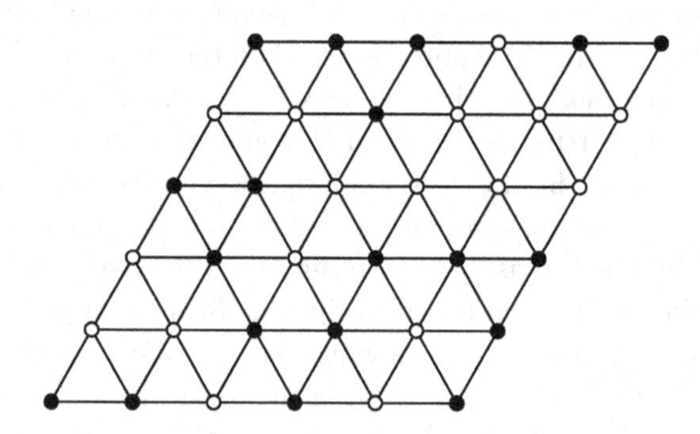

Figure 5. The rhombus R_6; solid circles represent open sites, and hollow circles closed sites.

Proof. Let $V^\star(R_n)$ be the event that there is a closed path in R_n joining the top of R_n to the bottom. Reflecting R_n in its long diagonal, and exchanging closed and open, we see that

$$\mathbb{P}_p\big(H(R_n)\big) = \mathbb{P}_{1-p}\big(V^\star(R_n)\big)$$

for any p and any n. In particular, $\mathbb{P}_{1/2}\big(H(R_n)\big) = \mathbb{P}_{1/2}\big(V^\star(R_n)\big)$. It thus suffices to prove that

$$\mathbb{P}_{1/2}\big(H(R_n)\big) + \mathbb{P}_{1/2}\big(V^\star(R_n)\big) = 1.$$

As in Chapter 3, we shall prove the much more detailed result that, whatever the states of the sites in R_n, exactly one of the events $H(R_n)$ and $V^\star(R_n)$ holds.

The proof is essentially the same as that of Lemma 1 of Chapter 3, although the picture is somewhat simpler. One can replace each site of T with a regular hexagon to obtain a tiling of the plane. Thus, what we have to show is that in the game of Hex, no draw is possible: if all the hexagons corresponding to R_n are coloured black or white, then either there is a black path from left to right, or a white path from top to bottom, but not both. (On a symmetric board, it follows easily that the first player has a winning strategy.)

To see this, we shall consider face percolation on the hexagonal lattice, which, as noted in Chapter 1, is equivalent to site percolation on T. More precisely, let us replace each open site of R_n by a black hexagon, and each closed site of R_n by a white hexagon, and consider additional black hexagons to the left and right of R_n, and white hexagons above and below R_n, as in Figure 6.

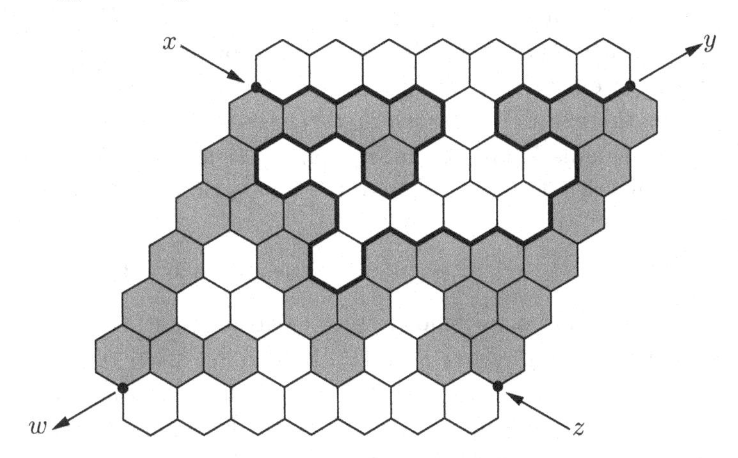

Figure 6. A partial tiling of the plane corresponding to Figure 5, obtained by replacing each open site in R_6 by a black hexagon and each closed site by a white hexagon, with additional black and white hexagons around the outside. The thick line is a path separating black and white hexagons, starting at x, with black hexagons on the right. This path must end at y (as shown) or at w.

The rest of the proof is exactly as for Lemma 1 of Chapter 3: let I be the *interface graph* formed by those edges of hexagons that separate a black region from a white region, with the endpoints of these edges as the vertices. Then every vertex of I has degree exactly 2, except for the four vertices x, y, z and w of degree 1 shown in Figure 6. The component of I containing x is thus a path. Following the path starting at x, there is always a black hexagon on the right and a white one on the left, so the path ends either at y or at w. In the former case, the black hexagons on the right contain a path in T witnessing $H(R_n)$, in the latter case, the white hexagons on the left witness $V^\star(R_n)$. As before, $H(R_n)$ and $V^\star(R_n)$ cannot both hold as otherwise K_5 could be drawn in the plane. □

Using the results of Menshikov and of Aizenman, Kesten and Newman, it is easy to deduce that the critical probability for site percolation on T is $1/2$, a result due to Kesten [1982]. Menshikov's Theorem (Theorem 7 of Chapter 4) tells us that $p_T = p_H$ in this context; from now on, we write p_c for their common value.

Theorem 8. *Let T be the equilateral triangular lattice in the plane. Then $p_c^s(T) = 1/2$.*

Proof. By Theorem 7 of Chapter 4 we have $p_H^s(T) = p_T^s(T) = p_c^s(T)$. Suppose first that $p_c^s(T) = p_T^s(T) > 1/2$. Then, by Theorem 9 of Chapter 4, we have exponential decay of the radius of an open cluster at $p = 1/2$, i.e., there is an $\alpha > 0$ such that $\mathbb{P}_{1/2}(0 \xrightarrow{n}) \le \exp(-\alpha n)$. Defining R_n as in Lemma 7, any of the sites on the right-hand side of R_n is at distance at least $n-1$ from any of the n sites on the left-hand side, so

$$\mathbb{P}_{1/2}\big(H(R_n)\big) \le n\mathbb{P}_{1/2}\big(0 \xrightarrow{n-1}\big) \le n\exp\big(-\alpha(n-1)\big).$$

As $n \to \infty$, the final bound tends to zero, contradicting Lemma 7.

Suppose next that $p_c^s(T) = p_H^s(T) < 1/2$, so $\theta(1/2) > 0$. Then, by Theorem 4, in the $p = 1/2$ site percolation on T there is with probability 1 a unique infinite open cluster. Let H_n be the hexagon centred at the origin with n sites on each side shown in Figure 7. As $\bigcup_n H_n = T$, if n is large enough then the $\mathbb{P}_{1/2}$-probability that some site in H_n is in an infinite open cluster is at least $1 - 10^{-6}$, say. Numbering the six sides of H_n in cyclic order, let L_i be the event that an infinite open path leaves H_n from side i. More precisely, L_i is the event that there is an infinite open path in T with initial site on the ith side of H_n (we may include both corners), and all other sites outside H_n. Then $\bigcup_i L_i$ is exactly the event that there is an infinite open cluster meeting H_n, so $\mathbb{P}_{1/2}(\bigcup_i L_i) \ge 1 - 10^{-6}$. As the events L_i are increasing, and, by symmetry, each has the same probability, it follows form Harris's Lemma (Lemma 3 of Chapter 2) that $\mathbb{P}_{1/2}(L_i) = \mathbb{P}_{1/2}(L_1) \ge 1 - 1/10$ for each i.

Let $L_i^\star$ be the event that an infinite closed path leaves H_n from the ith side. Then $\mathbb{P}_p(L_i^\star) = \mathbb{P}_{1-p}(L_i)$, so

$$\mathbb{P}_{1/2}\big(L_i^\star\big) = \mathbb{P}_{1/2}(L_i) \ge 1 - 1/10.$$

Hence, with probability at least $1 - 4/10 = 6/10 > 0$, the event $E = L_1 \cap L_2^\star \cap L_4 \cap L_5^\star$ holds; this event is illustrated in Figure 7. Now E is independent of the states of the sites in H_{n-1}. Thus, with positive probability, E holds and every site in H_{n-1} is closed. But then the closed paths $P_2^\star$, $P_5^\star$ guaranteed by the events $L_2^\star$ and $L_5^\star$ may be joined

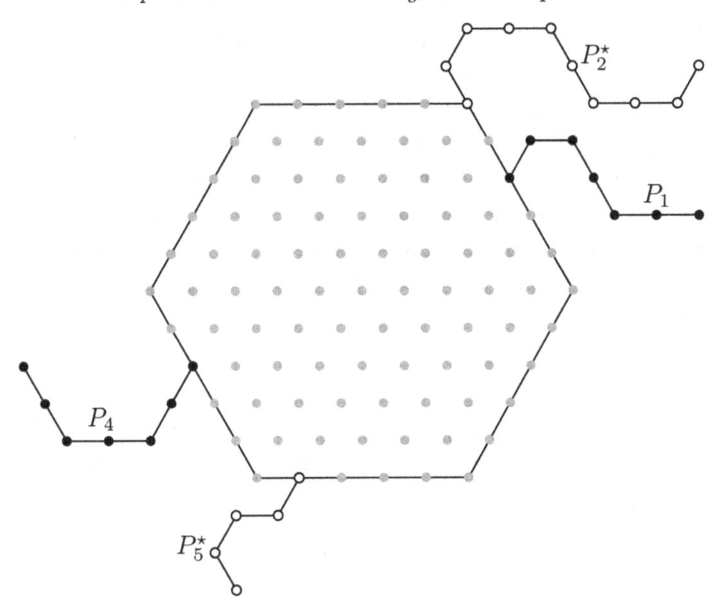

Figure 7. The hexagon H_6 with the initial segments of infinite open paths P_1 and P_4 leaving its 1st and 4th sides, and of infinite closed paths $P_2^\star$, $P_5^\star$ leaving the 2nd and 5th sides. Whatever the states of the grey sites (and the undrawn sites), the event $E = L_1 \cap L_2^\star \cap L_4 \cap L_5^\star$ holds.

to form a doubly infinite closed path separating the open paths P_1 and P_4 guaranteed by L_1 and L_4. It follows that, with positive probability, there are at least two infinite open clusters, contradicting Theorem 4. □

An alternative proof of Theorem 8 is given in Bollobás and Riordan [2006b], based on an RSW type theorem and the sharp-threshold results in Chapter 2.

The method used to prove Theorem 8 above may be applied to site percolation in the square lattice. This time, the critical probability cannot be obtained in this way, as the lattice is not self-dual. Indeed, let $\Lambda_\square = \mathbb{Z}^2$ be the planar square lattice, and let $\Lambda_\boxtimes$ be the graph with vertex set $\mathbb{Z}^2$ in which any two vertices at Euclidean distance 1 or $\sqrt{2}$ are adjacent. Thus $\Lambda_\boxtimes$ is obtained from $\Lambda_\square$ by adding both diagonals to each face of $\Lambda_\square$. It is easy to see that a finite open cluster in the site percolation on $\Lambda_\square$ is bounded by a set of closed sites that form a path in $\Lambda_\boxtimes$, and vice versa. Also, a path in $\Lambda_\square$ cannot cross a path in $\Lambda_\boxtimes$ without the two sharing a vertex. These observations are the starting point for the proof of the 'duality' result for $\Lambda_\square$ and $\Lambda_\boxtimes$, Theorem 10 below.

Let $\mathbb{P}_p$ be the product probability measure in which each site of $\mathbb{Z}^2$ is open with probability p, and let R be a rectangle in $\mathbb{Z}^2$. For $\Lambda = \Lambda_\square$ or $\Lambda = \Lambda_\boxtimes$, let $H_\Lambda(R)$ be the event that there is an open Λ-path crossing R horizontally, i.e., a set of open sites of R that form a path in the graph Λ crossing R horizontally. Similarly, let $V_\Lambda(R)$ be the event that there is an open Λ-path crossing R vertically. The following result corresponds to Lemma 1 of Chapter 3.

Lemma 9. *Let Λ be one of $\Lambda_\square$ and $\Lambda_\boxtimes$, let $\Lambda^\star$ be the other, and let R be a rectangle in $\mathbb{Z}^2$. Whatever the states of the sites in R, there is either an open Λ-path crossing R from left to right, or a closed $\Lambda^\star$-path crossing R from top to bottom, but not both. In particular,*

$$\mathbb{P}_p\big(H_\Lambda(R)\big) + \mathbb{P}_{1-p}\big(V_{\Lambda^\star}(R)\big) = 1. \tag{6}$$

Lemma 9 says that, if we colour the squares of an n by m chess board black and white in an arbitrary manner, then either a rook can move from the left side to the right passing only over black squares, or a king can move from top to bottom using only white squares, but not both. Bollobás and Riordan [2006b] gave a very simple proof of this result; this proof is very similar to the corresponding arguments for bond percolation on $\mathbb{Z}^2$ and for site percolation on the triangular lattice presented here. Figure 8 below, reproduced from Bollobás and Riordan [2006b], is essentially the complete proof. Although this fact is not needed for the proof, let us note that the tiling in the picture is a finite part of the lattice $(4, 8^2)$ shown in Figure 18. The same lattice was used in a different way in the proof of Lemma 1 of Chapter 3.

Theorem 10. *The critical probabilities for site percolation on the lattices $\Lambda_\square$ and $\Lambda_\boxtimes$ obey the relation $p_c^s(\Lambda_\square) + p_c^s(\Lambda_\boxtimes) = 1$.*

This result was first proved by Russo [1981] (see also Russo [1982]), by adapting the original arguments for bond percolation on $\mathbb{Z}^2$, in particular, the RSW Theorem. An alternative presentation of this approach is given in Bollobás and Riordan [2006b]. Once again, Theorem 10 is easy to deduce from the general results of Menshikov and of Aizenman, Kesten and Newman; we shall describe briefly the steps in such a deduction.

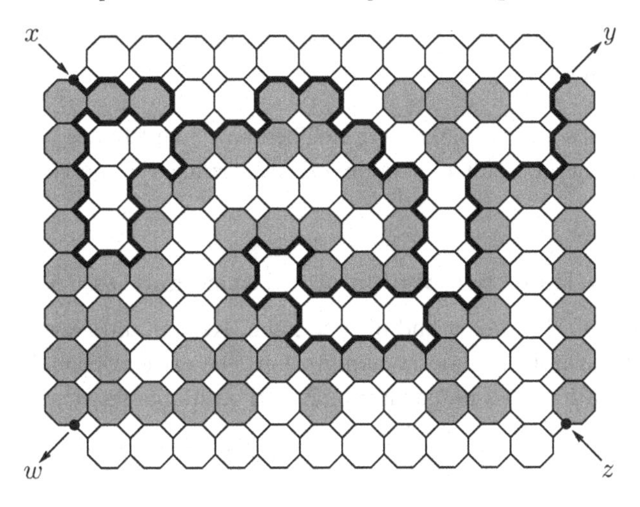

Figure 8. A rectangle R in $\mathbb{Z}^2$ with each site drawn as an octagon, with an additional row/column of sites on each side. 'Black' (shaded) octagons are open. Either there is a black path from left to right, or a white path (which may use the squares) from top to bottom. Following the interface between black and white regions starting at x, one emerges either at y or at w. In the first case (shown), the event $H_{\Lambda_\square}(R)$ holds. Otherwise $V_{\Lambda_\boxtimes}(R)$ holds.

Proof. Once again, by Menshikov's Theorem, we have $p_{\mathrm{H}}^{\mathrm{s}} = p_{\mathrm{T}}^{\mathrm{s}}$ for $\Lambda_\square$ and for $\Lambda_\boxtimes$, so it is legitimate to write $p_{\mathrm{c}}^{\mathrm{s}}$ for their common value.

Given an assignment of states to the sites of $\mathbb{Z}^2$, by a Λ-*cluster* we mean a maximal connected open subgraph of Λ, where $\Lambda = \Lambda_\square$ or $\Lambda_\boxtimes$.

Suppose first that $p_{\mathrm{c}}^{\mathrm{s}}(\Lambda_\square) + p_{\mathrm{c}}^{\mathrm{s}}(\Lambda_\boxtimes) > 1$. Then we may choose a $p \in (0,1)$ with $1 - p_{\mathrm{c}}^{\mathrm{s}}(\Lambda_\boxtimes) < p < p_{\mathrm{c}}^{\mathrm{s}}(\Lambda_\square)$. Note that $1 - p < p_{\mathrm{c}}^{\mathrm{s}}(\Lambda_\boxtimes)$. Taking each site of $\mathbb{Z}^2$ to be open independently with probability p, by Theorem 9 of Chapter 4 we have exponential decay of the radius of the open $\Lambda_\square$-cluster containing the origin, and exponential decay of the radius of the closed $\Lambda_\boxtimes$-cluster containing the origin. For large enough n, this contradicts Lemma 9 applied to an n by n square.

Suppose next that $p_{\mathrm{c}}^{\mathrm{s}}(\Lambda_\square) + p_{\mathrm{c}}^{\mathrm{s}}(\Lambda_\boxtimes) < 1$. Then there is a p with $p > p_{\mathrm{c}}^{\mathrm{s}}(\Lambda_\square)$ and $1 - p > p_{\mathrm{c}}^{\mathrm{s}}(\Lambda_\boxtimes)$. Taking each site open independently with this probability, by Theorem 4 there is, with probability 1, a unique infinite open $\Lambda_\square$-cluster, and a unique infinite closed $\Lambda_\boxtimes$-cluster. It follows as before that with positive probability there are infinite open $\Lambda_\square$-paths leaving a large square S from two opposite sides, and infinite closed $\Lambda_\boxtimes$-paths leaving S from the remaining sides. Hence, with positive probability, there are at least two infinite open $\Lambda_\square$-clusters, a contradiction. □

As we shall see in the next section, Theorems 8 and 10 are special cases of a more general result (Theorem 14) concerning symmetric lattices.

5.4 Bond percolation on a lattice and its dual

The results of Menshikov and of Aizenman, Kesten and Newman imply that, under a mild symmetry assumption, the critical probabilities for bond percolation on a planar lattice Λ and on its planar dual $\Lambda^\star$ satisfy

$$p_c^b(\Lambda) + p_c^b(\Lambda^\star) = 1. \tag{7}$$

When $\Lambda = \mathbb{Z}^2$, the relation above is exactly the Harris–Kesten Theorem. Later, we shall prove (7) in some generality (Theorem 13); first, we illustrate it with another simple example. As before, let T be the (equilateral) triangular lattice in the plane. Let H be the planar dual of T, defined in the usual way. Taking the sites of H to be the centres of the faces of T, then H is the (regular) hexagonal lattice, or *honeycomb*; see Figure 9.

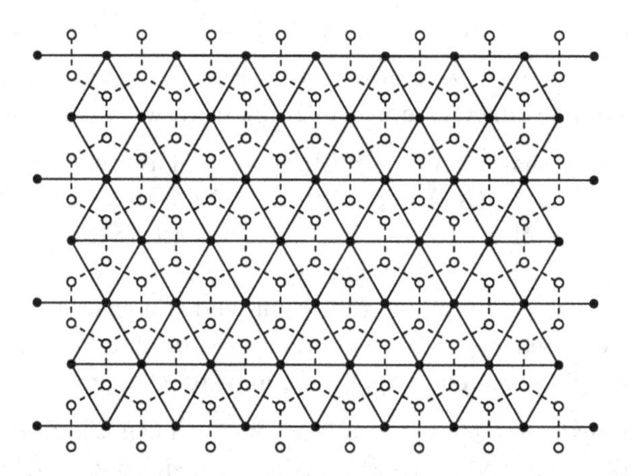

Figure 9. Portions of the triangular lattice T and its dual H, the hexagonal or *honeycomb* lattice

As we shall see later, the critical probabilities for bond percolation on

T and on H have been determined exactly by Wierman [1981], confirming a conjecture of Sykes and Essam [1963; 1964]. For the moment, we show only that these critical probabilities sum to 1.

Theorem 11. *The triangular lattice T and honeycomb lattice H satisfy*

$$p_c^b(T) + p_c^b(H) = 1. \tag{8}$$

Proof. The result follows easily from the general results of Menshikov and of Aizenman, Kesten and Newman. Suppose first that $p_c^b(T) + p_c^b(H) > 1$. Then we may chose $p \in (0,1)$ with $p < p_c^b(T)$ and $1-p < p_c^b(H)$. Let us take the bonds $e \in E(T)$ to be open independently with probability p, and each dual bond $e^\star \in E(H)$ to be open if and only if e is closed. Then by Menshikov's Theorem we have exponential decay of the radius of open clusters both in T and in H. Hence, taking a large enough 'rectangle' R as in Figure 9, with probability 99% there is neither an open path in T crossing R from left to right, nor an open path in H crossing R from top to bottom. But by planar duality, there is always a path of one of these two types: this is a special case of Lemma 2 of Chapter 3, whose proof is the same as that of Lemma 1 of that chapter. In this case, the figure obtained by replacing each degree d site of H or its dual by a $2d$-gon, and each bond-dual bond pair by a square, is the $(4,6,12)$ lattice shown in Figure 18.

To complete the proof of (8), it suffices to show that for any p, at most one of the percolation probabilities $\theta(T;p)$ and $\theta(H;1-p)$ is strictly positive. This follows from Theorem 4 as above: if both $\theta(T;p)$ and $\theta(H;1-p)$ are strictly positive then, taking bonds of T open independently with probability p, with probability 1 there is a unique infinite open cluster in T, and a unique infinite open cluster in H. Considering a large enough hexagon in T, it follows as before that with positive probability there are infinite open paths in T leaving the hexagon from the 1st and 3rd sides, and infinite open paths in H leaving from the 2nd and 4th sides. If the bonds of H inside the hexagon are also open, we find a doubly infinite open path in H separating two infinite open components in T, a contradiction. □

Having proved (7) in two special cases, for $\Lambda = \mathbb{Z}^2$, and for $\Lambda = T$, we turn to a considerably more general result. The arguments we have given so far used the fact that Λ had a suitable rotational symmetry, of order 4 in the first case, and order 6 in the second. In fact, the weaker assumption of order 2 rotational symmetry is enough, although one has to work a little harder to obtain (7) in this case. Also, there is a natural generalization of (7) to certain settings in which bonds of different 'types' may be open with different probabilities. In this context, it is convenient to work with a *weighted graph* $(\Lambda, \mathbf{p})$, i.e., a graph Λ together with an assignment of a weight $p_e \in [0,1]$ to every bond e of Λ. For each weighted graph there is a corresponding independent bond percolation model $M = M(\Lambda, \mathbf{p})$, in which the bonds of Λ are open independently, and each bond e is open with probability p_e.

To state a formal result, by a *planar lattice* we mean a connected, locally finite plane graph Λ (i.e., a planar graph with a given drawing in the plane), with $V(\Lambda)$ a discrete subset of $\mathbb{R}^2$, such that there are translations T_{v_1} and T_{v_2} of $\mathbb{R}^2$ through two independent vectors v_1 and v_2 each of which acts on Λ as a graph isomorphism. In particular, all the Archimedean lattices are planar lattices. Recall that two sites x and x', or two bonds e and e', are *equivalent* in a graph Λ if there is an automorphism φ of Λ mapping x to x', or e to e'. Note that any lattice is a finite-type graph, in the sense that there are finitely many equivalence classes of sites and of bonds under this relation.

To allow for models in which edges have different probabilities of being open, we define a *weighted planar lattice* $(\Lambda, \mathbf{p})$ as above: Λ is a planar lattice, and there are two translations T_{v_1} and T_{v_2} as above acting as automorphisms of $(\Lambda, \mathbf{p})$ as a weighted graph, i.e., preserving the edge weights. Perhaps the simplest non-trivial example is the square lattice, with $p_e = p$ for every horizontal bond and $p_e = q$ for every vertical bond, where $0 < p, q < 1$. Kesten [1982] showed that in this case, percolation occurs if and only if $p + q > 1$; see Theorem 15 below. Another simple example is shown in Figure 10. Note that in a weighted planar lattice, there can only be finitely many distinct edge weights.

We say that a graph Λ drawn in the plane is *centrally symmetric*, or simply *symmetric*, if the map $x \mapsto -x$ from $\mathbb{R}^2$ to itself acts on Λ as a graph isomorphism. For a weighted graph, this map should also preserve the weights. For example, the (weighted) planar lattice shown in in Figure 10 is symmetric if one takes the origin to be the centre of an appropriate face. If Λ is a planar lattice then, taking the vertices of the planar dual $\Lambda^\star$ to be the centroids of the faces of Λ, say, one can draw $\Lambda^\star$ as a planar lattice, as in Figure 10. We assume throughout that the bonds of both Λ and $\Lambda^\star$ are drawn with piecewise linear curves in the plane. If Λ is symmetric, then we may draw $\Lambda^\star$ so that it is also symmetric. The dual of a weighted planar lattice $(\Lambda, \mathbf{p})$ is the weighted planar lattice $(\Lambda^\star, \mathbf{q})$ in which the dual $e^\star$ of a bond e

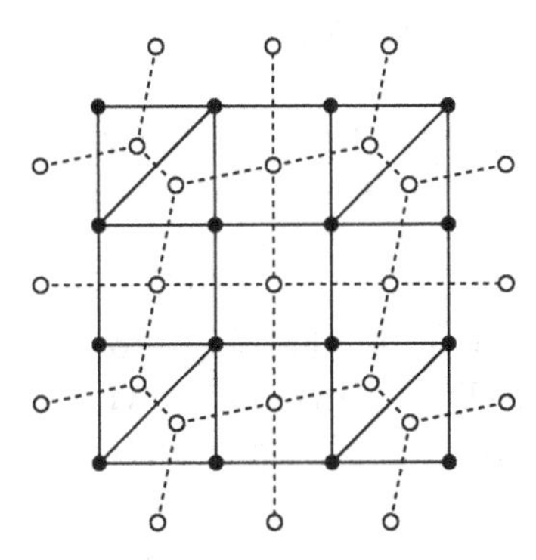

Figure 10. The planar lattice Λ (solid lines and filled circles) obtained by adding diagonals to every fourth face of $\mathbb{Z}^2$. If the horizontal, vertical and diagonal bonds are assigned weights p, q, and r respectively, then Λ becomes a weighted planar lattice. The dual $\Lambda^\star$ of Λ is drawn with hollow circles, at the centroids of the faces of Λ, and dashed lines.

has weight $q_{e^\star} = 1 - p_e$. As shown by Bollobás and Riordan [2006d], percolation cannot occur simultaneously on a symmetric planar weighted lattice and on its dual.

Theorem 12. *Let* $(\Lambda, \mathbf{p})$ *be a symmetric weighted planar lattice, with* $0 < p_e < 1$ *for every bond* e*. Then either* $\theta(\Lambda; \mathbf{p}) = 0$ *or* $\theta(\Lambda^\star; \mathbf{q}) = 0$*, where* $(\Lambda^\star, \mathbf{q})$ *is the dual weighted lattice.*

Proof. Suppose for a contradiction that $\theta(\Lambda; \mathbf{p}) > 0$ and $\theta(\Lambda^\star; \mathbf{q}) > 0$. In the proof of Theorem 4 it was not relevant that all bonds were open with equal probability. Thus, writing M and $M^\star$ for the independent bond percolation models associated to $(\Lambda, \mathbf{p})$ and to $(\Lambda^\star, \mathbf{q})$, we see that in each of M and $M^\star$ there is a unique infinite open cluster with probability 1. As usual, we realize the bond percolation models M and $M^\star$ simultaneously on the same probability space, by taking the dual $e^\star$ of a bond $e \in \Lambda$ to be open if and only if e is closed. Throughout the proof we write $\mathbb{P}$ for the probability measure on this probability space.

The basic idea of the proof is as follows: as before, any large square S is very likely to meet the unique infinite open clusters in Λ and in $\Lambda^\star$. If we had four-fold symmetry then, using the 'nth-root trick', we could deduce that for each side of S, with high probability there are infinite open paths in Λ and in $\Lambda^\star$ leaving S from that side. With only central symmetry, all we can conclude immediately is that there is *some* pair of opposite sides of S from which infinite open paths in Λ leave S with high probability.

The key idea is to move the 'corners' of S while keeping S the same. More precisely, instead of a square, we take S to be a circle whose boundary is divided into four arcs A_1, A_2, A_3, A_4, and consider infinite open paths leaving S from each A_i. If we move the dividing point between two arcs, then paths leaving one become more likely, and paths leaving the other less likely. If we move the dividing point gradually, then the probabilities will change in a roughly continuous manner, so at some point they will be roughly equal. By moving two opposite division points while preserving symmetry, we can find a symmetric decomposition of the boundary of S into four arcs so that open paths of Λ leaving the four arcs are roughly equally likely. Now, using the 'fourth-root trick', for *every* arc A_i it is very likely that there is an infinite open path in Λ leaving S from this arc. We cannot say that the same applies to $\Lambda^\star$, as we have chosen the arcs for Λ and not for $\Lambda^\star$. We observe, however, that among our four arcs, there is *some* pair of opposite arcs of S from which infinite open paths of $\Lambda^\star$ leave with high probability. Indeed, this follows from symmetry and the square-root trick. This gives us infinite paths in Λ, $\Lambda^\star$, Λ and $\Lambda^\star$ leaving the arcs of S in order and, as before, we can deduce a contradiction by showing that the two paths in $\Lambda^\star$ may be joined within S, giving two infinite open clusters in Λ. We shall now make this argument precise.

Let $S = S_r$ be the circle centred at the origin with radius r. Let $E(S)$ be the event that some site of Λ inside S is in an infinite open cluster in Λ, and let $E^\star(S)$ be the event that some site of $\Lambda^\star$ inside S lies in an infinite open cluster in $\Lambda^\star$. Writing D_r for the disc bounded by S_r, we have $\bigcup_r D_r = \mathbb{R}^2$, so the union $\bigcup_r E(S_r)$ is simply the event that there is an infinite open cluster somewhere in Λ, and we have $\lim_{r\to\infty} \mathbb{P}\big(E(S_r)\big) = 1$, and similarly for $E^\star(S_r)$. Let $\varepsilon = \varepsilon(\Lambda, \mathbf{p})$ be a positive constant that we shall specify later. Choosing r large enough, we have

$$\mathbb{P}\big(E(S_r)\big) \geq 1 - \varepsilon^4 \quad \textit{and} \quad \mathbb{P}\big(E^\star(S_r)\big) \geq 1 - \varepsilon^4. \tag{9}$$

For simplicity, we shall assume throughout this proof that no site of Λ or $\Lambda^\star$ lies on S_r, and that no bond of Λ or $\Lambda^\star$ is tangent to S_r. (More precisely, recalling that bonds of Λ and $\Lambda^\star$ are drawn as sequences of line segments, we assume that none of these segments is tangent to S_r.) This assumption is satisfied for all but a countable set of values of r. For the rest of the proof we fix such an r large enough that (9) holds, and write S for S_r.

Let c_i, $1 \le i \le 4$, be four distinct points on the boundary of S, numbered in anticlockwise order. We write $\mathbf{c}$ for the quadruple (c_1, c_2, c_3, c_4). We shall always choose these points so that no c_i is on a bond of Λ or $\Lambda^\star$. We write $A_i = A_i(\mathbf{c})$ for the boundary *arc* of S from c_i to c_{i+1}, taking $c_5 = c_1$. If vw is a bond of Λ with v inside S and w outside, then vw *leaves S from the arc A_i* if, travelling along the (piecewise linear) bond vw from v to w, the last point of S lies on the arc A_i. Let $E_i = E_i(\mathbf{c})$ be the event that there is an infinite open path in Λ *leaving S from the arc A_i*, i.e., an open path $P = v_0v_1v_2,\ldots$ with v_0 inside S and v_j outside S for all $j \ge 1$, such that v_0v_1 leaves S from the arc A_i; see Figure 11.

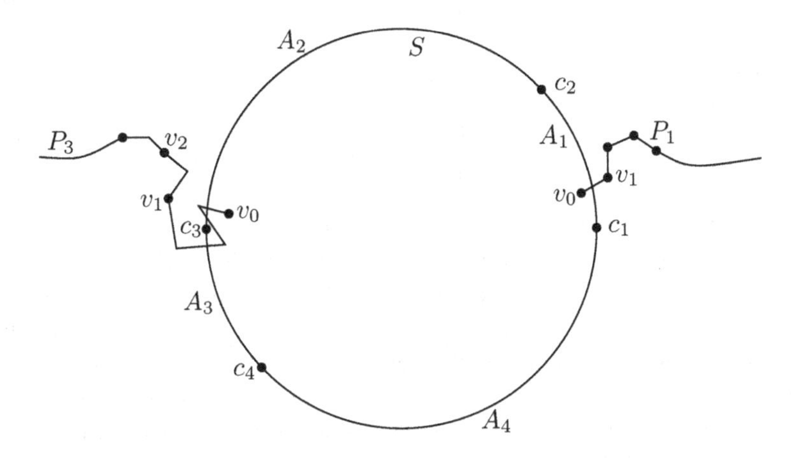

Figure 11. Possible open paths P_1 and P_3 witnessing the events $E_1 = E_1(\mathbf{c})$ and $E_3 = E_3(\mathbf{c})$. Usually, the bonds are straight line segments, as in P_1, but they need not be.

The precise details of the definition are not that important: the 'soft' arguments we shall present go through with many minor variants. For example, we could consider the last time the whole path leaves S, even if this is on a bond v_0v_1 with $v_0 \notin S$. For 'nice' drawings of 'nice' lattices, a bond typically crosses S at most once, so the condition is essentially that v_0v_1 crosses A_i.

Set

$$f_i(\mathbf{c}) = \mathbb{P}\big(E_i(\mathbf{c})\big)$$

and

$$g_i(\mathbf{c}) = 1 - f_i(\mathbf{c}) = \mathbb{P}\big(E_i(\mathbf{c})^c\big),$$

and define $f_i^\star$ and $g_i^\star$ similarly, using the dual lattice. As any infinite open path starting inside S must leave S somewhere, $\bigcup_i E_i(S)$ is exactly the event $E(S)$. The events $E_i(S)$ are increasing, so their complements $E_i(S)^c$ are decreasing. Thus, by Harris's Lemma (Lemma 3 of Chapter 2), for any $\mathbf{c}$ we have

$$\mathbb{P}(E(S)^c) = \mathbb{P}\left(\bigcap_{i=1}^{4} E_i(\mathbf{c})^c\right) \geq \prod_{i=1}^{4} \mathbb{P}(E_i(\mathbf{c})^c) = \prod_{i=1}^{4} g_i(\mathbf{c}).$$

From (9) it follows that

$$\prod_{i=1}^{4} g_i(\mathbf{c}) \leq \varepsilon^4. \tag{10}$$

Similarly,

$$\prod_{i=1}^{4} g_i^\star(\mathbf{c}) \leq \varepsilon^4. \tag{11}$$

The key observation is that, as we move one point, c_2, say, the probabilities $f_i(\mathbf{c})$ change in a 'continuous' manner. For a precise statement, it is more convenient to work with $g_i(\mathbf{c})$. The only properties of $\mathbf{c}$ that the event $E_i(\mathbf{c})$ depends on are which bonds of Λ leave S from which arcs A_j. Thus, as we move c_2, the probabilities $f_i(\mathbf{c})$ and $g_i(\mathbf{c})$ can only change when c_2 moves across a bond of Λ. Of course, $g_i(\mathbf{c})$ does jump at these points. Our claim is that there is a constant $C = C(\Lambda, \mathbf{p})$ such that, at any such jump, no $g_i(\mathbf{c})$ increases or decreases by more than a factor of C.

Let $\mathbf{c}$ and $\mathbf{c}' = (c_1, c_2', c_3, c_4)$ be such that exactly one bond e leaves S from the arc c_2c_2'. Without loss of generality we may suppose that c_1, c_2, c_2', c_3 lie in this order around S; see Figure 12. Thus, defining arcs A_i using the division points $\mathbf{c}$, the bond e leaves S across the arc $A_1 = c_1c_2$ while, defining arcs A_i' using the points $\mathbf{c}'$, the bond e leaves across A_2'. All other bonds leaving S do so across corresponding arcs A_i, A_i'. Thus, for $i = 3, 4$ the events $E_i(\mathbf{c})$ and $E_i(\mathbf{c}')$ coincide, so $f_i(\mathbf{c}) = f_i(\mathbf{c}')$.

Let E_e be the event that the bond e is open. The event $E_i(\mathbf{c})$ is defined in terms of *open* paths leaving S across the arc A_i. If e is closed, then no open path leaves S along the bond e, so which arc e crosses is irrelevant. Thus, the symmetric difference of $E_i(\mathbf{c})$ and $E_i(\mathbf{c}')$ is contained in the event E_e. In other words,

$$E_i(\mathbf{c})^c \cap E_e^c = E_i(\mathbf{c}')^c \cap E_e^c$$

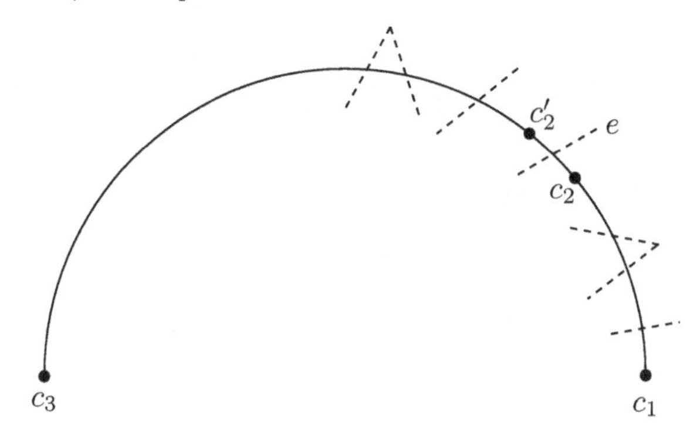

Figure 12. The effect of moving c_2 slightly to a new point c_2': various bonds of Λ are shown as dashed lines. We choose c_2' so that a unique bond e leaves S between c_2 and c_2'. The arcs A_i and A_i' are determined by $A_1 = c_1c_2$, $A_1' = c_1c_2'$, $A_2 = c_2c_3$ and $A_2' = c_2'c_3$.

holds for each i. Now $E_i(\mathbf{c})^c$ and E_e^c are decreasing events, so, by Harris's Lemma,

$$\mathbb{P}\big(E_i(\mathbf{c})^c \cap E_e^c\big) \geq \mathbb{P}\big(E_i(\mathbf{c})^c\big)\mathbb{P}\big(E_e^c\big).$$

Thus

$$\begin{aligned} g_i(\mathbf{c}') = \mathbb{P}\big(E_i(\mathbf{c}')^c\big) &\geq \mathbb{P}\big(E_i(\mathbf{c}')^c \cap E_e^c\big) \\ &= \mathbb{P}\big(E_i(\mathbf{c})^c \cap E_e^c\big) \geq \mathbb{P}\big(E_i(\mathbf{c})^c\big)\mathbb{P}\big(E_e^c\big) \geq c g_i(\mathbf{c}), \end{aligned}$$

where

$$c = c(\Lambda, \mathbf{p}) = \inf_e \mathbb{P}\big(E_e^c\big) = \inf_e (1 - p_e) > 0,$$

as Λ has finite type and each $p_e < 1$. Similarly, $g_i(\mathbf{c}) \geq c g_i(\mathbf{c}')$, establishing the claim.

Set $C = 1/c > 1$. Let us fix c_1 and c_3 as opposite points of S. Consider moving c_2 from very close to c_1 to very close to c_3. At the start of this process, no bonds cross A_1, so $E_1(\mathbf{c})$ cannot hold, and $g_1(\mathbf{c}) = 1 \geq g_2(\mathbf{c})$. Similarly, at the end, $g_2(\mathbf{c}) = 1 \geq g_1(\mathbf{c})$. Each time c_2 crosses a bond, $g_1(\mathbf{c})$ decreases by at most a factor C, and $g_2(\mathbf{c})$ increases by at most a factor C. It follows that we may choose c_2 so that

$$1/C \leq g_1(\mathbf{c})/g_2(\mathbf{c}) \leq C.$$

Let c_4 be the opposite point to c_2. Then by central symmetry we have

$$g_i(\mathbf{c}) = g_{i+2}(\mathbf{c}) \quad \textit{and} \quad g_i^\star(\mathbf{c}) = g_{i+2}^\star(\mathbf{c}). \tag{12}$$

Thus,

$$\prod_i g_i(\mathbf{c}) = g_1(\mathbf{c})^2 g_2(\mathbf{c})^2 \geq g_1(\mathbf{c})^4/C^2.$$

Using (10), it follows that $g_1(\mathbf{c}) \leq C^{1/2}\varepsilon$, and hence that

$$g_i(\mathbf{c}) \leq C^{3/2}\varepsilon \tag{13}$$

for *every* i. From (11), there is *some* j with $g_j^\star(\mathbf{c}) \leq \varepsilon$. As (13) holds for *every* i, we may assume without loss of generality that $j = 1$. Thus, using (12) again,

$$g_3^\star(\mathbf{c}) = g_1^\star(\mathbf{c}) \leq \varepsilon \quad \text{and} \quad g_4(\mathbf{c}) = g_2(\mathbf{c}) \leq C^{3/2}\varepsilon. \tag{14}$$

It is now easy to complete the proof of Theorem 12, although, to avoid the need to consider exactly how the bonds of Λ and $\Lambda^\star$ leave S, especially near the division points c_i, we shall introduce one more technicality.

Let d be a constant (much) larger than the length of any bond in Λ or in $\Lambda^\star$. Let $F_i = F_i(\mathbf{c})$ be the event that there is an infinite open path P in Λ leaving S across the arc A_i, such that no point of P lies within distance d of any c_j. This event is illustrated in Figure 13. Let $D(\mathbf{c})$ be the event that all bonds passing within distance d of any c_j are closed. There is a maximum number of bonds of Λ that any disc of radius d can meet, so there is a constant $c_1 = c_1(\Lambda, \mathbf{p}) > 0$ such that $\mathbb{P}\big(D(\mathbf{c})\big) \geq c_1$ for any $\mathbf{c}$. Clearly, if $D(\mathbf{c})$ holds, then $E_i(\mathbf{c})$ holds if and only if $F_i(\mathbf{c})$ holds. Using Harris's Lemma as above, it follows that

$$h_i(\mathbf{c}) = \mathbb{P}\big(F_i(\mathbf{c})^c\big) \leq C_1 g_i(\mathbf{c}), \tag{15}$$

where $C_1 = 1/c_1$. Replacing C_1 by $1/\min\{c_1(\Lambda, \mathbf{p}), c_1(\Lambda^\star, \mathbf{q})\}$, then both (15) and the corresponding equation

$$h_i^\star(\mathbf{c}) = \mathbb{P}\big(F_i^\star(\mathbf{c})^c\big) \leq C_1 g_i^\star(\mathbf{c}) \tag{16}$$

hold for any $\mathbf{c}$.

Let $\varepsilon = 1/(10C^{3/2}C_1)$, noting that this quantity depends on Λ and $\mathbf{p}$ only, not on r or $\mathbf{c}$. From (14) and (15) we have

$$h_2(\mathbf{c}), h_4(\mathbf{c}) \leq C_1 C^{3/2}\varepsilon \leq 1/10,$$

while from (14) and (16) we have

$$h_1^\star(\mathbf{c}), h_3^\star(\mathbf{c}) \leq C_1\varepsilon \leq 1/10.$$

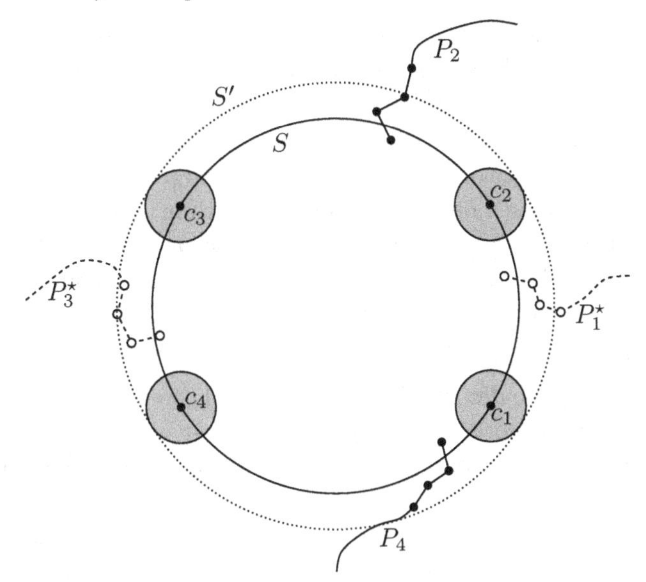

Figure 13. Open paths P_2 and P_4 witnessing the events $F_2(\mathbf{c})$ and $F_4(\mathbf{c})$. The open dual paths $P_1^\star$ and $P_3^\star$ witness $F_1^\star(\mathbf{c})$ and $F_3^\star(\mathbf{c})$. No site or bond of any of these paths meets the shaded circles. It follows that P_2, for example, leaves the larger circle S' through the arc corresponding to A_2.

Hence, with with probability at least $1 - 4/10 > 0$, the event

$$F = F_1^\star(\mathbf{c}) \cap F_2(\mathbf{c}) \cap F_3^\star(\mathbf{c}) \cap F_4(\mathbf{c})$$

holds, i.e., there are infinite open paths P_2 and P_4 in Λ leaving S from the arcs A_2 and A_4, and infinite open paths $P_1^\star$ and $P_3^\star$ in $\Lambda^\star$ leaving S from the arcs A_1 and A_3, with no P_i or $P_i^\star$ passing with distance d of the endpoint of any arc A_j. This event is illustrated in Figure 13.

When F holds, there are sub-paths P_2' and P_4' of P_2 and P_4 leaving the larger circle S' with radius $r+d$ from the arcs corresponding to A_2 and A_4. We may connect $P_1^\star$ and $P_3^\star$ by changing the states of all dual bonds $e^\star$ that meet S to open. The corresponding bonds e lie entirely within S', so after this change the paths P_2' and P_4' are still open. But then we have, with positive probability, two infinite open paths separated by a doubly infinite open dual path. This implies that there are two infinite open clusters in Λ, contradicting Theorem 4. □

As an immediate corollary of Theorem 12 we obtain the desired relationship between the critical probabilities for bond percolation on a planar lattice Λ and on its dual, assuming only central symmetry.

Theorem 13. *Let Λ be a symmetric planar lattice, and let $\Lambda^\star$ be its planar dual. Then $p_c^b(\Lambda) + p_c^b(\Lambda^\star) = 1$.*

Proof. As before, the inequality $p_c^b(\Lambda) + p_c^b(\Lambda^\star) \leq 1$ follows easily from Menshikov's Theorem, by considering a large region in the plane which must be crossed one way by an open path in Λ, or the other way by an open path in $\Lambda^\star$. In the other direction, $p_c^b(\Lambda) + p_c^b(\Lambda^\star) \geq 1$ is immediate from Theorem 12: if this inequality does not hold, then there is a $p \in (0,1)$ with $p > p_c^b(\Lambda)$ and $1-p > p_c^b(\Lambda^\star)$. But then $\theta(\Lambda; p)$ and $\theta(\Lambda^\star; 1-p)$ are strictly positive, contradicting Theorem 12. □

Theorem 13 includes the Harris–Kesten Theorem, Theorem 13 of Chapter 3, and Theorem 11 as special cases. It also applies to many other lattices, for example, all the Archimedean lattices shown in Figure 18.

Turning to site percolation, Kesten [1982] pointed out that the site percolation models on certain pairs of graphs are related in a way that is analogous to the connection between bond percolation on a planar graph and its dual; he called such graphs *matching* pairs, and noted that any planar lattice matches some graph. To see this, let Λ be a planar lattice, and let $\Lambda^\times$ be the graph on the same vertex set obtained from Λ by adding all diagonals to all faces. For example, if $\Lambda = \Lambda_\square$, then $\Lambda^\times = \Lambda_\boxtimes$. The proof of Lemma 9 extends immediately to show that, for a suitably chosen 'rectangle' in Λ, whatever the states of the sites of $V(\Lambda) = V(\Lambda^\times)$, either there is an open Λ-crossing from the left to the right, or a closed $\Lambda^\times$-crossing from the top to the bottom: to obtain the picture corresponding to Figure 8, replace each site v of Λ with degree d by a $2d$-gon that is black if v is open and white if v is closed, and each f-sided face of Λ by a white f-gon.

If Λ is symmetric, then trivial modifications of the proof of Theorem 12 show that $\theta_p(\Lambda)$ and $\theta_{1-p}(\Lambda^\times)$ cannot both be strictly positive, while $p_c^s(\Lambda) + p_c^s(\Lambda^\times) \leq 1$ is again immediate from Menshikov's Theorem, giving the following analogue of Theorem 13.

Theorem 14. *Let Λ be a symmetric planar lattice, and let $\Lambda^\times$ be the graph obtained from Λ by adding all diagonals to all faces of Λ. Then*

$$p_c^s(\Lambda) + p_c^s(\Lambda^\times) = 1. \qquad \square$$

As noted above, $\Lambda_\square^\times = \Lambda_\boxtimes$, so Theorem 14 implies Theorem 10. Since every face of the triangular lattice T is a triangle, $T^\times = T$, so Theorem 14 implies Theorem 8 as well.

We conclude this section with an application of Theorem 12 to a weighted graph. Let $(\mathbb{Z}^2, p_x, p_y)$ be the graph $\mathbb{Z}^2$, in which each horizontal bond has weight p_x and each vertical bond weight p_y. Kesten [1982, p. 82] showed that the 'critical line' for this model is given by $p_x+p_y = 1$.

Theorem 15. *Let $\theta(p_x, p_y)$ denote the probability that the origin is in an infinite open cluster in the independent bond percolation on $(\mathbb{Z}^2, p_x, p_y)$. For $0 < p_x, p_y < 1$, we have $\theta(p_x, p_y) > 0$ if and only if $p_x + p_y > 1$.*

Proof. The planar dual of the weighted graph $\Lambda = (\mathbb{Z}^2, p_x, p_y)$ is $\Lambda^\star = (\mathbb{Z}^2+(1/2, 1/2), 1-p_y, 1-p_x)$, the usual dual of $\mathbb{Z}^2$ with weight $1-p_y$ on each horizontal bond, and weight $1-p_x$ on each vertical bond. Rotating and translating, $\Lambda^\star$ is isomorphic to $(\mathbb{Z}^2, 1-p_x, 1-p_y)$.

Suppose first that $\theta(p_x, p_y) = 0$ for some p_x, p_y with $p_x + p_y > 1$, and fix $0 < p'_x < p_x$ and $0 < p'_y < p_y$ with $p'_x + p'_y > 1$. By the weighted version of Menshikov's Theorem, the radius of the open cluster containing a given site of $\Lambda' = (\mathbb{Z}^2, p'_x, p'_y)$ decays exponentially. As $1 - p'_y < p'_x$ and $1 - p'_x < p'_y$, the same is true in the dual, $(\mathbb{Z}^2 + (1/2, 1/2), 1 - p'_y, 1 - p'_x)$. But then the probability that a large square has either an open horizontal crossing or an open dual vertical crossing tends to zero, contradicting Lemma 1 of Chapter 3.

We have shown that the condition $p_x+p_y > 1$ is sufficient for $\theta(p_x, p_y)$ to be non-zero. To show that it is necessary, it suffices to show that $\theta(p, 1-p) = 0$ for every $0 < p < 1$. Since $(\mathbb{Z}^2, p, 1-p)$ is symmetric and self-dual, this follows from Theorem 12. □

As it happens, one does not need Theorem 12 to prove Theorem 15; the proof of the Harris–Kesten Theorem given in this chapter adapts immediately. Indeed, suppose that $\theta(p, 1-p) > 0$. Considering a large square S, it follows from the 'fourth-root' trick that there is *some* side of S from which an infinite open path leaves with high probability. Of course, the same holds for the opposite side. The dual weighted lattice is isomorphic to the original lattice rotated through 90 degrees, so infinite open dual paths leave the remaining two sides of S with high probability, and one can complete the proof as before. In the next section we shall apply Theorem 12 to prove a more difficult result, that a certain analogue of Theorem 15 holds for the triangular lattice.

5.5 The star-delta transformation

Sykes and Essam [1963] noticed a second connection between bond percolation on the hexagonal and triangular lattices H and T, other than that given by duality. This connection involves the *star-triangle transformation* or *star-delta* transformation, a basic transformation in the theory of electrical networks. To describe this, let G_1 and G_2 be the two graphs shown in Figure 14. Suppose that the bonds of G_1 are open

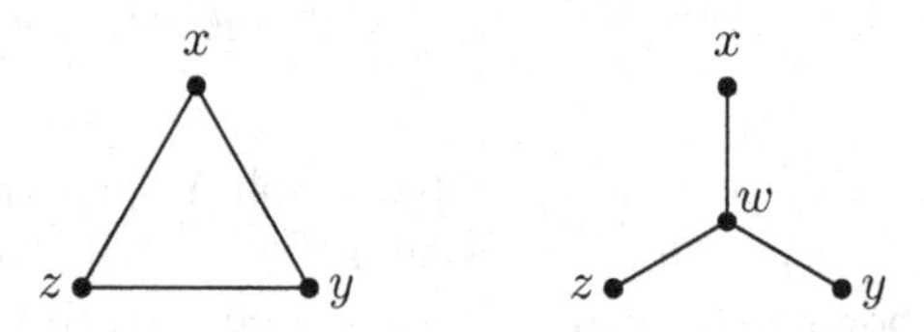

Figure 14. A triangle, G_1, and a star, G_2, with the same 'attachment' sites x, y and z.

independently with probability p_1, and those of G_2 with probability p_2. In either graph, there are five possibilities for which sites among $\{x, y, z\}$ are connected to each other by open paths: all three may be connected, none may be connected, and some pair may be connected to each other but not to the third. In other words, the partition of $\{x, y, z\}$ induced by any subgraph of G_1 or of G_2 is $\{\{x, y, z\}\}$, or $\{\{x\}, \{y\}, \{z\}\}$, or one of the three partitions isomorphic to $\{\{x, y\}, \{z\}\}$. These cases have the probabilities shown below:

pairs connected	probability in G_1	probability in G_2
all	$p_1^3 + 3p_1^2(1-p_1)$	p_2^3
none	$(1-p_1)^3$	$(1-p_2)^3 + 3p_2(1-p_2)^2$
$\{x, y\}$	$p_1(1-p_1)^2$	$p_2^2(1-p_2)$

Serendipitously, there is a solution to the three equations suggested by the table above, i.e., to

$$\begin{aligned} p_1^3 + 3p_1^2(1-p_1) &= p_2^3, \\ (1-p_1)^3 &= (1-p_2)^3 + 3p_2(1-p_2)^2, \text{ and} \\ p_1(1-p_1)^2 &= p_2^2(1-p_2). \end{aligned} \tag{17}$$

Indeed, the last equation is satisfied whenever $p_2 = 1 - p_1$. Substituting $p_2 = 1 - p_1$ into either of the first two equations gives the *same* equation,

$$p_1^3 - 3p_1 + 1 = 0.$$

This equation has a unique solution in $(0, 1)$, namely

$$p_0 = 2\sin(\pi/18) = 0.3472\ldots.$$

Let $\mathcal{O}_i$, $i = 1, 2$, be the random (open) subgraph of G_i obtained by selecting each bond independently with probability p_i, where $p_1 = p_0$ and $p_2 = 1 - p_0$. As all three equations in (17) are satisfied, the random graphs $\mathcal{O}_1$ and $\mathcal{O}_2$ are *equivalent* with respect to the sites x, y, and z: we may couple $\mathcal{O}_1$ and $\mathcal{O}_2$ so that exactly the same pairs of sites from $\{x, y, z\}$ are connected in $\mathcal{O}_1$ as in $\mathcal{O}_2$.

In the context of independent bond percolation, each bond of a (usually infinite) graph is open independently with a certain probability, which we may think of as a weight. The observations above mean that if a (finite or infinite) weighted graph Λ has G_1 as a subgraph, with bond weights p_0, then we may replace G_1 by G_2, with weights $1 - p_0$. Ignoring the 'internal' site of G_2, this operation does not change the distribution of open clusters. This is a simple example of the 'substitution method' that we shall return to in the next chapter.

Using the star-triangle transformation, it is easy to deduce from Theorem 11 that $p_c^b(T) = p_0$. This result was derived by Sykes and Essam [1963; 1964] without rigorous proof. Wierman [1981] gave the first rigorous proof, based on the star-triangle transformation and a Russo–Seymour–Welsh type theorem.

Theorem 16. *Let T be the triangular lattice in the plane, and H the hexagonal or honeycomb lattice. Then*

$$p_c^b(T) = 2\sin(\pi/18)$$

and

$$p_c^b(H) = 1 - 2\sin(\pi/18).$$

Proof. As before, it follows from Menshikov's Theorem that the two critical probabilities associated to each lattice are equal, so it is legitimate to write p_c^b for their common value.

Let H' be the graph obtained by replacing every second triangle in T by a star with the same attachment sites; see Figure 15. Then H' is isomorphic to H; we shall keep the notation separate to indicate the different relationships to T, reserving H for the planar dual of T. Informally, bond percolation on T with parameter p_0 is equivalent to bond percolation on H' with parameter $1 - p_0$, so both are supercritical, both are

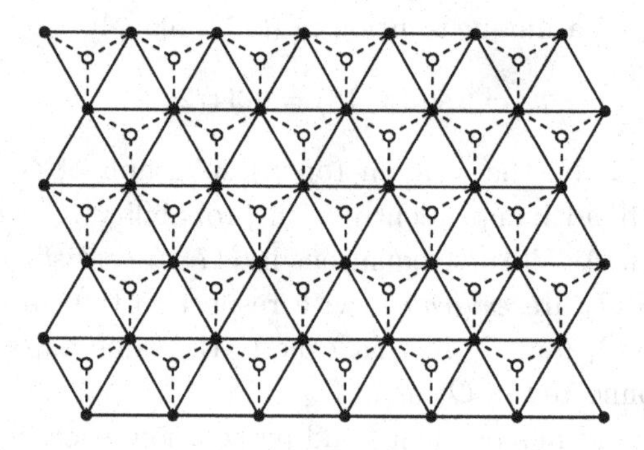

Figure 15. The triangular lattice T, and the graph H' obtained by replacing each downward pointing face of T by a star. The sites of H' are the circles (solid and hollow); its bonds are the dashed lines.

subcritical, or both are critical. By Theorem 11, it follows that both are critical, giving the result.

More formally, by a *domain* in T, or in H', we shall mean one of the triangles to which we applied the star-triangle transformation, or the resulting star in H'. Consider the probability measures $\mathbb{P}^{\mathrm{b}}_{T,p_0}$ in which bonds of T are open independently with probability p_0, and $\mathbb{P}^{\mathrm{b}}_{H',1-p_0}$, in which bonds of H' are open independently with probability $1-p_0$. We have shown above that the restrictions of these measures to a single domain D may be coupled so that the same attachment sites of D are joined by open paths within D in the two measures. We may extend the coupling to all domains simultaneously by independence. Any path in T or in H' between two sites of T may be split into a sequence of paths P_i within domains D_i, with the ends of P_i being attachment sites of D_i. It follows that, under our coupling, two sites x, y of T are joined by an open path in T if and only if they are joined by an open path in H'.

Recalling that 0 is a site of T, let C_0 be the open cluster of T containing 0, and let C'_0 be the open cluster of H' containing 0. Under our coupling, we have $C'_0 \cap V(T) = C_0$. As C'_0 is a connected subgraph of H', and sites of $V(H') \setminus V(T)$ are joined only to sites in $V(T)$, which have degree three, we have $|C'_0| \le 4|C'_0 \cap V(T)|$. Thus

$$|C_0| \le |C'_0| \le 4|C_0|$$

holds always in the given coupling, so

$$\mathbb{P}(|C_0| \ge n) \le \mathbb{P}(|C'_0| \ge n) \le \mathbb{P}(|C_0| \ge n/4)$$

for every n. Letting $n \to \infty$, we see that $\theta(H'; 1-p_0) = \theta(T; p_0)$. In other words, as H' is isomorphic to the hexagonal lattice H, we have $\theta(H; 1-p_0) = \theta(T; p_0)$. In proving Theorem 11, we showed that, for any p, at most one of $\theta(H; 1-p)$ and $\theta(T; p)$ can be strictly positive. Thus,

$$\theta(H; 1-p_0) = \theta(T; p_0) = 0,$$

which gives $p_c^b(T) \geq p_0$ and $p_c^b(H) \geq 1-p_0$. Since $p_c^b(T) + p_c^b(H) = 1$ by Theorem 11, it follows that $p_c^b(T) = p_0$ and $p_c^b(H) = 1-p_0$. □

Let us summarize what the results above, the Harris–Kesten Theorem, Theorem 8 and Theorem 16, tell us about the critical probabilities associated to the three regular planar lattices.

Theorem 17. *For the square lattice $\mathbb{Z}^2$, we have*

$$p_c^b(\mathbb{Z}^2) = 1/2,$$

for the triangular lattice T,

$$p_c^s(T) = 1/2 \quad \textit{and} \quad p_c^b(T) = 2\sin(\pi/18),$$

and for the hexagonal or honeycomb lattice H,

$$p_c^b(H) = 1 - 2\sin(\pi/18).$$ □

In a sense, the summary above is a little misleading: for these three lattices, four of the six critical probabilities are known exactly, but there are very few other natural lattices for which even one critical probability is known exactly.

The observation of Sykes and Essam [1963] concerning the star-delta transformation is a little more general: let G_1 be a triangle in which the bonds have probabilities p_a, p_b and p_c of being open, and G_2 a star in which the corresponding (i.e., opposite) bonds have probabilities r_a, r_b and r_c of being open. Then G_1 and G_2 are equivalent if and only if

$$p_i(1-p_j)(1-p_k) = (1-r_i)r_j r_k \tag{18}$$

for $\{i, j, k\} = \{a, b, c\}$,

$$p_a p_b p_c + (1-p_a)p_b p_c + p_a(1-p_b)p_c + p_a p_b(1-p_c) = r_a r_b r_c,$$

and

$$(1-p_a)(1-p_b)(1-p_c) = (1-r_a)(1-r_b)(1-r_c) + r_a(1-r_b)(1-r_c) + (1-r_a)r_b(1-r_c) + (1-r_a)(1-r_b)r_c.$$

The equations (18) are satisfied by taking $r_i = 1 - p_i$ for each i, and then both the remaining equations reduce to

$$p_a p_b p_c - p_a - p_b - p_c + 1 = 0. \tag{19}$$

This more general star-triangle transformation was used by Sykes and Essam [1964] to study percolation on a triangular lattice in which the states of the bonds are independent, but the probability that a bond is open depends on its orientation. They derive (non-rigorously) equation (19) for the 'critical surface' in this three-parameter model.

Using Theorem 13 in place of Theorem 11, the proof of Theorem 16 given above adapts immediately to this weighted model, to give the following result.

Theorem 18. *Let $\Lambda = (T, p_x, p_y, p_z)$ be the weighted triangular lattice in which bonds in the three directions have weights p_x, p_y and p_z, respectively, where $0 < p_x, p_y, p_z < 1$. Let $\theta(p_x, p_y, p_z)$ be the probability that the origin is in an infinite open cluster in the independent bond percolation model corresponding to Λ. Then $\theta(p_x, p_y, p_z) > 0$ if and only if*

$$p_x + p_y + p_z - p_x p_y p_z > 1. \qquad \square$$

This result was proved by Grimmett [1999], using ideas of Kesten [1982; 1988]. In the light of Menshikov's Theorem, the hard part is to show that percolation does not occur when $p_x + p_y + p_z - p_x p_y p_z \le 1$. Kesten [1982] deduced this result from a theorem that he stated without proof. A version of this theorem that is in many ways more general was later proved by Gandolfi, Keane and Russo [1988], but their result assumes symmetry under reflections in the coordinate axes, which this model does not have.

A star-delta transformation related to that discussed above is important in the theory of electrical networks, where it has a much longer history. The operation on the graph is the same, but the weights (resistances) transform differently, to satisfy the different notion of equivalence (that the response, i.e., net current at each attachment vertex, to each input, i.e., set of potentials at the attachment vertices, is the same). In electrical networks, it turns out that every star is equivalent to a triangle, and vice versa; see Bollobás [1998, pp. 43–44].

Remarkably, Wierman [1984] was able to use the star-triangle transformation with unequal edge weights to obtain the exact critical probability for a certain lattice, where each bond is open with the same probability. Let S^+ be the square lattice with one diagonal added to every second face, shown (rotated) on the left of Figure 16, together with its dual D.

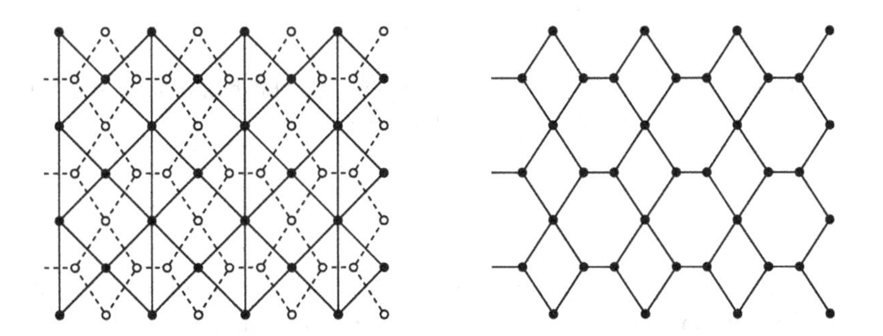

Figure 16. The lattice S^+ obtained from the square lattice by adding a diagonal to every other face, shown on the left (solid circles and lines) together with its dual D (dashed lines and hollow circles). For clarity, D is drawn separately on the right.

Let S' be the lattice shown on the left in Figure 17 below, obtained from S^+ by replacing each of the diagonal bonds by a double bond. Then S^+, with every bond open with probability p, is equivalent to S', with the original bonds of S^+ open with probability p, and the new bonds open with probability $p' = 1 - (1-p)^{1/2}$. (As usual, the states of different bonds are independent.) Applying the star-triangle transformation, one obtains the lattice D' formed from D by subdividing certain bonds; see Figure 17.

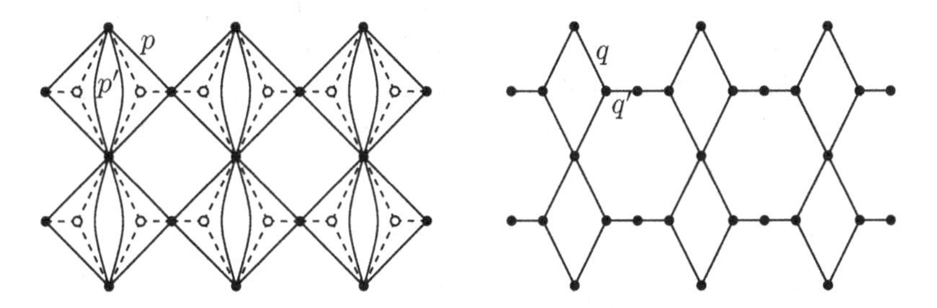

Figure 17. If certain bonds of S^+ are replaced by double bonds, and the triangle-star transformation is applied as shown, the resulting lattice is D with certain bonds subdivided.

Taking the undivided bonds to be open with probability $q = 1 - p$, and the divided bonds with probability $q' = \sqrt{q} = 1 - p'$, we see that D' is equivalent to D, with every bond open with probability q. The conditions for equivalence in the star-triangle transformation are satisfied provided (19) holds with $p_a = p_b = p$, $p_c = p'$. This condition reduces to

$$1 - p - 6p^2 + 6p^3 - p^5. \tag{20}$$

Using these transformations and arguing as for the hexagonal and triangular lattices above, Wierman deduces that

$$p_c^b(S^+) = 0.404518\ldots,$$

a root of (20).

An *Archimedean lattice* is a tiling of the plane by regular polygons in which all vertices are equivalent, i.e., the automorphism group of the tiling acts transitively on the vertices. The square, triangular and hexagonal lattices are all Archimedean, the lattice S^+ and its dual are not. The complete set of Archimedean lattices is shown in Figure 18. The notation, which is that of Grünbaum and Shephard [1987], is self-explanatory: it gives us the orders of the faces when we go round a vertex. At this point we have essentially exhausted the list of Archimedean lattices whose exact critical probability is known; there are two further examples that may be easily derived from those above. Let K be the Kagomé lattice, shown in Figure 18. Then K is the line-graph of the honeycomb H, so we have

$$p_c^s(K) = p_c^b(H) = 1 - 2\sin(\pi/18).$$

Also, let K^+ be the $(3, 12^2)$ or *extended Kagomé* lattice shown in Figure 18. Then K^+ is the line graph of the lattice H_2 obtained by subdividing each bond of H exactly once. As noted in Chapter 1, the relation $p_c^b(H_2) = p_c^b(H)^{1/2}$ is immediate: an open bond in the subdivided graph is only 'useful' if its partner bond is also open. Thus, as noted by Suding and Ziff [1999], among others,

$$p_c^s(K^+) = p_c^b(H_2) = p_c^b(H)^{1/2} = \big(1 - 2\sin(\pi/18)\big)^{1/2}. \tag{21}$$

In the next chapter we shall review some of the upper and lower bounds for the critical probabilities of Archimedean lattices.

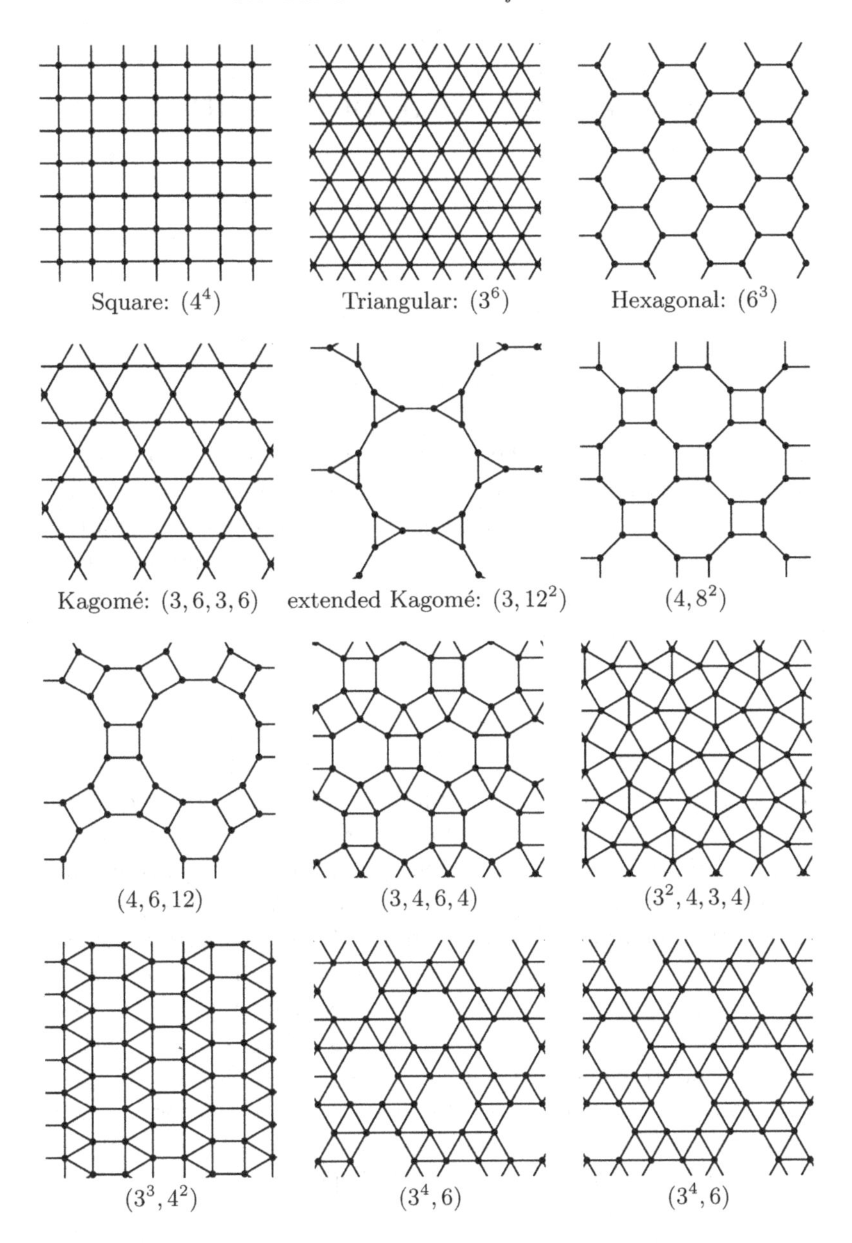

Figure 18. The 11 Archimedean lattices, i.e., tilings of the plane with regular convex polygons in which all vertices are equivalent. The notation for the unnamed lattices is that of Grünbaum and Shephard [1987]. 10 of the lattices are equivalent under rotation and translation to their mirror images. The final lattice, $(3^4, 6)$, is not, and is shown in both forms.

6

Estimating critical probabilities

In general, there is no hope of determining the exact critical probabilities $p_c^s(\Lambda)$ and $p_c^b(\Lambda)$ for a general graph Λ, even if Λ is a planar lattice. Nevertheless, there are many ways of proving rigorous bounds on these critical probabilities. In this chapter we shall describe several of these, starting with the *substitution method*, a special case of which we saw in the previous chapter.

6.1 The substitution method

To describe the substitution method, we shall use the terminology of *weighted graphs*: all our graphs will have a weight p_e associated to each bond e, with $0 \leq p_e \leq 1$. We shall consider independent bond percolation on a weighted graph $(\Lambda, \mathbf{p})$, where each bond e of Λ is open with probability p_e independently of the other bonds. We often suppress the weights in the notation. A weighted graph $(G, \mathbf{p})$ with a specified set A of *attachment* sites generates a random partition Π of A: two sites in A are in the same class of Π if they are joined by an open path in G. As in the previous chapter, we say that two weighted graphs G_1, G_2 with the same set A of attachment sites are *equivalent* if the associated random partitions Π_1, Π_2 have the same distribution, i.e., if the corresponding percolation measures can be coupled so that $\Pi_1 = \Pi_2$ always holds. In general, exact equivalence is too much to hope for.

Let us say that a partition π_2 of A is *coarser* than π_1, and write $\pi_2 \geq \pi_1$, if any two sites of A that are in the same class in π_1 are also in the same class in π_2. In other words, π_2 is coarser than π_1 if and only if π_1 is a refinement of π_2; in this context, a coarser partition is 'better', as it will correspond to more connections in the percolation model; this is the reason for our notation. We say that a weighted graph G_2 is *stronger* than G_1, and write $G_2 \geq G_1$, if we may couple the corresponding percolation measures so that Π_2 is always coarser than Π_1. In this case Π_2 *stochastically dominates* Π_1. Note that G_1 and G_2 are equivalent if and only if each is stronger than the other.

Let Λ_1 and Λ_2 be two infinite weighted graphs. Suppose that Λ_1 may be decomposed into edge-disjoint *domains* $D_{1,i}$, $i = 1, 2, \ldots$, each having a specified set A_i of *attachment sites.* We assume that each $D_{1,i}$ is a subgraph of Λ_1, that their union is Λ_1, and that two domains may meet only in sites that are attachment sites of both. Typically, the graphs $D_{1,i}$ are all isomorphic. Suppose that Λ_2 has a decomposition into domains $D_{2,i}$, where each $D_{2,i}$ has the same attachment sites as $D_{1,i}$. We have seen an example of such a decomposition already, in connection with the star-triangle transformation. Indeed, referring to Figure 15 of Chapter 5, we may take Λ_1 to be the triangular lattice, Λ_2 the hexagonal lattice H', the domains $D_{1,i}$ to be every second triangle in Λ_1 and each $D_{2,i}$ to be the corresponding star in Λ_2. Arguing as in the proof of Theorem 16 of Chapter 5, since any path from one domain to another must pass through an attachment site, the only property of $D_{1,i}$ that is relevant for percolation on Λ_1 is the induced partition $\Pi_{1,i}$ of the set A_i. Hence, if $D_{1,i}$ is equivalent to $D_{2,i}$ for every i, then percolation occurs on Λ_1 if and only if it occurs on Λ_2: for any fixed attachment site x, the percolation probabilities $\theta(\Lambda_1; x)$ and $\theta(\Lambda_2; x)$ are exactly equal.

Similarly, if $D_{2,i} \geq D_{1,i}$ for every i, then $\theta(\Lambda_2; x) \geq \theta(\Lambda_1; x)$: one can couple the percolation measures on Λ_1 and Λ_2 so that any pair of attachment sites that are joined in Λ_1 are also joined in Λ_2. This fact allows us to derive relationships between the critical probabilities of two lattices; if the critical probability of one is known, then we can bound the critical probability of the other. This technique is known as the *substitution method*, and is due to Wierman [1990].

At first sight, it is not clear how one can tell whether a given weighted graph is stronger than another, but there is a simple algorithm. An *up-set* in the partition lattice on a set A is simply a set $\mathcal{U}$ of partitions of A such that whenever $\pi_1 \in \mathcal{U}$ and $\pi_2 \geq \pi_1$, then $\pi_2 \in \mathcal{U}$. It is not too hard to show that $G_2 \geq G_1$ if and only if, for every up-set, we have $\mathbb{P}(\Pi_2 \in \mathcal{U}) \geq \mathbb{P}(\Pi_1 \in \mathcal{U})$; in fact, this is an easy consequence of Hall's Matching Theorem [1935] (see also Bollobás [1998, p. 77]). Thus, the condition $G_2 \geq G_1$ is equivalent to a finite set of polynomial inequalities on the weights of the bonds. Let us illustrate this with a simple example, giving bounds on the critical probability $p_{\mathrm{c}}^{\mathrm{b}}(K) = p_{\mathrm{H}}^{\mathrm{b}}(K) = p_{\mathrm{T}}^{\mathrm{b}}(K)$, where K is the *Kagomé* lattice shown on the right of Figure 1. In the

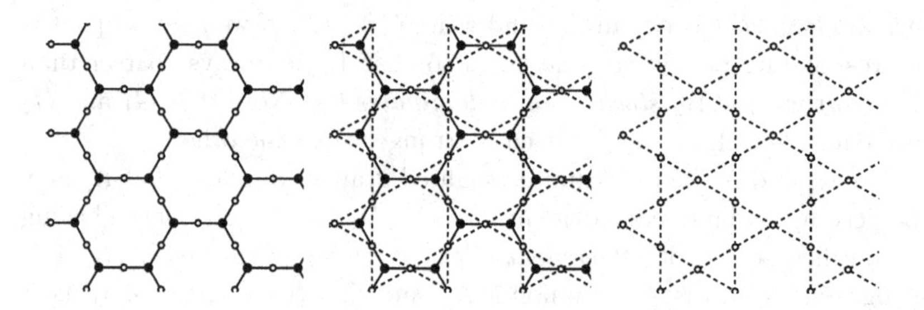

Figure 1. A two-step transformation from the hexagonal lattice H to the Kagomé lattice K: first subdivide the bonds of H to obtain the lattice H_2 shown on the left. Then apply the star-triangle transformation to the original sites of H (middle figure). The result is the Kagomé lattice (right-hand figure).

notation of Grünbaum and Shephard [1987] for Archimedean lattices, K is the $(3,6,3,6)$ lattice.

Ottavi [1979] noticed that the Kagomé lattice may be obtained from the honeycomb or hexagonal lattice H by first subdividing every bond, and then applying the star-triangle transformation to the (now edge-disjoint) stars centred at the original sites of H; see Figure 1. (A more primitive sequence of such 'equivalent' graphs was shown in Figure 6 of Chapter 1.) Let S_s denote the star with attachment sites $\{x, y, z\}$ in which each bond has weight s, and T_t the triangle on $\{x, y, z\}$ in which each bond has weight t. We have seen that S_s is equivalent to T_t if and only if $t = p_0$ and $s = 1 - p_0$, where $p_0 = 2\sin(\pi/18)$. We would like to know for which pairs (s, t) we have $S_s \geq T_t$, and for which pairs we have $S_s \leq T_t$.

Recall that there are 5 partitions of $\{x, y, z\}$: one in which all three are connected (are in the same part), one in which none are connected, and three in which exactly one pair is connected. We shall refer to these as the partitions of *type* 3, 0 and 1 respectively, so the type of a partition is the number of connected pairs. Repeating the calculation in the last section of the previous chapter, the probability that S_s induces a given partition of type i is $f_i(s)$, where

$$f_3(s) = s^3, \quad f_1(s) = s^2(1-s), \text{ and } f_0(s) = (1-s)^3 + 3s(1-s)^2.$$

The corresponding probabilities for T_t are given by $g_i(t)$, with

$$g_3(t) = t^3 + 3t^2(1-t), \quad g_1(t) = t(1-t)^2, \text{ and } g_0(t) = (1-t)^3.$$

There are 10 up-sets in the partition lattice on a three-element set: two are *trivial*: the empty up-set, and the up-set consisting of all partitions. Any other up-set must contain the type-3 partition, cannot contain the type-0 partition, and may contain any subset of the three type-1 partitions. Let us write $\mathcal{U}_j$ for one of the non-trivial up-sets containing j partitions of type 1. In this symmetric setting, we have $S_s \geq T_t$ if and only if four inequalities hold: each $\mathcal{U}_j$ must be at least as likely in the partition induced by S_s as in that induced by T_t.

In the partition induced by S_s we have $\mathbb{P}(\mathcal{U}_j) = f_3(s) + jf_1(s)$, while in that induced by T_t we have $\mathbb{P}(\mathcal{U}_j) = g_3(t) + jg_1(t)$, so $S_s \geq T_t$ if and only if

$$f_3(s) + jf_1(s) \geq g_3(t) + jg_1(t) \tag{1}$$

holds for $j = 0, 1, 2, 3$. Similarly, $T_t \geq S_s$ if and only if the reverse inequalities hold. Of course, if (1) holds for $j = 0$ and for $j = 3$ then it also holds for $j = 1$ and $j = 2$, so there are only two conditions to verify.

We know the critical probability for H_2, the left-hand lattice in Figure 1. Indeed, writing p_{c} for p_{H} or p_{T} (which are equal for any lattice by Menshikov's Theorem), from equation (21) of Chapter 5 we have

$$p_{\mathrm{c}}^{\mathrm{b}}(H_2) = p_{\mathrm{c}}^{\mathrm{b}}(H)^{1/2} = \big(1 - 2\sin(\pi/18)\big)^{1/2} = s_0,$$

say. As shown in Figure 1, this weighted graph H_2, with bond weights s, has a partition into weighted stars S_s. Replacing each star with a triangle T_t, we obtain the Kagomé lattice with bond weights t. It follows that, if $\theta(H_2; s) > 0$ and $T_t \geq S_s$, then $\theta(K; t) > 0$. As $\theta(H_2; s) > 0$ for any $s > s_0$, we thus have

$$p_{\mathrm{c}}^{\mathrm{b}}(K) \leq \inf\{t : T_t \geq S_s \text{ for some } s > s_0\} = \inf\{t : T_t \geq S_{s_0}\}.$$

Similarly,

$$p_{\mathrm{c}}^{\mathrm{b}}(K) \geq \sup\{t : S_{s_0} \geq T_t\}.$$

Solving the simple polynomial inequalities (1) for t with $s = s_0$, this method gives the bounds

$$0.51822 \leq p_{\mathrm{c}}^{\mathrm{b}}(K) \leq 0.54128.$$

These inequalities and the proof we have just given are due to Wierman [1990]. In the same paper, he obtains the stronger upper bound $p_{\mathrm{c}}^{\mathrm{b}}(K) \leq 0.5335$ by considering a larger substitution – replacing the union of two adjacent stars in H_2 by the union of the corresponding triangles in K. In this case there are four attachment sites, so the partition lattice is more complicated. By using larger and larger substitutions, better and better bounds may be obtained. However, the calculations quickly become impractical if carried out in a naive way. Using various methods of simplifying the calculations, and a substitution with six attachment sites, Wierman [2003b] sharpened these bounds considerably.

Theorem 1. *Let K be the Kagomé or* $(3,6,3,6)$ *lattice. Then*

$$0.5209 \leq p_c^b(K) \leq 0.5291. \qquad \square$$

Returning to the star-triangle transformation, the condition $S_s \geq T_t$ is, for some purposes, unnecessarily strong. Suppose that we have a weighted graph Λ_1 with S_s as a subgraph (joined only at the attachment sites), and we replace S_s by T_t to obtain Λ_2. We would like conditions on s and t that allow us to deduce that 'percolation is more likely' in Λ_1 than in Λ_2. More precisely, we should like conditions that ensure that the event $\{0 \to \infty\}$, that a particular site (the origin) is in an infinite open cluster, is at least as likely in Λ_1 as in Λ_2. We write $S_s \succeq T_t$ if this holds for all pairs of weighted graphs (Λ_1, Λ_2) related in the way we have described. Ottavi [1979] found the set of pairs (s,t) for which $S_s \succeq T_t$.

To present this result, let us first decide the states of the bonds in E, the set of all bonds of Λ_1 outside S_s. Note that E is also the set of bonds of Λ_2 outside T_t. For each attachment site $v \in \{x,y,z\}$, there may or may not be an open path from 0 to v in $\Lambda_1 \setminus S_s$, and there may or may not be an infinite open cluster of $\Lambda_1 \setminus S_s$ meeting v. In other words, the events $\{0 \to v\}$ and $\{v \to \infty\}$ may or may not hold in $\Lambda_1 \setminus S_s$. Given the states of the bonds in E, the conditional probability that $0 \to \infty$ depends only on which of the events $\{0 \to v\}$ and $\{v \to \infty\}$ hold in $\Lambda_1 \setminus S_s$. Indeed, $0 \to \infty$ in Λ_1 if and only if there are sites v_1, $v_2 \in \{x,y,z\}$ that are connected within S_s, with $0 \to v_1$ and $v_2 \to \infty$ in $\Lambda_1 \setminus S_s$. Here, $v_1 = v_2$ is allowed.

Much of the time, this conditional probability is 0 or 1: if $0 \to v$ and $v \to \infty$ for some $v \in \{x,y,z\}$, then $0 \to \infty$ even if all bonds in S_s are closed. Similarly, if $0 \not\to v$ for all $v \in \{x,y,z\}$, or $v \not\to \infty$ for all v, then $0 \to \infty$ cannot hold, whatever the states of the bonds in S_s. This leaves only three non-trivial cases: we must have an attachment site, x, say, with $0 \to x$, and another site, y, say with $y \to \infty$. For the third site, we may have $0 \to z$, $z \to \infty$, or $0 \not\to z \not\to \infty$. In the last case, $0 \to \infty$ in Λ_1 if and only if x and y are connected in S_s. If $0 \to z$ and $z \not\to \infty$, then $0 \to \infty$ in Λ_1 if and only if one of x and z is connected to

y in S_s. Similarly, if $z \to \infty$, then $0 \to \infty$ in Λ_1 if and only if one of y and z is connected to x in S_s. The relevant events in S_s have respective probabilities $f_3(s) + f_1(s)$, $f_3(s) + 2f_1(s)$ and $f_3(s) + 2f_1(s)$.

We have shown that $\mathbb{P}_{\Lambda_1}(0 \to \infty)$ is a weighted average of the quantities 1, 0, $f_3(s) + f_1(s)$, and $f_3(s) + 2f_1(s)$, where the weights depend on $\Lambda_1 \setminus S_s$. Furthermore, $\mathbb{P}_{\Lambda_2}(0 \to \infty)$ is a weighted average of 1, 0, $g_3(t) + g_1(t)$, and $g_3(t) + 2g_1(t)$ with the *same* weights, determined by $\Lambda_2 \setminus T_t = \Lambda_1 \setminus S_s$. Thus $S_s \succeq T_t$ holds precisely when the two inequalities

$$f_3(s) + f_1(s) \geq g_3(t) + g_1(t) \text{ and } f_3(s) + 2f_1(s) \geq g_3(t) + 2g_1(t) \quad (2)$$

hold, i.e., when (1) holds for $j = 1$ and $j = 2$. This is a much weaker condition than $S_s \geq T_t$. The reverse relation $T_t \succeq S_s$ holds provided the reverse inequalities to (2) hold.

Ottavi showed that if $s = s_0$ is the critical probability for H_2, and $t = 0.52893$, then $T_t \succeq S_s$, so the subgraph S_s of any weighted graph Λ may be replaced by T_t, and the probability that a given site is in an infinite cluster (or that a given pair of sites are connected by an open path) will not decrease. It might seem that the inequality $p_c^b(K) \leq 0.52893$ follows easily; unfortunately, this is not the case. As the natural argument is so close to working, let us examine it in detail, to see where it fails.

One would hope that, if $T_t \succeq S_s$, then percolation is at least as likely in a graph formed by gluing together copies of T_t at their attachment sites, as in the corresponding graph obtained from copies of S_s. However, the relation $T_t \succeq S_s$ allows us to replace *one* copy of T_t by a copy of S_s in an 'outside' graph which is the same before and after the substitution, but it does not allow us to continue and replace a second copy. After replacing the first copy, the outside graphs E_1 and E_2 are different: in E_2 we have already replaced a copy of S_s by T_t. One might hope that, as E_2 is 'better' than E_1, there is no real problem. But what does 'better' mean? It could mean that any connection in the outside graph E_1 between an attachment site of our second substitution and 0 or ∞ is also present in E_2. Until we have looked at the second substitution, we do not know which connections we will require, so we should impose the condition that the first substitution preserves *all* such connections (while perhaps adding new ones). The condition $T_t \succeq S_s$ does not allow us to do this, only to preserve a connection chosen in advance.

The arguments above illustrate the power of the notion of stochastic domination: if $T_t \geq S_s$, then it is very easy to prove that we may replace as many copies of S_s by copies of T_t as we like. The key to the application of the substitution method is to find suitable weighted graphs G_1, G_2 with $G_1 \geq G_2$.

Wierman [2002] used the substitution method to obtain bounds on $p_c^b(\Lambda)$ for other Archimedean lattices Λ, obtaining the following result.

Theorem 2. *Among the Archimedean lattices Λ, the extended Kagomé, or $(3, 12^2)$, lattice K^+ maximizes $p_c^b(\Lambda)$, with*

$$0.7385 < p_c^b(K^+) < 0.7449. \qquad \square$$

Although we have described the substitution method only for bond percolation, essentially the same method can be used to study site percolation. For example, Wierman [1995] adapted his method to obtain an upper bound on $p_c^s(\mathbb{Z}^2)$.

Theorem 3. *The critical probability $p_c^s(\mathbb{Z}^2)$ for site percolation on the square lattice satisfies $p_c^s(\mathbb{Z}^2) \leq 0.679492$.* $\square$

For a list of other rigorous bounds on the critical probabilities for lattices, see Wierman and Parviainen [2003].

6.2 Comparison with dependent percolation

Another method of bounding critical probabilities is implicit in the use of dependent percolation in Chapter 3. Let us illustrate this by giving an upper bound for $p_c^s(\mathbb{Z}^2) = p_H^s(\mathbb{Z}^2) = p_T^s(\mathbb{Z}^2)$; we shall mention a much more sophisticated version of this idea later. As usual, we write $\mathbb{P}_p = \mathbb{P}^s_{\mathbb{Z}^2,p}$ for the independent site percolation measure on $\mathbb{Z}^2$, where each site is open with probability p; we denote this model by M. Let ℓ be a parameter to be chosen later, and partition the vertex set of $\mathbb{Z}^2$ into ℓ by ℓ squares S_v, $v \in \mathbb{Z}^2$. Thus, for $v = (a, b)$,

$$S_v = S_{(a,b)} = \{(x, y) \in \mathbb{Z}^2 : a\ell \leq x < (a + 1)\ell,\ b\ell \leq y < (b + 1)\ell\}.$$

The set of squares S_v has the structure of $\mathbb{Z}^2$ in a natural sense. To make use of this, for each bond $e = uv$ of $\mathbb{Z}^2$, let $R_e = S_u \cup S_v$, so R_e is a 2ℓ by ℓ (or ℓ by 2ℓ) rectangle. If e and f are vertex-disjoint, then the rectangles R_e and R_f are vertex-disjoint. Thus, if we define a bond percolation model $\widetilde{M}$ on $\mathbb{Z}^2$ in which the state of a bond $e \in E(\mathbb{Z}^2)$ is determined by the states of the sites in R_e, then the associated probability measure $\widetilde{\mathbb{P}}$ on $2^{E(\mathbb{Z}^2)}$ will be 1-independent. The idea is to define $\widetilde{M}$ so that an infinite path in $\widetilde{M}$ guarantees an infinite path in the original site percolation M.

We have seen a way of doing this in Chapter 3, using 3 by 1 rectangles for clarity in the figures. We use the same idea here. Recall that $H(R)$ denotes the events that a given rectangle R is crossed horizontally by an open path, and $V(R)$ the event that R is crossed vertically by an open path. For a horizontal bond $e = ((a,b),(a+1,b))$ of $\mathbb{Z}^2$, let $G(R_e)$ be the event $H(R_e) \cap V(S_{(a,b)})$ illustrated in Figure 2. For a vertical bond

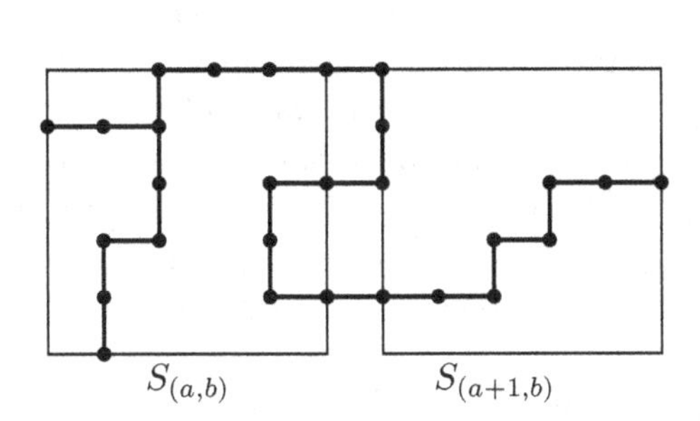

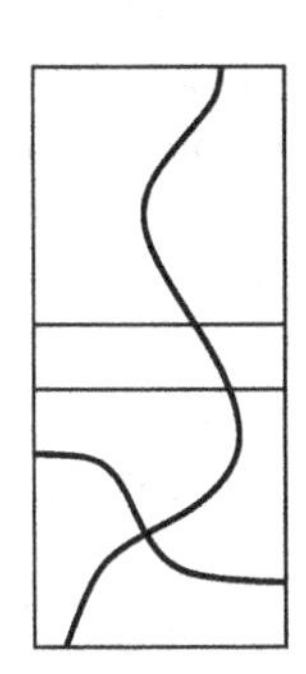

Figure 2. The figure on the left shows open paths guaranteeing that the event $G(R_e)$ holds, where $e = ((a,b),(a+1,b))$ is a horizontal bond of $\mathbb{Z}^2$; the sites shown are open. The corresponding event for a vertical bond e is shown schematically on the right.

$e = ((a,b),(a,b+1))$ of $\mathbb{Z}^2$, let $G(R_e) = V(R_e) \cap H(S_{(a,b)})$. In either case, let e be open in $\widetilde{M}$ if and only if $G(R_e)$ holds. Since horizontal and vertical crossings of the same square must meet, if there is an infinite open path in $\widetilde{M}$ then there is an infinite open cluster in M. Note that the probability $\mathbb{P}_p\big(G(R_e)\big)$ is the same for all bonds e.

Let $p_1 = p_1(\mathbb{Z}^2)$ be the infimum of the set of p such that, in any 1-independent bond percolation measure on $\mathbb{Z}^2$ in which each bond is open with probability at least p, the origin is in an infinite open cluster with positive probability. As we saw in Chapter 3, it is very easy to show that $p_1 < 1$. Here, the value of p_1 is important; we shall use the result of Balister, Bollobás and Walters [2005] that $p_1 < 0.8639$; see Lemma 15 of Chapter 3.

Suppose that for some parameters s and p we can show that

$$f_\ell(p) = \mathbb{P}_p\big(G(R_e)\big) > p_1.$$

Then $\widetilde{\mathbb{P}}$ is a 1-independent measure on $\mathbb{Z}^2$ in which each bond is open with some probability $p > p_1$, so with positive probability there is an infinite open path in $\widetilde{M}$, and hence in M. Thus $p \geq p_c^s(\mathbb{Z}^2)$.

Suppose that $p > p_c^s(\mathbb{Z}^2)$ is fixed. Then, by Theorem 10 of Chapter 5, we have $1 - p < p_c^s(\Lambda_\boxtimes)$, where $\Lambda_\boxtimes$ is the square lattice with both diagonals added to every face. Exponential decay of closed clusters in $\Lambda_\boxtimes$ follows by Menshikov's Theorem, so the probability that a 2ℓ by ℓ rectangle R is crossed the short way by a closed $\Lambda_\boxtimes$-path tends to zero as $\ell \to \infty$. Hence, by Lemma 9 of Chapter 5, the probability that R is crossed the long way by an open path in $\Lambda_\square$ tends to 1. It follows that $f_\ell(p) \to 1$, so there is some ℓ such that $f_\ell(p) \geq 0.8639 > p_1$. Thus, in principle, the method above gives arbitrarily good upper bounds: for each ℓ, find the minimal (or an almost minimal) value of p_ℓ of p such that $f_\ell(p) \geq 0.8639$. Then each p_ℓ is an upper bound on $p_c^s(\mathbb{Z}^2)$, and the sequence p_ℓ converges to $p_c^s(\mathbb{Z}^2)$ from above. It is easy to check that $f_2(p) = 6p^5q^3 + 16p^6q^2 + 8p^7q^1 + p^8$, where $q = 1 - p$, giving the bound $p_c^s(\mathbb{Z}^2) \leq 0.8798$. With a computer, it is easy to show that

$$\begin{aligned} f_3(p) = {} & 37p^8q^{10} + 399p^9q^9 + 1737p^{10}q^8 + 4027p^{11}q^7 + 5466p^{12}q^6 \\ & + 4527p^{13}q^5 + 2335p^{14}q^4 + 757p^{15}q^3 + 153p^{16}q^2 + 18p^{17}q^1 + p^{18}, \end{aligned}$$

giving $p_c^s(\mathbb{Z}^2) \leq 0.845$. An exact evaluation of $f_4(p)$ gives $p_c^s(\mathbb{Z}^2) \leq 0.817$. Unfortunately, the straightforward method of evaluating $f_\ell(p)$, namely counting for which of the $2^{2\ell^2}$ configurations in a 2ℓ by ℓ rectangle R_e the event $G(R_e)$ holds, quickly becomes impractical, at around $\ell = 5$.

When the substitution method can be applied, it tends to give better bounds for the same computational effort. Note, however, that the method described here is much more robust. For example, irregularities in the graph are not a problem, provided we can show that $\mathbb{P}_p\big(G(R_e)\big) \geq p_1' > p_1$ for every bond e. While one can perform different substitutions in different parts of the graph, the substitutions have to fit together exactly to give the structure of a graph with known critical probability. This will often not be possible.

Another very important advantage of the 1-independent approach is that, if one is prepared to accept an error probability of 1 in a million, say, or 1 in a billion, then good bounds can easily be obtained by this method. Indeed, there is a very easy way to estimate the numerical value of $f_\ell(p)$ very precisely: generate N configurations in a 2ℓ by ℓ rectangle R_e at random, using the measure $\mathbb{P}_p$, and count the number M of configurations for which $G(R_e)$ holds. The random number M has a binomial distribution with parameters N and $f_\ell(p)$, so if N is large, then with very high probability M/N will be close to $f_\ell(p)$. They key point is that one can bound the error probability, even without knowing $f_\ell(p)$.

Given any $\varepsilon > 0$, one can give a simple procedure for producing upper bounds on $p_c^s(\mathbb{Z}^2)$ which, *provably*, has probability at most ε of giving an incorrect bound: first (by trial and error, or guesswork) decide on parameters ℓ and p for which $f_\ell(p)$ seems likely to be large enough. Then calculate numbers N and M_0 such that $\mathbb{P}(X \geq M_0) \leq \varepsilon$, where $X \sim \mathrm{Bi}(N, 0.8639)$. Then generate N samples as above, and if $M \geq M_0$ of these have the property $G(R_e)$, assert $p_c^s(\mathbb{Z}^2) \leq p$ as a bound. If, in fact, $p_c^s(\mathbb{Z}^2) > p$, then $f_\ell(p) \leq 0.8639$, and the probability that the sampling procedure generates at least M_0 successful trials is at most ε.

This method works very well in practice, and the dependence of the running time on ε is very modest. For example, using the parameters $\ell = 10,000$, $p = 0.594$, $N = 1000$ and $M_0 = 935$, with a moderate computational effort we obtain the bound $p_c^s(\mathbb{Z}^2) \leq 0.594$ with an error probability of $\varepsilon < 10^{-12}$.

Similarly, for the matching lattice $\Lambda_\boxtimes$ with vertex set $\mathbb{Z}^2$ and bonds between each pair of vertices at Euclidean distance 1 or $\sqrt{2}$, using the parameters $\ell = 20,000$, $p = 0.408$, $N = 1000$ and $M_0 = 935$ we obtain $p_c^s(\Lambda_\boxtimes) \leq 0.408$ with a confidence of $1 - 10^{-12}$. Together these bounds give

$$p_c^s(\mathbb{Z}^2) \in [0.592, 0.594]$$

with extremely high confidence. A little more computational effort ($\ell = 80,000$, $N = 1000$, $M_0 = 915$) gives the result $p_c^s(\mathbb{Z}^2) \in [0.592, 0.593]$ with 99.9999% confidence.

Note that we cannot say that the probability that $p_c^s(\mathbb{Z}^2)$ lies in the interval $[0.592, 0.593]$ is at least 99.9999%: the critical probability is not a random quantity, so this statement either holds or it does not. In the language of statistics, we have given a confidence interval for the unknown deterministic quantity $p_c^s(\mathbb{Z}^2)$. As described, the procedure produces a confidence interval whose upper limit is either the bound 0.593, say, that we aim for, or infinity, if $M < M_0$.

Narrow confidence intervals for many other critical probabilities have been obtained by Riordan and Walters [2006]; for example, $p_c^s(H) \in [0.6965, 0.6975]$ with 99.9999% confidence, where H is the hexagonal lattice.

As usual in statistics, we must be a little careful in generating confidence intervals: for example, it is not legitimate to perform various runs of the sampling procedure with various parameters, and only use one result. In practice, this is not a problem for two reasons: firstly, we can perform as many 'dummy' runs as we like to get an idea of parameters that are very likely to work, and then one real run with these parameters. Also, it is easy to get a failure probability of 10^{-9}, say, for each run. It is then legitimate to perform 1000 runs with different parameters, and take the best bounds obtained: as long as the probability that each run gives an incorrect bound is at most 10^{-9}, the final bounds obtained still give a 99.9999% confidence interval for the critical probability.

There is another pitfall to bear in mind when implementing the probabilistic procedure described above: we have implicitly assumed that a source of random numbers is available. In practice, for this kind of simulation one usually uses a pseudo-random number generator. Not all the widely used ones are sufficiently good for this purpose. Indeed, the current standard generator 'random()' used with the programming language C is not.

For example, using this generator we obtained estimates for $f_{10}(0.731)$ of 0.8661 ± 0.0002 and 0.8631 ± 0.0002, depending on the order in which the random states of the sites were assigned. (The uncertainties given are two standard deviations.) Using the much better 'Mersenne Twister' generator MT19337 written by Matsumoto and Nishimura, we obtain $f_{10}(0.731) = 0.8630 \pm 0.0002$. Note that this (presumably) true value is smaller than 0.8639, while one of the values obtained using random() is larger. In generating the confidence intervals given above, we used the Mersenne Twister. Of course, one can re-run the same procedure with a different generator, or with 'true' random numbers obtained from, for example, quantum noise in a diode.

The methods described above can be easily adapted to give good upper and lower bounds on $p_{\mathrm{c}}^{\mathrm{b}}(\Lambda)$ and $p_{\mathrm{c}}^{\mathrm{s}}(\Lambda)$ for any of the Archimedean lattices Λ. For lower bounds, to prove that percolation does not occur at a particular value of p one can use 1-independence as in the proof of exponential decay of the volume in Chapter 3.

For another approach to lower bounds, let $S_n(0)$ be the set of sites at distance n from the origin, and let N_n be the number of sites in $S_n(0)$ that may be reached from 0 by open paths using only sites within distance n of 0. Lemma 8 of Chapter 4 (or its equivalent for bond percolation) states that if, for some n, we have $\mathbb{E}_p(N_n) < 1$, then there is exponential decay of the radius of the open cluster C_0, which certainly implies that $p \leq p_{\mathrm{c}}$. In any lattice, for any $p < p_{\mathrm{c}}$, Menshikov's Theorem implies that $\mathbb{E}_p(N_n) \to 0$ as $n \to \infty$, so arbitrarily good bounds may in principle be obtained in this way. As before, exact calculations of $\mathbb{E}_p(N_n)$ are impractical except for very small n, but estimates with rigorous error bounds may be obtained for larger n. This observation is applicable to any finite-type graph, including any lattice in any dimension.

For bond percolation on a planar lattice Λ with inversion symmetry, to find a lower bound on $p_c^b(\Lambda) = p_H^b(\Lambda) = p_T^b(\Lambda)$, one may find an upper bound on $p_c^b(\Lambda^\star)$ and use Theorem 13 of Chapter 5, which states that $p_c^b(\Lambda) + p_c^b(\Lambda^\star) = 1$. Similarly, for site percolation, we may find an upper bound on $p_c^s(\Lambda^\times)$ and apply Theorem 14 of Chapter 5. This approach seems to give better results in practice. The reason is that it is easier to estimate the probability of an event by sampling than to estimate the expectation of a random variable that might in principle take quite large values occasionally.

6.3 Oriented percolation on $\mathbb{Z}^2$

The study of oriented percolation, and, in particular, bond percolation on the oriented graph $\vec{\mathbb{Z}}^2$, is a major topic in its own right; see Durrett [1984] for a survey of the early results in this area. Oriented percolation is in general much harder to work with than unoriented percolation, a fact that is reflected in the difficulty of obtaining good bounds on $p_H^b(\vec{\mathbb{Z}}^2)$. Indeed, Durrett described the question of finding a sequence of rigorous upper bounds on $p_H^b(\vec{\mathbb{Z}}^2)$ that decrease to the true value as an important open problem. His bound, $p_H^b(\vec{\mathbb{Z}}^2) \leq 0.84$, was very far from the lower bound of 0.6298 obtained by Dhar [1982]. In contrast, for lower bounds, it is easy to produce bounds that do tend up to the true value. Indeed, let N_n be the number of sites on the line $x + y = n$ that may by reached from the origin. If $\mathbb{E}_p^b(N_n) < 1$ for some n, then Lemma 8 of Chapter 4 shows that $p \leq p_T^b(\vec{\mathbb{Z}}^2) \leq p_H^b(\vec{\mathbb{Z}}^2)$. Indeed, Hammersley [1957b] deduced that $p_H^b(\vec{\mathbb{Z}}^2) \geq 0.5176$ from the fact that $\mathbb{E}_p^b(N_2) = 4p^2 - p^4$. Of course, by Menshikov's Theorem, $p_H^b(\vec{\mathbb{Z}}^2) = p_T^b(\vec{\mathbb{Z}}^2) = p_c^b(\vec{\mathbb{Z}}^2)$, say.

There have been many Monte Carlo estimates of $p_c^b(\vec{\mathbb{Z}}^2)$; such results are not our main focus, so we shall give only a few examples, rather than attempt a complete list: Kertész and Vicsek [1980] gave the estimate $p_c^b(\vec{\mathbb{Z}}^2) = 0.632 \pm 0.004$, Dhar and Barma [1981] reported $p_c^b(\vec{\mathbb{Z}}^2) = 0.6445 \pm 0.0005$, and Essam, Guttmann and De'Bell [1988] $p_c^b(\vec{\mathbb{Z}}^2) = 0.644701 \pm 0.000001$, for example.

Balister, Bollobás and Stacey [1993; 1994] gave better upper bounds on $p_c^b(\vec{\mathbb{Z}}^2)$, using the basic strategy of comparison with 1-independent percolation, but in a much more complicated way than that described in the previous section. Their approach does give a sequence of rigorous upper bounds tending down to the true value. As the argument is rather involved, we shall give only an outline.

When considering oriented percolation on $\vec{\mathbb{Z}}^2$, our aim is to decide for which p the percolation probability $\theta(p) = \theta_0(p)$ is non-zero. In doing so, we may of course restrict our attention to that part of $\vec{\mathbb{Z}}^2$ that may conceivably be reached from the origin, namely the positive quadrant $\vec{Q}$. The basic idea is to use independent bond percolation on $\vec{Q}$ to define a 're-scaled' 1-independent bond percolation on $\vec{Q}$. To do this, we choose parameters b and h, and arrange rhombi with $b+1$ sites along the bottom-left and top-right sides as in Figure 3. More precisely,

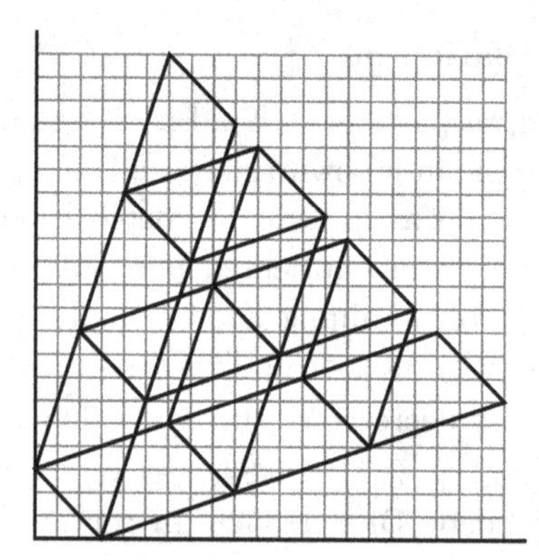
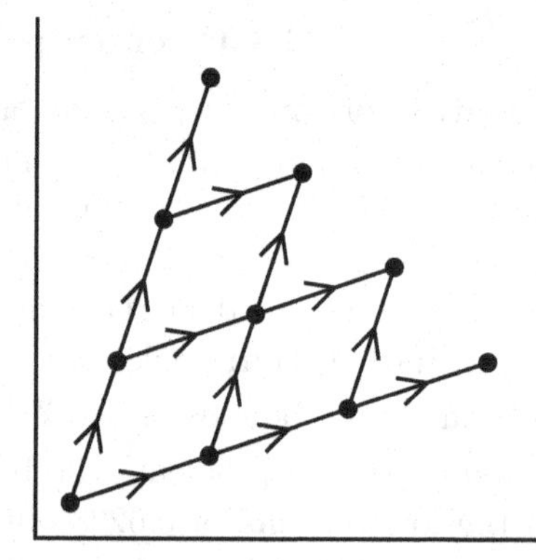

Figure 3. Rhombi with 4 sites along the base (bottom left) and the top (upper right). If each rhombus is replaced by an oriented edge, and the top and bottom of each rhombus by a vertex, the resulting oriented graph is isomorphic to $\vec{\mathbb{Z}}^2$. In reality the rhombi are 'jagged': each contains exactly 4 sites from each of the 9 layers of $\vec{\mathbb{Z}}^2$ that it meets.

writing $L_t = \{(x, y) \in \mathbb{Z}^2 : x, y \geq 0, x + y = t\}$ for *layer* t of $\vec{Q}$, for each site v in L_t we choose a set S_v of consecutive sites in $L_{h(t+1)}$, so that the sets S_v are disjoint. Also, for each bond $\vec{e} \in \vec{Q}$ we choose a (roughly rhombic) subgraph $R_{\vec{e}}$ of $\vec{Q}$ in such a way that $R_{\vec{e}}$ and $R_{\vec{f}}$ are disjoint whenever $\vec{e}$ and $\vec{f}$ are bonds of $\vec{Q}$ that do not share a site. We require that if $\vec{e} = \vec{uv}$, then the *base* of $R_{\vec{e}}$, defined in the natural way, is exactly S_u, and the *top* of $R_{\vec{e}}$ is S_v.

As before, we write $C_0 = C_0^+$ for the *open out-cluster of the origin*, i.e., the set of sites that may be reached from the origin by open (oriented) paths. It is easy to show that there is a $p_1 = p_1(\vec{\mathbb{Z}}^2) < 1$ such that, for any 1-independent bond percolation measure $\widetilde{\mathbb{P}}$ on $\vec{\mathbb{Z}}^2$ in which each bond is open with probability at least p_1, we have $\widetilde{\mathbb{P}}(|C_0| = \infty) > 0$; the argument is very similar to the proof of Lemma 14 in Chapter 3. Indeed, if C_0 is finite, then there is a dual cycle S surrounding the origin, such that every oriented bond starting inside S and ending outside is closed. As shown in Chapter 3 there are, very crudely, at most $4\ell 3^{2\ell-2}$ dual cycles S of length 2ℓ surrounding the origin. For each, there is a set $\vec{E}$ of exactly ℓ oriented bonds starting inside S and ending outside. Two horizontal bonds in $\vec{E}$ cannot share a site, and nor can two vertical bonds in $\vec{E}$, so there is a set $\vec{E}' \subset \vec{E}$ consisting of $\lceil \ell/2 \rceil$ vertex-disjoint bonds. The $\widetilde{\mathbb{P}}$-probability that all bonds in $\vec{E}'$ are closed is at most $(1-p_1)^{\ell/2}$, so the expected number of cycles with the required property is at most

$$\sum_{\ell \geq 2} \frac{4\ell}{9} 3^{2\ell}(1-p_1)^{\ell/2},$$

which is less than 1 if $p_1 = 0.997$, say. Note that the expression above is exactly the same as that appearing in the proof of Lemma 14 of Chapter 3, not by coincidence. By counting more carefully the cycles S that may actually arise as the boundary of C_0, as in Balister, Bollobás and Stacey [1999], one can obtain a better bound on p_1.

If we define an event $G(R_{\vec{e}})$ depending on the states of the bonds in $R_{\vec{e}}$ so that an infinite oriented path of bonds $\vec{e}$ for which $G(R_{\vec{e}})$ holds guarantees an infinite open oriented path in $\vec{Q}$, and if, for some p, we can show that $\mathbb{P}_p(G(R_{\vec{e}})) \geq p_1$ for every $\vec{e}$, then it follows easily that $p_{\mathrm{H}}(\vec{\mathbb{Z}}^2) \leq p$. However, it is not easy to see how to define $G(R_{\vec{e}})$: piecing together paths in this context is much harder than in the unoriented case.

The actual argument of Balister, Bollobás and Stacey is much more subtle. Choosing the regions $R_{\vec{e}}$ so that they are isomorphic to one another, they seek a non-trivial up-set $\mathcal{A} \subset \mathcal{P}([b+1])$ with the following property: let $\vec{e} = \vec{uv}$, and consider the random set U of sites in S_u that may be reached from the origin by open oriented paths. This set may be naturally identified with a subset of $[b+1]$. Let V be the set of sites in S_v that may be reached from U by open paths in $R_{\vec{e}}$; again this set may be identified with a subset of $[b+1]$. The condition required is that, for some p, $\rho \in (0,1)$, whenever $U \in \mathcal{A}$, then

$$\mathbb{P}_p(V \in \mathcal{A} \mid U) \geq \rho. \tag{3}$$

There is a further restriction, that $\mathcal{A}$ be symmetric under the operation $i \mapsto b+1-i$, so that the identifications described above may be made consistently.

Balister, Bollobás and Stacey [1993] show that, under these assumptions, one can construct a re-scaled oriented bond percolation on $\vec{\mathbb{Z}}^2$ in which each bond is open with probability at least ρ, and which is effectively 1-independent. (More precisely, given the states of the bonds below some layer, the bonds in that layer are 1-independent.) They show that, if $p \geq (3-\sqrt{5})/2$ and $\rho \geq 1-(1-p)^2$, then such a process dominates the original independent process. Using Menshikov's Theorem, one can then deduce that $p \geq p_{\mathrm{c}}^{\mathrm{b}}(\vec{\mathbb{Z}}^2)$.

In summary, if one can find a suitable region R, an up-set $\mathcal{A}$, and a $p \geq (3-\sqrt{5})/2$ such that (3) holds with $\rho = 1-(1-p)^2$ for all $U \in \mathcal{A}$, then p is an upper bound for $p_{\mathrm{c}}^{\mathrm{b}}(\vec{\mathbb{Z}}^2)$. Note that this does not correspond to a direct construction of a 1-independent percolation model as in the unoriented case. Such a direct construction could be achieved by defining $G(R_{\vec{e}})$ so that if $G(R_{\vec{uv}})$ holds and $U \in \mathcal{A}$ then $V \in \mathcal{A}$; indeed, one could take this as the definition of $G(R_{\vec{e}})$. But then the condition that $G(R_{\vec{e}})$ have probability at least p_1 amounts to

$$\mathbb{P}_p\big(U \in \mathcal{A} \implies V \in \mathcal{A}\big) \geq p_1,$$

which is infinitely stronger than (3) with $\rho = p_1$.

In order to apply the method above, it seems that we have to make a large number of choices: first, we have to choose a suitable region R, and a probability p. Then, we must also choose one of the $2^{\Theta\left(2^b/\sqrt{b}\right)}$ possible up-sets to test. Fortunately, there is an algorithm given by Balister, Bollobás and Stacey [1993] that, for a given R and p, tests whether there is *any* up-set $\mathcal{A}$ with the required property (and finds the maximal one if there is). Using this algorithm, together with the arguments outlined above, they proved that $p_{\mathrm{c}}^{\mathrm{b}}(\vec{\mathbb{Z}}^2) \leq 0.6863$. They also showed that, as in the case of unoriented percolation, the method gives arbitrarily good bounds with sufficient computational effort. This is a much harder result than the (almost trivial) equivalent for unoriented percolation described in the previous section.

Using a more sophisticated version of the argument, involving a more complicated reduction from the dependent percolation to independent percolation, Balister, Bollobás and Stacey [1994] obtained the following bounds.

Theorem 4. *The critical probabilities for oriented bond and site percolation on the square lattice satisfy*

$$p_c^b(\vec{\mathbb{Z}}^2) \le 0.6735 \quad and \quad p_c^s(\vec{\mathbb{Z}}^2) \le 0.7491. \qquad \square$$

As in the case of unoriented percolation, a (simpler) version of the argument just described may be adapted to give bounds with 99.99% confidence, and far stronger bounds can be obtained in this way with much less computational effort. Indeed, Bollobás and Stacey [1997] showed that $p_c^b(\vec{\mathbb{Z}}^2) \le 0.647$ with 99.999967% confidence. This bound is very close to the value of approximately 0.6445 suggested by simulations.

Liggett [1995] used a beautiful and totally different approach to bound $p_c^b(\vec{\mathbb{Z}}^2)$ and $p_c^s(\vec{\mathbb{Z}}^2)$, based on an idea of David Williams, who conjectured that $p_c^b(\vec{\mathbb{Z}}^2) \le 2/3$. To describe this, note that oriented percolation on $\vec{\mathbb{Z}}^2$ corresponds to a Markov chain in a natural way. Recall that the tth *layer* is the set $L_t = \{(x, y) \in \mathbb{Z}^2) : x, y \ge 0,\ x + y = t\}$. The set R_t of sites in L_t that may be reached from the origin by open paths depends only on R_{t-1}, and on the states of the bonds between L_{t-1} and L_t (for oriented bond percolation), or the states of the sites in L_t (for site percolation). More explicitly, let us code R_t by the x-coordinates of the points it contains, setting

$$A_t = \{x : (x, t - x) \text{ may be reached from } 0 \text{ by an open path}\}.$$

Then $A_0 = \{0\}$. For bond percolation, given A_t, the probability that $x \in A_{t+1}$ is 0, $p_1 = p$ or $p_2 = 2p - p^2$ according to whether $|A_t \cap \{x-1, x\}|$ is 0, 1 or 2. Furthermore, given A_t, the events $\{x \in A_{t+1}\}$ are independent for different x. For site percolation, the Markov chain is the same, except that $p_1 = p_2 = p$. Thus, it is natural to extend the definition of the Markov chain (A_t) to any parameters $0 \le p_1 \le 1$ and $0 \le p_2 \le 1$. Note that this Markov chain has stationary transition probabilities: the distribution of A_{t+1} given that $A_t = A$ is the same for any t. Of course, this Markov chain underlies any analysis of oriented percolation; for the approach described above, we did not need to define it explicitly. Liggett [1995] proved the following result.

Theorem 5. *If the parameters p_1 and p_2 satisfy the inequalities*

$$1/2 < p_1 \le 1 \quad and \quad 4p_1(1 - p_1) \le p_2 \le 1, \tag{4}$$

then the Markov chain (A_t) *defined above satisfies* $\mathbb{P}(\forall t : A_t \neq \emptyset) > 0$.

For bond percolation, where $p_1 = p$ and $p_2 = 2p - p^2$, the conditions of Theorem 5 are satisfied for $p \geq 2/3$. For site percolation, where $p_1 = p_2 = p$, the relations (4) hold for any $p \geq 3/4$. Thus Liggett's result immediately implies the following bounds.

Theorem 6. *The critical probabilities for oriented bond and site percolation on the square lattice satisfy*

$$p_{\mathrm{c}}^{\mathrm{b}}(\vec{\mathbb{Z}}^2) \leq 2/3 \quad \textit{and} \quad p_{\mathrm{c}}^{\mathrm{s}}(\vec{\mathbb{Z}}^2) \leq 3/4. \qquad \square$$

Given a finite set $A \subset \mathbb{Z}$, let A^k denote the random set obtained by running the Markov chain for k steps starting with the set A. Thus, if $A = \{0\}$, then A^n has the distribution of A_n. The basic idea of Liggett's proof is as follows: we hope to define a function H on the finite subsets A of $\mathbb{Z}$ so that $H(\emptyset) = 1$, and for all $A \neq \emptyset$ we have $H(A) < 1$ and

$$\mathbb{E}\big(H(A^1)\big) \leq H(A). \tag{5}$$

Of course, whether such an H exists depends on the parameters p_1 and p_2 of the Markov chain. If such an H does exist, then from the Markov property and (5) we have

$$\mathbb{E}\big(H(A_n)\big) = \mathbb{E}\Big(\mathbb{E}\big(H(A_n \mid A_{n-1})\big)\Big) \leq \mathbb{E}\big(H(A_{n-1})\big) \leq \\ \cdots \leq \mathbb{E}\big(H(A_0)\big) = H(\{0\}) < 1$$

for every n. This implies that the percolation process is supercritical: otherwise, we would have $\mathbb{P}(A_n = \emptyset) \to 1$ as $n \to \infty$. As $H(\emptyset) = 1$ and H is bounded, this would imply $\mathbb{E}\big(H(A_n)\big) \to 1$, a contradiction.

Liggett [1995] showed that, if the conditions (4) hold, then an H with the required properties can be found. The weight function H used by Liggett is rather complicated.

To describe Liggett's argument, let T be a random variable taking positive integer values, with $\mathbb{E}(T) < \infty$, and let ν be the stationary renewal measure on sequences $\eta \in \{0,1\}^{\mathbb{Z}}$ associated to T. Thus, the distribution of the random sequence η is invariant under translation, and the gaps between successive 1s are independent and have the distribution of T. The weight $H(A)$ is defined as the ν-probability that $\eta(x) = 0$ for every $x \in A$. Note that this certainly satisfies $H(\emptyset) = 1$ and $H(A) < 1$ for $A \neq \emptyset$. Of course, whether the key equation (5) is satisfied depends on the choice of T.

Writing $F(n)$ for $\mathbb{P}(T \geq n)$, Liggett shows that (5) is satisfied in the special case that A is an interval if and only if

$$p_1^2 F(n+1) = (1-p_2) \sum_{k=1}^{n-1} F(k)F(n+1-k) + (1-p_1)^2 F(n) \qquad (6)$$

for every $n \geq 1$. The derivation of this condition is relatively simple; most of the work lies ahead. First, one must show that, if (4) holds, then the solution to (6) with $F(1) = 1$ makes sense, i.e., corresponds to $F(n) = \mathbb{P}(T \geq n)$ for some T. This amounts to showing that $F(n)$ is decreasing and that $\sum_n F(n) = \mathbb{E}(T) < \infty$. Second, one must show that (5) holds not only for intervals, but also for arbitrary sets A. Even though the Markov chain operates on disjoint intervals separately, in some sense, due to the complex definition of H this is by no means a simple consequence of the interval case. Indeed, the arguments of Liggett [1995] for both steps are far from easy.

Let us remark that oriented percolation on $\vec{\mathbb{Z}}^2$ may be thought of as a discrete-time version of the contact process on $\mathbb{Z}$. Indeed, Liggett's proof described above is similar in outline to an argument of Holley and Liggett [1978] giving an upper bound for the critical parameter of this contact process. As noted by Liggett [1995], it is much harder to turn the outline into a proof for oriented percolation than for the contact process, due to the discrete nature of the process.

At the time of writing, Liggett's bound $p_c^b(\vec{\mathbb{Z}}^2) \leq 2/3$ is the best rigorous upper bound on $p_c^b(\vec{\mathbb{Z}}^2)$. Liggett notes, however, that his method cannot be pushed further: the solution $F(n)$ to (6) above has the required properties if *and only if* the conditions (4) are satisfied. In contrast, the method of Balister, Bollobás and Stacey [1993; 1994] gives arbitrarily good bounds with increasing computational effort. Thus, at some point it will become feasible to obtain a better bound in this way. Indeed, this may have already happened: the amount of computing power available now is much larger than it was when the bound $p_c^b(\vec{\mathbb{Z}}^2) \leq 0.6735$ was obtained in 1994. For site percolation, the bound of Balister, Bollobás and Stacey [1994] given in Theorem 4 is already better than 3/4.

As Liggett remarks, Lincoln Chayes pointed out to him that Theorem 5 gives bounds on the critical probabilities for percolation on oriented lattices other than $\vec{\mathbb{Z}}^2$. Indeed, let $\vec{H}$ be the oriented hexagonal lattice shown in Figure 4. Taking every second column of sites as a layer L_t as in the figure, a given site in L_{t+1} may be reached only from one

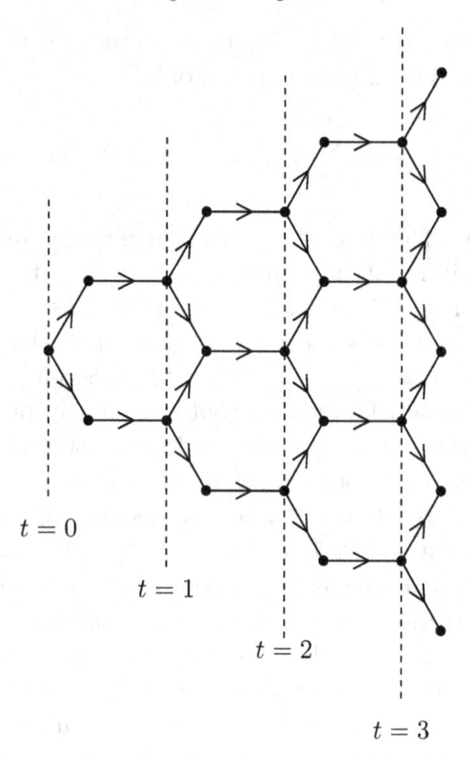

Figure 4. The portion of the oriented hexagonal lattice that may be reached from the origin. Considering the sets of points in every second column that may be reached from the origin by open paths leads to a simple Markov chain.

of two consecutive sites in L_t. Furthermore, all routes from L_t to a site $v \in L_{t+1}$ are disjoint from all routes from L_t to any other site $w \in L_{t+1}$. Hence, both for site and for bond percolation, given the set R_t of sites in L_t that may be reached from the origin, each site $x \in L_{t+1}$ is in R_{t+1} independently of the other sites. Thus, the sets R_t may be described by the same Markov chain used in the case of $\vec{\mathbb{Z}}^2$, with

$$p_1 = p^2 \quad \text{and} \quad p_2 = p(2p - p^2)$$

for bond percolation on $\vec{H}$, and

$$p_1 = p_2 = p^2$$

for site percolation on $\vec{H}$. Consequently, Theorem 5 implies the bounds

$$p_c^b(\vec{H}) \le \frac{1 + \sqrt{33}}{8} \quad \text{and} \quad p_c^s(\vec{H}) \le \frac{\sqrt{3}}{2}.$$

6.4 Non-rigorous bounds

It seems that the critical probabilities for even very simple graphs can be determined exactly only in a small number of exceptional cases. In the light of this, it is not surprising that a huge amount of work has gone in to estimating such critical probabilities. Here we shall describe briefly some of the techniques used, and mention a few of the older and a few of the more recent papers on this topic: a complete survey of this field is beyond the scope of this book.

Most non-rigorous estimates of critical probabilities for lattices are based on Monte Carlo techniques, i.e., on simulating the behaviour of percolation on some finite lattice. However, there are many different ways of extrapolating from such finite simulations to predict the critical probabilities.

For simplicity, we shall consider site percolation on the square lattice $\mathbb{Z}^2$. Let Λ_n be the finite portion of this lattice consisting of a square with n sites on each side. If A_n is any event depending on the states of the sites in Λ_n, then we may estimate $\mathbb{P}_p(A_n) = \mathbb{P}^{\mathrm{s}}_{\mathbb{Z}^2,p}(A_n)$ in the obvious way: for each of N trials, generate the states of the sites in Λ_n at random, according to the measure $\mathbb{P}_p$, and count the number M of trials in which A_n holds.

The problem is this: how do we estimate $p_{\mathrm{c}} = p^{\mathrm{s}}_{\mathrm{H}}(\mathbb{Z}^2) = p^{\mathrm{s}}_{\mathrm{T}}(\mathbb{Z}^2)$ from the probabilities of events depending only on the states of finitely many sites? One possible approach is the following. Let $H_n = H(\Lambda_n)$ be the event that Λ_n has an open horizontal crossing. From Menshikov's Theorem and Lemma 9 of Chapter 5, we know that $\lim_{n\to\infty} \mathbb{P}_p(H_n) = 0$ if $p < p_{\mathrm{c}}$, while $\lim_{n\to\infty} \mathbb{P}_p(H_n) = 1$ if $p > p_{\mathrm{c}}$. For each n, the function $\mathbb{P}_p(H_n)$ is increasing in p. Hence, as n gets larger and larger, this function increases from close to 0 to close to 1 in a smaller and smaller 'window' around p_{c}. In particular, the value $p(n)$ at which $\mathbb{P}_{p(n)}(H_n) = 1/2$, say, tends to p_{c}. By estimating $\mathbb{P}_p(H_n)$ for various values of p, we can estimate $p(n)$. By choosing n as large as is practical, or by considering several different values of n and extrapolating, one can then estimate $p_{\mathrm{c}} = \lim_{n\to\infty} p(n)$. A problem with this approach is that, even for a single value of n, we must estimate $\mathbb{P}_p(H_n)$ for many values of p.

A computationally efficient version of this method was developed by Newman and Ziff [2000; 2001], based on the following simple idea. Let Q_n be some random variable that depends on the states of the sites in Λ_n, whose expectation we wish to estimate. For example, Q_n could be the indicator function of the event H_n, so $\mathbb{E}_p(Q_n) = \mathbb{P}_p(H_n)$. If there are $N = N(n)$ sites in Λ_n, then, writing X for the random number of open sites in Λ_n, we have

$$\mathbb{E}_p(Q_n) = \sum_{k=0}^{N} \mathbb{E}(Q_n \mid X = k)\mathbb{P}_p(X = k) = \sum_{k=0}^{N} b(N, p, k)\mathbb{E}(Q_n \mid X = k),$$

where $b(N, p, k) = \binom{N}{k} p^k (1-p)^{N-k}$ is the probability that a binomial random variable with parameters N and p takes the value k. Thus, to obtain (non-independent) estimates of $\mathbb{E}_p(Q_n)$ for all values of p, it suffices to estimate $\mathbb{E}(Q_n \mid X = k)$ for each value of k. After conditioning on $X = k$, the set of open sites is a random subset $\mathcal{O}_k$ consisting of k sites of Λ_n, distributed uniformly over all $\binom{N}{k}$ such sets. We can efficiently generate a sequence $(\mathcal{O}_k)_0^N$ of random subsets with the right (marginal, i.e., individual) distributions by considering a *random set process*: start with $\mathcal{O}_0 = \emptyset$, and generate $\mathcal{O}_{t+1}$ from $\mathcal{O}_t$ by adding a site of $\Lambda_n \setminus \mathcal{O}_t$ chosen at random, with each of the $N - t$ sites equally likely. If we generate R such sequences independently, for each k we may estimate $\mathbb{E}(Q_n \mid X = k)$ as the average of Q_n over the R samples for $\mathcal{O}_k$ we have generated. Combining these estimates as above gives estimates of $\mathbb{E}_p(Q_n)$ for *all* p simultaneously.

Often, the behaviour of Q_n varies in a simple way as the state of one site is changed from closed to open, in which case this procedure is very fast. For example, taking Q_n to be the indicator function of the event $H_n = H(\Lambda_n)$, the values of Q_n for an entire sequence $(\mathcal{O}_k)$ can be calculated in *linear* time (in the number of sites), using a 'union/find' algorithm to keep track of the set of open clusters at each stage. This enables very accurate estimates of the curve $\mathbb{P}_p(H_n)$ to be made even for quite large n, which in turn means that $p(n)$ can be estimated accurately. Using a variant of this method, considering the event than some open cluster on a torus 'wraps around' the torus in a certain way, Newman and Ziff report the following estimate for $p_c = p_c^s(\mathbb{Z}^2)$:

$$p_c^s(\mathbb{Z}^2) = 0.59274621 \pm 0.00000013.$$

However, there are two problems with the approach just described. One is that the estimates of $\mathbb{E}_p(Q_n)$ for different p are not independent. This makes the statistical analysis somewhat harder. A much more serious problem is the effect of 'finite-size scaling': we know that $p(n)$, say, tends to p_c, but not at what rate. It is believed that

$$p(n) - p_c = n^{-\alpha + o(1)} \tag{7}$$

for some constant α; this would follow from the existence of a certain 'critical exponent'; see Chapter 7. The existence of this exponent is known only for site percolation on the triangular lattice, where p_c is in any case known exactly. Even if (7) were known, it would not help to give rigorous bounds: such a relation still tells us nothing about the relationship between p_c and the finite set of values $p(n)$ we have estimated, as the $o(1)$ term could be very large (or small) for small values of n.

In summary, Monte Carlo methods as described above produce estimates for p_c that do converge in probability to the true value, as the size of the finite lattice studied and the number of simulation runs increases. However, while the statistical error can be analyzed, the rate of convergence to p_c is unknown. Thus any error bounds given by these methods amount to an educated guess as to the final uncertainty. This contrasts with the rigorous 99.99% confidence intervals obtained by considering 1-independent percolation, where the only source of error is statistical, and the error probability for a given bound can be determined exactly.

To show that this problem of estimating the uncertainty is a real one, note that there is some disagreement about the rate of convergence for the Newman–Ziff method described above. Newman and Ziff [2001] say that the finite-size error decreases as $N^{-11/8}$, where N is the number of sites in the finite lattice studied. However, the numerical results of Parviainen [2005] strongly suggest that the rate is closer to $N^{-5/4}$. There are also many examples of reported results contradicted by later estimates, or even by rigorous results. An example is the estimate $p_H^s(\overrightarrow{\mathbb{Z}}^2) = 0.632 \pm 0.004$ given by Kertész and Vicsek [1980] we mentioned earlier. Nonetheless, these methods do give very accurate estimates of p_c; the trouble is that one cannot be sure how accurate!

7

Conformal invariance – Smirnov's Theorem

The celebrated 'conformal invariance' conjecture of Aizenman and Langlands, Pouliot and Saint-Aubin [1994] states, roughly, that if Λ is a planar lattice with suitable symmetry, and we consider percolation on Λ with probability $p = p_c(\Lambda)$, then as the lattice spacing tends to zero certain limiting probabilities are invariant under conformal maps of the plane $\mathbb{R}^2 \cong \mathbb{C}$. This conjecture has been proved for only one standard percolation model, namely independent site percolation on the triangular lattice. The aim of this chapter is to present this remarkable result of Smirnov [2001a; 2001b], and to discuss briefly some of its consequences.

In the next section we describe the conformal invariance conjecture, in terms of the limiting behaviour of crossing probabilities, and present Cardy's explicit prediction for these conformally invariant limits. In Section 2, we present Smirnov's Theorem and its proof; as we give full details of the proof, this section is rather lengthy. Finally, we shall very briefly describe some consequences of Smirnov's Theorem concerning the existence of certain 'critical exponents.'

7.1 Crossing probabilities and conformal invariance

Throughout this chapter we identify the plane $\mathbb{R}^2$ with the set $\mathbb{C}$ of complex numbers in the usual way. A *domain* $D \subset \mathbb{C}$ is a non-empty connected open subset of $\mathbb{C}$. If D and D' are domains, then a *conformal map* from D to D' is a bijection $f : D \to D'$ which is analytic on D, i.e., analytic at every point of D. Note that f^{-1} is then analytic on D'. Locally, a conformal map preserves angles: the images of two crossing line segments are curves crossing at the same angle; this is why the term conformal is used. By the Riemann Mapping Theorem (see, for example,

Duren [1983, p. 11] or Beardon [1979, p. 206]), if $D, D' \neq \mathbb{C}$ are simply connected domains, then there is a conformal map from D to D'.

Roughly speaking, conformal invariance of critical percolation means that certain (random) limiting objects can be defined whose distribution is unchanged by conformal maps. Here we shall only consider a more down-to-earth statement concerning crossing probabilities. For this, we shall need to consider domains whose boundaries are reasonably well behaved.

We write $\overline{D}$ for the closure of a domain D. Let us say that a simply connected domain D is a *Jordan domain* if the boundary $\overline{D} \setminus D$ of D is a Jordan curve, i.e., the image Γ of a continuous injection $\gamma : \mathbb{T} \to \mathbb{C}$, where $\mathbb{T} = \mathbb{R}/\mathbb{Z}$ is the *circle*, which we may view either as $[0, 1]$ with 0 and 1 identified, or as $\{z : |z| = 1\}$. We shall write $\Gamma(D)$ for the boundary of D, and $\gamma = \gamma(D)$ for a function γ as above, noting that γ is unique up to parametrization. By a *k-marked domain* $(D; P_1, \dots, P_k)$, we mean a Jordan domain D together with k points $P_1, P_2, \dots, P_k$ on the boundary of D. We always assume that as the boundary $\Gamma(D)$ is traversed anticlockwise, the points P_i appear in this order. We shall often suppress the marking in our notation, writing D_k for $(D; P_1, \dots, P_k)$.

Given a marked domain D_k, we write $A_i = A_i(D_k)$ for the boundary arc from P_i to P_{i+1} (in the anticlockwise direction), where we include both endpoints, and the indices are taken modulo k; see Figure 1.

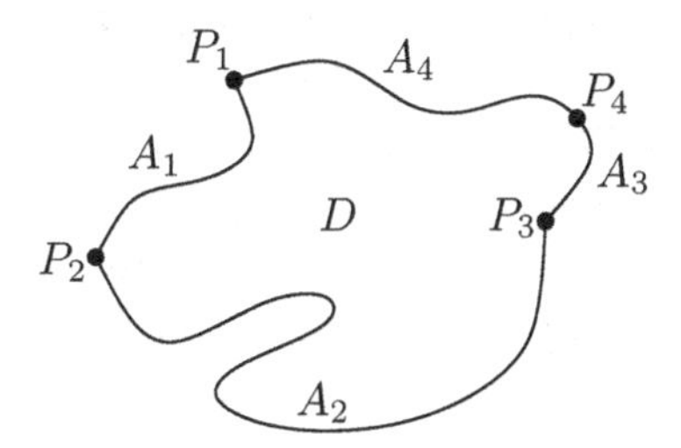

Figure 1. A 4-marked domain $D_4 = (D; P_1, P_2, P_3, P_4)$.

Let Λ be a planar lattice. Given a real number $\delta > 0$, by $\delta\Lambda$ we mean the lattice obtained by scaling Λ by δ about the origin. Thus, for example, $\delta\mathbb{Z}^2$ is the graph with vertex set $\{(\delta a, \delta b) : a, b \in \mathbb{Z}\}$ in which two vertices at distance δ are joined by an edge. Suppose that we have an assignment of states, open or closed, to the bonds or sites of $\delta\Lambda$. If $D_4 = (D; P_1, P_2, P_3, P_4)$ is a 4-marked domain, then by an *open crossing of D from A_1 to A_3 in $\delta\Lambda$*, we mean an open path $v_0 v_1 \dots v_t$ in $\delta\Lambda$

such that $v_1, \ldots, v_{t-1}$ lie inside D, v_0 and v_t are outside D, and the line segments v_0v_1 and $v_{t-1}v_t$ meet the arcs $A_1 = A_1(D_4)$ and $A_3 = A_3(D_4)$, respectively. When the context is clear, we may omit the references to the arcs A_1 and A_3 and to the lattice $\delta\Lambda$.

The quintessential example of a 4-marked domain is a rectangle with the corners marked. Let $D = (a, b) \times (c, d)$, and let P_i be the vertices of the rectangle $\overline{D}$, labelled as in Figure 2. If Λ is the square lattice

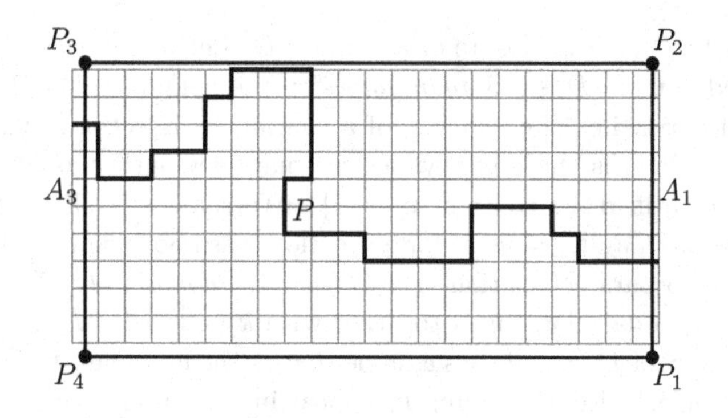

Figure 2. A path P in the lattice $\delta\mathbb{Z}^2$ crossing a rectangular 4-marked domain D_4. The path P is an open crossing if all sites or bonds of P are open. An open crossing of D_4 in site or bond percolation on $\delta\mathbb{Z}^2$ is exactly an open horizontal crossing of the rectangular subgraph R' of $\delta\mathbb{Z}^2$ shown in the figure.

$\mathbb{Z}^2$, then the subgraph R of $\delta\Lambda$ induced by the sites in D is a rectangle in $\delta\mathbb{Z}^2$ in the sense of Chapter 3. For $\delta < \min\{b - a, d - c\}$, an open crossing of $D_4 = (D; P_1, P_2, P_3, P_4)$ in the site or bond percolation on $\delta\Lambda$ is exactly an open horizontal crossing of the rectangle R' obtained from R by adding an extra column of points to each vertical side. We shall see later that the precise manner in which we treat the boundary is not important: we could have defined an open crossing of D_4 as an open path inside D starting at a site 'adjacent to' A_1 and ending at a site 'adjacent to' A_3, for example.

Let D_4 be a 4-marked domain, and Λ a planar lattice. Considering either site or bond percolation on Λ, for $\delta > 0$ and $0 < p < 1$, let

$$P_\delta(D_4, \Lambda, p) = \mathbb{P}_p\big(D_4 \text{ has an open crossing in } \delta\Lambda\big),$$

where $\mathbb{P}_p$ is the site or bond percolation measure on $\delta\Lambda$ in which sites or bonds are open independently with probability p. Of course, as usual, we should indicate in the notation $P_\delta(D_4, \Lambda, p)$ whether we consider site or bond percolation, but this will not be necessary: shortly, we shall restrict our attention to one specific model, site percolation on the triangular lattice.

If $p < p_c = p_H(\Lambda) = p_T(\Lambda)$ is fixed, then it follows from Menshikov's Theorem (see Theorems 6 and 7 of Chapter 4) that $P_\delta(D_4, \Lambda, p) \to 0$ as $\delta \to 0$. Indeed, the arcs $A_1 = A_1(D_4)$ and $A_3 = A_3(D_4)$ are separated by a distance $d > 0$. Roughly speaking, as $\delta \to 0$, if D_4 has an open crossing then one of the $O(1/\delta^2)$ sites near A_1 must be joined by an open path to a site at graph distance $\Theta(d/\delta) = \Theta(1/\delta)$. By Menshikov's Theorem and exponential decay below p_T (Theorem 9 of Chapter 4), the probability of this event is $O(1/\delta^2)\exp(-\Theta(1/\delta)) = o(1)$ as $\delta \to 0$. (The bound $O(1/\delta^2)$ on the number of sites in D 'adjacent to' A_1 is of course rather crude; note, however, that this number need not be $O(1/\delta)$, if A_1 has a fractal structure, for example.) Similarly, considering the dual lattice (for bond percolation) or matching lattice (for site percolation), it follows that $P_\delta(D_4, \Lambda, p) \to 1$ as $\delta \to 0$ with $p > p_c$ fixed. This begs the question of what happens when $p = p_c$.

From now on, we take p to be the critical probability $p_c = p_c(\Lambda)$, and write $P_\delta(D_4, \Lambda)$ for $P_\delta(D_4, \Lambda, p_c)$. To be pedantic, we should indicate whether we consider site or bond percolation, but we shall not do so. We have seen in Chapter 3 that if $D_4 = (a, b) \times (c, d)$ is a rectangle with the corners marked, then the Russo–Seymour–Welsh Theorem implies that there is a constant $\alpha(D_4) > 0$ such that

$$\alpha < P_\delta(D_4, \mathbb{Z}^2) < 1 - \alpha \tag{1}$$

holds for all sufficiently small δ. (In fact, $\delta < \min(b - a, d - c)$ will do, since this condition on δ ensures that D must contain points of $\delta\mathbb{Z}^2$.) A corresponding statement for any lattice with suitable symmetry can be proved along the same lines. Using Harris's Lemma, it is not hard to deduce that (1) holds for any 4-marked domain D_4.

In the light of (1), it is highly plausible that for any lattice Λ and any 4-marked domain D_4, the limit $\lim_{\delta\to 0} P_\delta(D_4, \Lambda)$ exists and lies in $(0, 1)$. Langlands, Pichet, Pouliot and Saint-Aubin [1992] studied the behaviour of this limit, assuming it exists, by performing numerical experiments with rectangular domains D_4, for site and bond percolation on the square, triangular and hexagonal lattices. These experiments suggested to them that the limiting crossing probabilities $P(D_4, \Lambda)$ are universal, i.e., independent of the lattice Λ. (This is an oversimplification: in general, one must first apply a linear transformation to the lattice Λ; for the cases listed, this is not necessary.) Aizenman then suggested that these crossing probabilities should be conformally invariant; supported by additional experimental data, this was stated as a 'hypothesis' by Langlands, Pouliot and Saint-Aubin [1994].

Conjecture 1. *Let Λ be a 'suitable' lattice in the plane, and let $D_4 = (D; P_1, P_2, P_3, P_4)$ a 4-marked domain. Then the limit*

$$P(D_4) = P(D_4, \Lambda) = \lim_{\delta \to 0} P_\delta(D_4, \Lambda)$$

exists, lies in $(0,1)$, and is independent of the lattice Λ. Furthermore, if D_4 and D_4' are conformally equivalent 4-marked domains, then

$$P(D_4) = P(D_4').$$

Let us spell out the meaning of conformal equivalence in this context. Carathéodory's Theorem (see, for example, Duren [1983, p.12], or Beardon [1979, p. 226] for a proof) states that if D and D' are Jordan domains, then any conformal map f from D to D' may be extended to a continuous map $\overline{f}$ from $\overline{D}$ to $\overline{D'}$. As f^{-1} may also be so extended, it follows that $\overline{f}$ maps the boundary of D bijectively onto that of D'. Let $D_4 = (D; (P_i))$ and $D_4' = (D'; (P_i'))$ be 4-marked domains. Then D_4 and D_4' are *conformally equivalent* if there is a conformal map f from D to D' whose continuous extension to $\overline{D}$ maps P_i to P_i' for $1 \leq i \leq 4$. Note that the crucial condition is that P_i is mapped into P_i': we always have conformal maps from D to D'.

To describe the scope of Conjecture 1, let us define what we mean by a lattice: a *two-dimensional lattice* Λ is a connected, locally finite graph Λ, with $V(\Lambda)$ a discrete subset of $\mathbb{R}^2$, such that there are translations T_{v_1} and T_{v_2} of $\mathbb{R}^2$ through two independent vectors v_1 and v_2 each of which acts on Λ as a graph isomorphism. Note that we do not require Λ to be a planar graph. Also, as in the case of planar lattices as defined in previous chapters, the vertex set $V(\Lambda)$ need not be a lattice in the algebraic sense: in general, $V(\Lambda)$ is a finite union of translates of an algebraic lattice. In the conjecture above, a 'suitable' lattice includes any two-dimensional lattice with rotational symmetry of order at least 3. This includes the square, triangular and hexagonal lattices, for example.

One cannot expect conformal invariance for site percolation, say, on an arbitrary lattice Λ. Suppose, for example, that Λ is obtained from $\mathbb{Z}^2$ by applying the shear described by a matrix M. Then conformal invariance for $\mathbb{Z}^2$ would imply that $P(D_1, \Lambda) = P(D_2, \Lambda)$ whenever the domains $M^{-1}D_1$ and $M^{-1}D_2$ are conformally equivalent. This is *not* the same as D_1 and D_2 being conformally equivalent. The conjecture of Aizenman and Langlands, Pouliot and Saint-Aubin [1994] states that for any two-dimensional lattice Λ, there is a non-singular matrix M such that $P(D, M\Lambda)$ is conformally invariant, and equal to $P(D, \mathbb{Z}^2)$, say. They also state that the same result should hold for many non-lattice percolation models, citing experimental results of Maennel for Gilbert's disc model (defined in Chapter 8) and Yonezawa, Sakamoto, Aoki, Nosé and Hori [1988] for percolation on an aperiodic Penrose tiling. Conjecture 1 is also believed to hold for random Voronoi percolation in the plane (also defined in the next chapter); see Aizenman [1998] and Benjamini and Schramm [1998].

As there are so many conformal maps, Conjecture 1 is extremely strong. Indeed, any simply connected domain $D \neq \mathbb{C}$ is conformally equivalent to the open unit disc $B_1(0)$. Thus, any 4-marked domain is conformally equivalent to the unit disc with some four points z_1, z_2, z_3, z_4 specified on its boundary. The conformal maps from $B_1(0)$ onto itself are the Möbius transformations. Given two triples of distinct points on the boundary of $B_1(0)$ in the same cyclic order, say (z_1, z_2, z_3) and (z_1', z_2', z_3'), there is a unique Möbius transformation mapping z_i to z_i'. Thus, any 4-marked domain D_4 is equivalent to the unit disc with the four marked points $1, \sqrt{-1}, -1$ and z, say, where $|z| = 1$ and $\mathrm{Im}(z) < 0$. Hence, the equivalence classes of 4-marked domains under conformal equivalence may be parametrized by a single 'degree of freedom'.

Although, throughout this chapter, we work in the complex plane $\mathbb{C} \cong \mathbb{R}^2$, we shall not often need to write complex numbers explicitly. Thus, we reserve the letter i for an integer (as in z_i above), and write $\sqrt{-1}$ rather than i.

A natural parametrization of 4-marked domains is in terms of the *cross-ratio*: given four distinct points z_1, z_2, z_3, z_4 appearing in this cyclic order around the boundary of $B_1(0)$, their cross-ratio η is the real number

$$\eta = \frac{(z_4 - z_3)(z_2 - z_1)}{(z_4 - z_2)(z_3 - z_1)} \in (0, 1). \tag{2}$$

Möbius transformations preserve cross-ratios; in fact, two markings of the domain $B_1(0)$ are conformally equivalent if and only if they have the same cross-ratio. Given a 4-marked domain $D_4 = (D; P_1, P_2, P_3, P_4)$, we may define the cross-ratio $\eta(D_4)$ as the cross-ratio of any marking of $B_1(0)$ conformally equivalent to D_4: this is unambiguous, as any two such markings are themselves conformally equivalent, so they are related by a Möbius transformation and have the same cross-ratio. With this definition, two 4-marked domains D_4, D_4' are conformally equivalent if and only if $\eta(D_4) = \eta(D_4')$. Thus the conjecture of Aizenman and Langlands, Pouliot and Saint-Aubin [1994] states that $P(D_4, \Lambda)$ is given by some function $p(\eta)$ of the cross-ratio $\eta = \eta(D_4)$.

Inspired by Aizenman's prediction of conformal invariance, Cardy [1992] proposed an exact conformally invariant formula $\pi(D_4)$ for the limiting crossing probability $P(D_4, \Lambda)$, i.e., a formula $\pi(\eta)$ for the function $p(\eta)$. Using methods of conformal field theory, which, as he stated, are not rigorously founded in this context, he obtained the formula

$$p(\eta) = \pi(\eta) = \frac{3\Gamma(2/3)}{\Gamma(1/3)^2}\eta^{1/3}{}_2F_1(1/3, 2/3; 4/3; \eta).$$

Here ${}_2F_1$ denotes the standard hypergeometric function, defined by

$${}_2F_1(a, b; c; z) = \sum_{n=0}^{\infty} \frac{a^{(n)}b^{(n)}}{c^{(n)}} \frac{z^n}{n!},$$

where $x^{(n)}$ is the *rising factorial* $x^{(n)} = x(x+1)(x+2)\cdots(x+n-1)$ (see Abramowitz and Stegun [1966, p. 556]). It is not clear whether Cardy's derivation can be made rigorous. In fact, there is (essentially) only one case where conformal invariance has been proved rigorously, namely site percolation on the triangular lattice. As we shall see in the next section, Smirnov's proof gives conformal invariance and Cardy's formula at the same time.

The case where D_4 is a rectangle with the corners marked is of special interest. Let $D_4(r)$ be the domain $(0, r) \times (0, 1)$, with the corners marked as in Figure 2. Then the *aspect ratio* of $D_4(r)$, i.e., the ratio of the width of the rectangle $D_4(r)$ to its height, is r; the cross-ratio $\eta(D_4(r))$ is given by some function $\eta(r)$, which is easily seen to be a decreasing function from $(0, \infty)$ to $(0, 1)$. In particular, there is one rectangle $D_4(r)$ for every cross-ratio $\eta \in (0, 1)$, so any 4-marked domain is conformally equivalent to some rectangle $D_4(r)$. For this reason, equivalence classes of 4-marked domains under conformal invariance are often known as *conformal rectangles.*

As every rectangle has an axis of symmetry, and any 4-marked domain is conformally equivalent to a rectangle, Conjecture 1 implies the invariance of $P(D_4)$ under reflections of the domain D_4; see Figure 3. Similarly, the cross-ratio $\eta(D_4)$, and hence Cardy's formula $\pi(D_4)$, are preserved by reflections, and hence by 'antiholomorphic' maps, i.e., bijections which preserve the magnitude but not the sign of angles.

Given a 4-marked Jordan domain $D_4 = (D; P_1, P_2, P_3, P_4)$, set $D_4^\star = (D; P_2, P_3, P_4, P_1)$. An open crossing of $D_4^\star$, i.e., an open crossing of $D_4^\star$

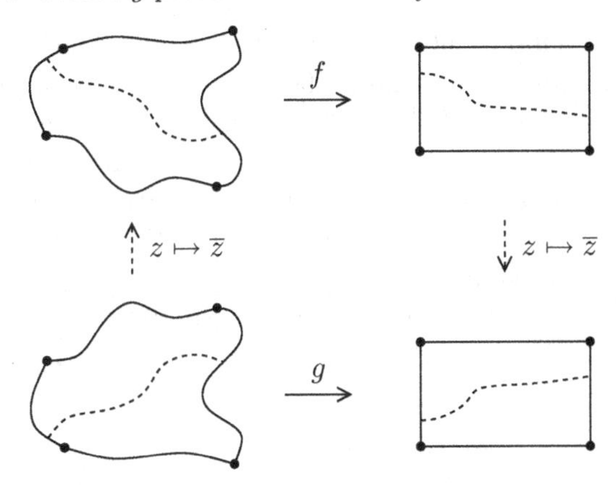

Figure 3. If f is a conformal map, then so is the map $g : z \mapsto \overline{f(\overline{z})}$. Hence, a 4-marked domain D_4 and its mirror image may be conformally mapped to the same rectangle.

from its first to its third arcs, is simply an open crossing of D_4 from A_2 to A_4. In the light of results such as Lemma 1 of Chapter 3, it is natural to expect that $P(D_4) + P(D_4^\star) = 1$. As $\eta(D_4^\star) = 1 - \eta(D_4)$, one can check that Cardy's formula does predict this, i.e., that

$$\pi(1 - \eta) = 1 - \pi(\eta).$$

By constructing an explicit conformal map from the circle to $D_4(r)$, one can use Cardy's formula to evaluate the value $\pi(D_4(r))$ predicted for $P(D_4(r))$. In fact, Cardy [1992] worked with the *upper half plane* $U = \{z : \mathrm{Im}(z) > 0\}$ instead of the open unit disc. We shall not consider unbounded domains here, but the definitions and results extend easily to such domains under suitable conditions. Marking four points z_1, z_2, z_3, z_4 on the real axis to obtain a 4-marked domain $U_4 = (U; (z_i))$, the cross-ratio $\eta(U_4)$ is also given by (2). For $0 < k < 1$, let $U_4(k)$ be the domain U with the four points $-k^{-1}, -1, 1, k^{-1}$ marked. Then $U_4(k)$ may be mapped by a Schwartz–Christoffel transformation

$$z \mapsto \int_0^z \frac{\mathrm{d}t}{\sqrt{(1-t^2)(1-k^2t^2)}}$$

to the interior of the rectangle R with corners $\pm K(k^2)$, $\pm K(k^2) + K(1-k^2)\sqrt{-1}$, where

$$K(u) = \int_0^1 \frac{\mathrm{d}t}{\sqrt{(1-t^2)(1-ut^2)}}$$

is the complete elliptic integral of the first kind. (Our notation here follows Abramowitz and Stegun [1966, p. 590]. The same notation is sometimes used for $K(u^2)$.) The aspect ratio of r is $r = r(k) = 2K(k^2)/K(1-k^2)$, so, to find $\pi(D_4(r))$, one can invert (numerically) this formula to find k; then $\pi(D_4(r)) = \pi(U_4(k)) = \pi(\eta)$, where $\eta = \eta(U_4(k)) = (1-k)^2/(1+k)^2$. (In particular, this shows that $\eta(r) = \eta(D_4(r))$ is monotone decreasing in r.)

Starting from Cardy's formula, Ziff [1995a; 1995b] gave a relatively simple formula not for $\pi(D_4(r))$ itself, but for its derivative with respect to r:

$$\frac{\mathrm{d}}{\mathrm{d}r}\pi(D_4(r)) = -\frac{2^{4/3}\pi\Gamma(2/3)}{\Gamma(1/3)^2}\left(\sum_{n=-\infty}^{\infty}(-1)^n e^{-3\pi r(n+1/6)^2}\right)^4.$$

The calculations outlined above illustrate an unfortunate property of conformal invariance: given two families of 4-marked domains, each parametrized by a real parameter, even if both families are 'nice', the conformal transformation from one to the other may transform the parameter in a rather ugly way. Thus, one cannot expect Cardy's formula to take a simple form for a given 'nice' family of domains.

There is, however, one family for which Cardy's formula can be written very simply, an observation made by Lennart Carleson in connection with joint work with Peter Jones. Let D be the equilateral triangle in $\mathbb{R}^2 \cong \mathbb{C}$ with vertices $P_1 = (1,0)$, $P_2 = (1/2, \sqrt{3}/2)$ and $P_3 = (0,0)$, and let $P_4 = (x,0)$, $0 < x < 1$, as in Figure 4.

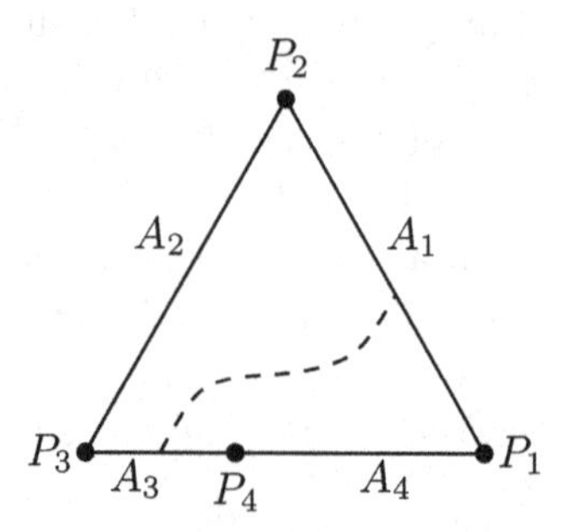

Figure 4. Carleson's 4-marked domain T_x.

Carleson showed that for the special 4-marked domain $T_x = (D; (P_i))$, Cardy's formula takes the extremely simple form

$$\pi(T_x) = x. \tag{3}$$

As we shall see in the next section, it is in this form that Smirnov proved Cardy's formula for site percolation on the triangular lattice. Note that the relation $\pi(\eta) + \pi(1-\eta) = 1$ is easy to verify from (3). Indeed, permuting the labels of the marked points to obtain the dual domain $T_x^\star$ as before, $T_x^\star$ is the mirror image of T_{1-x}. Writing $\eta = \eta(T_x)$ for the cross-ratio of T_x, we have $\eta(D_4^\star) = 1 - \eta(D_4)$ for any 4-marked domain, so

$$\pi(1-\eta) = \pi(T_x^\star) = \pi(T_{1-x}) = 1 - x = 1 - \pi(T_x) = 1 - \pi(\eta). \tag{4}$$

The hypothesis stated by Langlands, Pouliot and Saint-Aubin [1994] is a little more general than Conjecture 1. Their conjecture extends to events such as 'there is an open crossing from A_1 to A_3 and an open crossing from A_2 to A_4', and corresponding events involving any finite number of crossings of a domain whose boundary is split into a finite number of arcs. Conjecture 1 also has consequences that are at first sight unrelated, namely the existence of various critical exponents. We shall return to this briefly in Section 3.

7.2 Smirnov's Theorem

For the rest of this chapter we restrict our attention to site percolation on the triangular lattice T, and the re-scaled lattices δT. Throughout this section we consider only critical percolation, i.e., we study the probability measure $\mathbb{P}$ in which each site is open independently with probability $p = p_{\mathrm{H}}^{\mathrm{s}}(T) = 1/2$. As noted earlier, Smirnov [2001a] (see also [2001b]) proved the remarkable result that the conformal invariance suggested by Aizenman does indeed hold in this case.

Once its walls have been breached, one might hope that the castle would rapidly fall, i.e., that a proof for general lattices, or at least for other 'nice' lattices such as $\mathbb{Z}^2$, would follow quickly. Although this was widely expected, no proofs of other such results have emerged, except for an adaptation of Smirnov's argument to a somewhat unnatural model based on bond percolation on the triangular lattice given by Chayes and Lei [2006]. It seems that Smirnov's proof depends essentially on special properties of T.

In this section we shall present Smirnov's proof in detail. Let us note that even the expanded version of this proof in Smirnov [2001b] is only an outline. For the heart of the proof, showing that certain functions related to crossing probabilities are harmonic, it is very easy to dot the i's and cross the t's. However, one must also deal with certain boundary conditions. Smirnov [2001b] gives a suggestion as to how this should be done, but it does not seem to be at all easy to turn this suggestion into a proof. Beffara [2005], and independently Tóth, suggested considering certain symmetric combinations of Smirnov's harmonic functions; Beffara showed that this greatly simplifies the boundary conditions, eliminating the need to consider partial derivatives at the boundary. Ráth [2005] has used these ideas to give an expanded account of Smirnov's proof; nevertheless, even this presentation is far from giving all the details.

We shall prove Smirnov's Theorem in the precise form below; the original statement is somewhat more general in terms of the domains considered. The proof we shall present here is rather lengthy. We follow the strategy of Smirnov, as modified by Beffara. Along the way, however, we prove the many 'obvious' statements that are required.

Theorem 2. *Let $D \subset \mathbb{C}$ be a simply connected, bounded domain, whose boundary is a Jordan curve Γ. Let P_i, $1 \le i \le 4$, be distinct boundary points of D, appearing in this cyclic order as Γ is traversed, so $D_4 = (D; P_1, P_2, P_3, P_4)$ is a 4-marked domain. Let $A_i = A_i(D_4)$ be the arc of Γ from P_i to P_{i+1}, where P_5 is taken to be P_1. Then $P(D_4) = \lim_{\delta\to 0} P_\delta(D_4, T)$ exists and is given by Cardy's formula, where $P_\delta(D_4, T)$ is the probability that there is an open crossing of D from A_1 to A_3 in the critical site percolation on the lattice δT.*

The basic idea of the proof is to define certain functions f_δ^i that generalize crossing probabilities. We regard D_4 as a 3-marked domain D_3 by temporarily forgetting the fourth marked point P_4, replacing the boundary arcs A_3, A_4 by a single arc $A_3' = A_3 \cup A_4$. We then define functions $f_\delta^i(z)$ on $\overline{D}$, $i = 1, 2, 3$. Roughly speaking $f_\delta^1(z)$ is the probability that there is an open path in $\delta T \cap D$ from A_2 to A_3' separating z from A_1; the remaining f_δ^i are defined similarly. These functions generalize crossing probabilities: returning to the 4-marked domain D_4, an open crossing from A_1 to A_3 is the same as a crossing of D_3 from A_1 to A_3' separating the point P_4 from A_2; see Figure 5. Thus, the crossing probability $P_\delta(D_4, T)$ is just the value of $f_\delta^2(z)$ at the point $z = P_4$.

Smirnov proves a certain 'colour switching lemma' which implies an equality between certain discrete derivatives of the functions f_δ^i. It will follow that, as $\delta \to 0$, the functions f_δ^i converge uniformly to harmonic functions f^i. These functions satisfy boundary conditions that ensure that they transform in a conformally invariant way as D_3 is transformed,

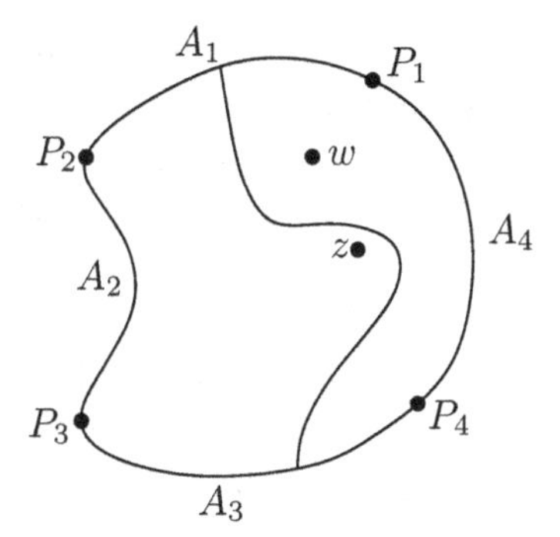

Figure 5. There is an open crossing of D from A_1 to A_3 if and only if there is an open path from the arc A_1 to the arc $A_3' = A_3 \cup A_4$ that separates P_4 from A_2. The indicated path separates P_4 from A_2, and w from A_2, but does not separate z from A_2. The function $f_\delta^2(z)$ is the probability that there is an open path in δT from A_1 to A_3' separating z from A_2.

and that if D_3 is an equilateral triangle with the corners marked, then each f^i is the linear function that takes the value 0 on the arc A_i and 1 on the opposite point. Cardy's formula in Carleson's form then follows.

7.2.1 A consequence of an RSW-type theorem

One of the key ingredients of Smirnov's proof is a consequence of the Russo–Seymour–Welsh (RSW) Theorem, concerning open paths crossing annuli with very different inner and outer radii. So far, we have proved an RSW-type theorem only for bond percolation on the square lattice. Of the many proofs of this result, most (perhaps all?) can be adapted fairly easily to site percolation on the triangular lattice T, to deduce the following result.

Theorem 3. *Let $\rho > 1$ be constant. There is a constant $c(\rho) > 0$ such that, if $n \geq 2$ and R is a ρn by n rectangle in $\mathbb{R}^2$ with any orientation, then the probability that the critical site percolation on T contains an open crossing of R joining the two short sides is at least $c(\rho)$.*

Here, the notion of an open crossing is that for 4-marked discrete domains, i.e., an open path $v_0v_1 \ldots v_t$ in T with $v_1, \ldots, v_{t-1}$ inside R, such that the line segments v_0v_1 and $v_{t-1}v_t$ meet opposite short sides of R. For a detailed proof of Theorem 3 based on the strategy used for bond percolation on $\mathbb{Z}^2$ in Section 1 of Chapter 3, see Bollobás and Riordan [2006b].

We shall use the following immediate consequence of Theorem 3 several times. Let A be an annulus, i.e., the region between two concentric circles C_1 and C_2. We say that A has an *open crossing in the lattice* δT if there is an open path from a site inside the inner circle C_1 to a site outside the outer circle C_2.

Lemma 4. *Let A be an annulus with inner radius r_- and outer radius r_+. If $r_+/r_- \geq 2$ and $r_- \geq 1000\delta$, then the probability that A has an open crossing in δT is at most $(r_-/r_+)^\alpha$, where $\alpha > 0$ is an absolute constant.*

Proof. Replacing α by $\alpha/2$, we may assume that $r_+ = 2^k r_-$ for some integer $k \geq 1$. In any annulus with inner and outer radii r and $2r$, we can find six rectangles as shown in Figure 6. If δ/r is small enough (and

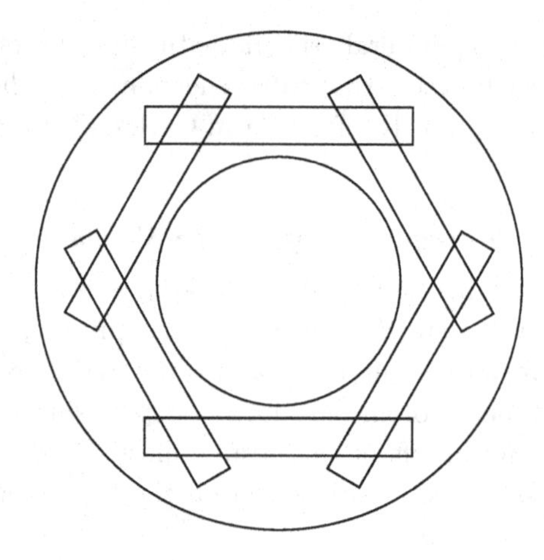

Figure 6. Six rectangles inside an annulus A of inner radius r and outer radius $2r$. If each rectangle is crossed the long way by a closed path in δT, then the annulus A cannot be crossed by an open path.

$\delta/r < 1/1000$ will certainly do), then the shorter side of each rectangle is at least 2δ, and the δ-neighbourhood of each rectangle is contained in the annulus, so the event that it has a closed crossing depends only on the states of sites in the annulus. As $p_c^s(T) = 1/2$, Theorem 3 applies equally well to closed crossings. Hence, for each of our six rectangles, the probability that it contains a closed path joining the two short sides is at least c, where $c > 0$ is an absolute constant. By Harris's Lemma

(Lemma 3 of Chapter 2), the probability that all six rectangles contain such paths is at least c^6. Hence, with probability at least c^6, there is a closed cycle in δT separating the inner and outer circles of the annulus.

Recall that $r_+ = 2^k r_-$. Thus, inside the given annulus A, we can find $k = \log(r_+/r_-)/\log 2$ disjoint annuli $A_i \subset A$ with inner and outer radii r_i and $2r_i$, respectively. Let E_i be the event that A_i contains a closed cycle separating its inside from its outside. If $r_- \geq 1000\delta$, then $\mathbb{P}(E_i) \geq c^6$ for each i. If A has an open crossing, then none of the events E_i can hold. But the events E_i are independent, so

$$\mathbb{P}(A \text{ has an open crossing}) \leq \mathbb{P}(\text{no } E_i \text{ holds}) = \prod_{i=1}^{k} \mathbb{P}(E_i^c) \leq (1-c^6)^k,$$

and the result follows with $\alpha = -\log(1-c^6)/(\log 2)$. □

Thinking of open sites as black and closed sites as white, a path P is *monochromatic* if all sites in P have the same colour. Lemma 4 applies equally well to closed crossings, and hence to monochromatic crossings.

Roughly speaking, Lemma 4 says that if we have a 'small' region of the plane then, when our mesh δ is fine enough, this region is unlikely to be joined either by an open path or by a closed path to any part of the plane 'far away'. To a certain extent, the form of the bound does not matter: any upper bound of the form $f(r_-/r_+)$ with $f(x) \to 0$ as $x \to 0$ suffices for the proof of Smirnov's Theorem. Note that even this weaker form of the lemma is much stronger than the fact that there is no percolation at the critical point, which implies only that the probability above tends to zero as $r_+/\delta \to \infty$ with r_-/δ fixed.

7.2.2 Discrete domains

The heart of Smirnov's proof is a lemma stating that certain probabilities associated to percolation on δT within a domain D are exactly equal. (Of course, this is just a statement concerning a subgraph of the triangular lattice T.) In presenting and proving this statement, we shall often consider the (re-scaled) triangular lattice δT together with its dual, the hexagonal lattice δH obtained by associating to each site $v \in \delta T$ a regular hexagon H_v in the natural way, to obtain a tiling of the plane.

By a *discrete domain with mesh* δ we mean a finite induced subgraph G_δ of δT such that the union of the (closed) hexagons H_v, $v \in G_\delta$, is simply connected. When considering G_δ as a graph, the mesh δ is irrelevant, so we may take $\delta = 1$ and view G_δ as a subgraph G of T.

In terms of the lattice T (or δT), the condition that G (or G_δ) be simply connected is equivalent to requiring that both G and its *outer boundary in* T, $\partial^+(G)$, are connected subgraphs of T, where $\partial^+(G)$ consists of the set of sites of $T \setminus G$ adjacent to some site in G. In fact, we shall impose the following additional restriction on our discrete domains G: we shall assume that neither G nor $\partial^+(G)$ has a cut-vertex. This restriction is irrelevant to the mechanics of the arguments that follow, but simplifies the presentation slightly.

The *inner boundary* $\partial^-(G)$ of a discrete domain $G \subset T$ is just the set of sites of G that are adjacent to some site of $T \setminus G$. Our additional assumptions ensure that both $\partial^-(G)$ and $\partial^+(G)$ are the vertex sets of simple cycles in T: indeed, viewing G as a union of hexagons, its topological boundary ∂G is a simple cycle in the hexagonal lattice H. Following this cycle in an anticlockwise direction, say, the hexagons (i.e., sites of T) seen on the left form $\partial^-(G)$, and those on the right form $\partial^+(G)$; see Figure 7. The condition that neither G nor $\partial^+(G)$ has a

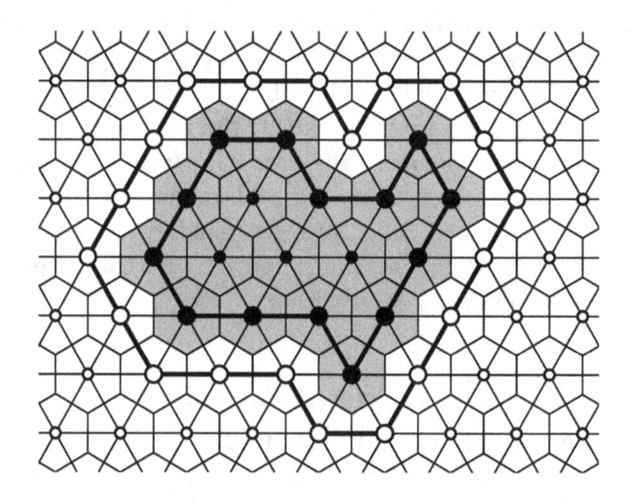

Figure 7. A discrete domain G (filled circles and the lines joining them), drawn with the corresponding hexagons shaded. The two thick lines are the cycles in T corresponding to the outer and inner boundaries $\partial^+(G)$ and $\partial^-(G)$ of G. The topological boundary ∂G of (the set of hexagons corresponding to) G is the cycle in H separating shaded and unshaded hexagons.

cut-vertex ensures that we do not visit the same vertex of $\partial^-(G)$ or of $\partial^+(G)$ more than once. From now on we shall view $\partial^-(G)$ and $\partial^+(G)$ as cycles in the graph T.

In preparation for Smirnov's key lemma, we need a further definition. A *k-marked discrete domain* is a discrete domain G (or G_δ) together with k distinct sites $v_1, \ldots, v_k$ of its boundary $\partial^-(G)$, appearing in this order as $\partial^-(G)$ is traversed anticlockwise. We shall abuse notation by writing simply G for a k-marked discrete domain $(G; v_1, \ldots, v_k)$. Purely for convenience, we impose the additional condition that each marked site v_i is adjacent to at least two sites of $T \setminus G$. (This condition is very mild: any site v of $\partial^-(G)$ that does not have two neighbours in $T \setminus G$ is adjacent to one that does.)

Given a k-marked discrete domain, we define the *arc* $A_i = A_i(G)$ to be the set of sites of $\partial^-(G)$ appearing between v_i and v_{i+1} (with $v_{k+1} = v_1$) as $\partial^-(G)$ is traversed anticlockwise. We include both v_i and v_{i+1} into A_i. In this discrete context, an *open crossing* of G from A_i to A_j is simply an open path in G starting at a site of A_i and ending at a site of A_j.

Given a k-marked discrete domain $G = (G; (v_i))$, as every marked site v_i has two neighbours outside G, we can partition the outer boundary $\partial^+(G)$ of G into vertex-disjoint paths $A_i^+ = A_i^+(G)$, $1 \le i \le k$, so that each A_i^+ starts at a site adjacent to v_i and ends at a site adjacent to v_{i+1}. Indeed, traversing $\partial^+(G)$ anticlockwise, we may take A_i^+ to run from the second neighbour of v_i to the first neighbour of v_{i+1}; see Figure 8. Note that a site $v \in G$ is adjacent to some site of A_i^+ if and only if $v \in A_i$.

Recall from Lemma 7 of Chapter 5 that a rhombus in the triangular lattice always contains either a horizontal open crossing, or a vertical closed crossing, but not both. This statement, and its proof, extends immediately to 4-marked discrete domains. Although the proof for the general case contains nothing new, we give it in full, since we shall soon use similar ideas in a more complicated way.

Lemma 5. *Let G be a 4-marked discrete domain, and let $A_i = A_i(G)$, $1 \le i \le 4$. Whatever the states of the sites in G, this graph contains either an open crossing from A_1 to A_3, or a closed crossing from A_2 to A_4, but not both. In particular, the probability that G has an open crossing from A_1 to A_3 and the probability that G has an open crossing from A_2 to A_4 sum to 1.*

Proof. As above, let $\partial^+(G)$ denote the outer boundary of G, which is partitioned into arcs A_i^+. Consider the partial tiling of the plane by hexagons, consisting of one hexagon for each site of $G \cup \partial^+(G)$. Colour the hexagon H_v corresponding to $v \in G$ black if v is open, and white

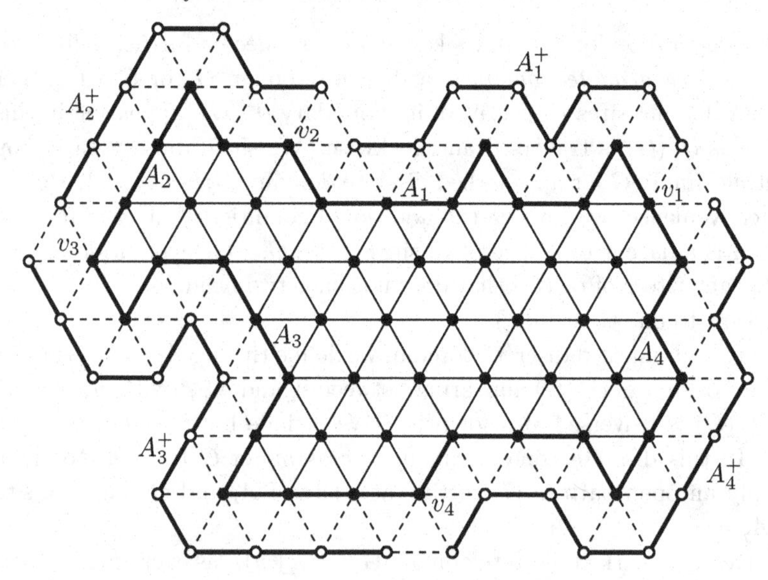

Figure 8. A 4-marked discrete domain $(G; v_1, v_2, v_3, v_4)$ (filled circles and lines joining them). The outer boundary $\partial^+(G)$, a cycle in T, consists of the hollow circles and the lines joining them. The thick lines show the inner boundary $\partial^-(G)$ of G, a union of arcs A_i, $1 \le i \le 4$, sharing endpoints, and the corresponding disjoint arcs $A_i^+ \subset \partial^+(G)$.

if v is closed. Colour the hexagons H_v corresponding to $v \in A_1^+ \cup A_3^+$ black, and those corresponding to $v \in A_2^+ \cup A_4^+$ white, as in Figure 9.

Let I be the interface graph formed by those edges of the hexagonal lattice separating black and white hexagons, together with their endpoints as the vertices. Then every vertex of I has degree two except for four vertices y_i, $1 \le i \le 4$, shown in Figure 9, which have degree 1. Orienting each edge of I so that the hexagon on its right is black, the component of I starting at y_1 is thus a path P ending either at y_2 or at y_4. Suppose that P ends at y_4, as in Figure 9. Then the black hexagons on the right of P form a connected subgraph of $G \cup A_1^+ \cup A_3^+$ joining A_1^+ to A_3^+. Any such subgraph contains a path within G joining a site v adjacent to A_1^+ to a site w adjacent to A_3^+. But then $v \in A_1$ and $w \in A_3$, so G has an open crossing from A_1 to A_3. Similarly, if P ends at y_2, then the white hexagons on the left of P give a closed crossing of G from A_2 to A_4. Crossings of both kinds cannot exist simultaneously, as otherwise K_5 could be drawn in the plane.

The second statement follows immediately: as each site is open independently with probability $1/2$, the probability of an open crossing

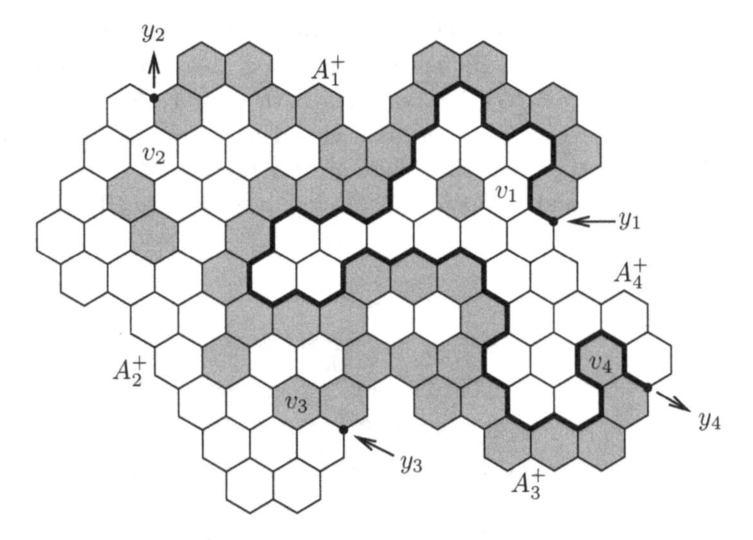

Figure 9. The hexagons H_v corresponding to $G \cup \partial^+(G)$, with those corresponding to the marked vertices v_1, v_2, v_3, v_4 labelled. A hexagon H_v, $v \in A_i^+$, is coloured black for $i = 1, 3$ and white for $i = 2, 4$. A hexagon H_v, $v \in G$, is black if v is open and white if v is closed. There is an open path in G from A_1 to A_3 if and only if there is a black path in the figure from A_1^+ to A_3^+, and hence if and only if the interface between black and white hexagons joins y_1 to y_4.

from A_2 to A_4 is the same as the probability of a closed crossing from A_2 to A_4. □

This lemma completes our brief review of the basic properties of critical site percolation on the triangular lattice T. In the next subsection we present the first step in Smirnov's proof of conformal invariance.

7.2.3 Colour switching

We are now ready to present the key lemma in Smirnov's proof: this 'colour switching' lemma states that the probabilities of certain events involving paths in a discrete domain are *exactly* equal. We shall shall often identify a site of T or δT with the corresponding hexagon. In particular, in the figures that follow, rather than drawing site percolation on the triangular lattice, we draw 'face percolation' on the hexagonal lattice, since it is easier to shade hexagons than points.

Let G be a 3-marked discrete domain, and let x_1, x_2, x_3 be three sites in G forming the vertices of a triangle in G, labelled in anticlockwise order around this triangle. Thinking of open sites as black and closed sites as white, we write B_i for the event that there is an open path joining x_i to $A_i = A_i(G)$, i.e., an open path from x_i to a site in A_i, and W_i for the event that there is a closed path from x_i to A_i. Let $B_1B_2W_3$ denote the event that there are *vertex-disjoint* paths P_i from x_i to A_i, with P_1 and P_2 open, and P_3 closed; see Figure 10. Note that

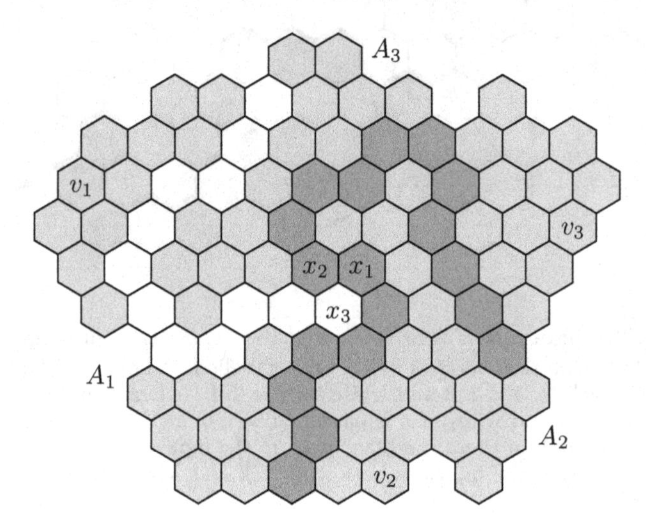

Figure 10. The hexagons corresponding to a 3-marked discrete domain $(G; v_1, v_2, v_3)$, this time without its outer boundary. If the vertices corresponding to the heavily shaded hexagons are open, and those corresponding to the unshaded hexagons are closed, then $B_1B_2W_3$ holds.

$B_1B_2W_3 = B_1 \square B_2 \square W_3$ is just the box product of the events B_1, B_2 and W_3, as defined in Chapter 2. Define $B_1W_2B_3 = B_1 \square W_2 \square B_3$ similarly, and so on.

The probability distribution associated to critical percolation on the triangular lattice is of course preserved if we change the state of every open site to closed, and vice versa. Thus,

$$\mathbb{P}(B_1W_2W_3) = \mathbb{P}(W_1B_2B_3), \tag{5}$$

and so on, so there are four potentially distinct probabilities associated to events of the form $X_1Y_2Z_3$, $X, Y, Z \in \{B, W\}$. Smirnov's Colour-Switching Lemma states that three of these are equal.

Lemma 6. *Let G be a 3-marked discrete domain, and let x_1, x_2, x_3 be sites of G forming a triangle in G, labelled in anticlockwise order around* *this triangle. Then*

$$\mathbb{P}(B_1B_2W_3) = \mathbb{P}(B_1W_2B_3) = \mathbb{P}(W_1B_2B_3). \tag{6}$$

In contrast to (5), there is no symmetry of the overall setup that implies (6). Lemma 6 does *not* say that when looking for disjoint paths, the colours are irrelevant, as it makes no statement about $\mathbb{P}(B_1B_2B_3)$.

To prove Lemma 6, we shall show that

$$\mathbb{P}(B_1W_2B_3) = \mathbb{P}(B_1W_2W_3). \tag{7}$$

Applying (5), or the relation $\mathbb{P}(B_1W_2W_3) = \mathbb{P}(B_1B_2W_3)$, which is essentially equivalent to (7), one of the equalities in (6) follows immediately. The other inequality follows similarly, or by relabelling.

In turn, (7) is equivalent to

$$\mathbb{P}\big(B_1W_2B_3 \mid B_1W_2\big) = \mathbb{P}\big(B_1W_2W_3 \mid B_1W_2\big). \tag{8}$$

The idea of the proof is as follows. Whenever B_1W_2 holds, one can find 'innermost' open and closed paths Q_1 and Q_2 witnessing B_1W_2. We condition not only on B_1W_2, but also on the precise values of the paths Q_1 and Q_2. We can find these paths without examining the states of sites 'outside' them.

Next we shall show that if $B_1W_2B_3$ holds, then there is an open path P_3 from x_3 to A_3 outside the innnermost paths Q_1 and Q_2. Thus, the conditional probability that $B_1W_2B_3$ holds is just the probability that the 'outside' contains an open path from x_3 to A_3. As we have not yet examined the states of any sites 'outside' Q_1 and Q_2, this is the same as the probability that the domain 'outside' Q_1 and Q_2 contains a closed path from x_3 to A_3, which is the conditional probability that $B_1W_2W_3$ holds.

At the risk of seeming too pedantic, we shall present a detailed proof of Lemma 6 using the ideas above. Note that a little caution is needed: on the level of the vague outline just given, it might seem that the same argument shows that $\mathbb{P}(B_1B_2B_3 \mid B_1B_2) = \mathbb{P}(B_1B_2W_3 \mid B_1B_2)$. In fact, $\mathbb{P}(B_1B_2W_3)$ and $\mathbb{P}(B_1B_2B_3)$ are not in general equal.

To find the innermost paths witnessing B_1W_2, we shall follow a certain interface between hexagons. Consider the partial tiling of the plane by hexagons, with one hexagon for each site of $G \cup \partial^+(G)$. As before, we colour a hexagon H_v corresponding to a site v of G black if v is open, and white if v is closed. This time, we colour the hexagons corresponding to A_1^+ black, those corresponding to A_2^+ white, and those to A_3^+ grey. As before, let I be the subgraph of the hexagonal lattice consisting of all edges between black and white hexagons, with their endpoints as the vertices. This time, every vertex of I has degree 2 except for a vertex y where A_1^+ meets A_2^+, and one or more vertices y_i incident with grey hexagons; see Figure 11. Let P be the component of I containing y. Then P is a path starting at y and ending at a vertex y_i where hexagons of all three colours meet.

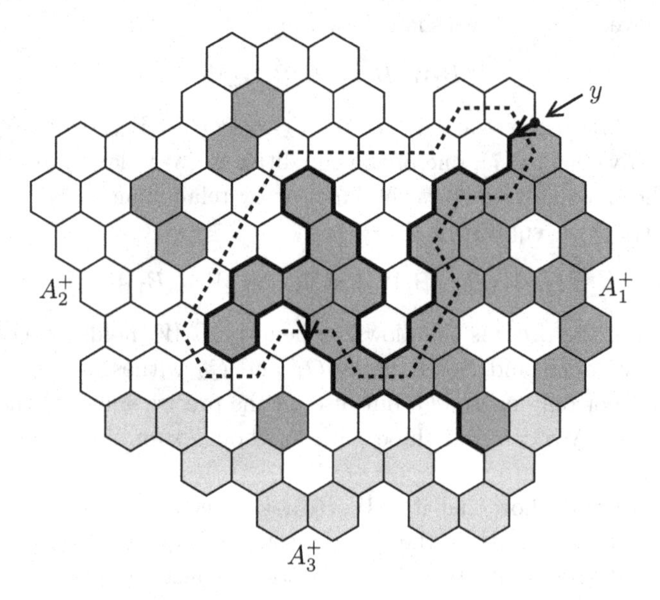

Figure 11. A 3-marked discrete domain with its outer boundary. Hexagons corresponding to A_1^+ are black, those corresponding to A_2^+ white, and those to A_3^+ grey. Internal hexagons are black if the corresponding site is open and white if it is closed. The interface path P starting at y ends at a grey hexagon.

Let w be the centroid of $x_1x_2x_3$, so w is a vertex of the hexagonal lattice. Let $\overrightarrow{e} = \overrightarrow{zw}$ be the edge of the hexagonal lattice separating x_1 from x_2, oriented towards w; see Figure 12.

We shall prove Lemma 6 via a sequence of three simple claims. In the statements of these, the assumptions of Lemma 6 are to be understood.

Claim 7. *If B_1W_2 holds, then the interface path P starting at y traverses the edge $\overrightarrow{e}$ in the positive direction.*

Proof. Suppose that the event B_1W_2 holds, and let P_1 be any open path from x_1 to A_1, and P_2 any closed path from x_2 to A_2. Note that P_1 and P_2 are necessarily disjoint. As any site in A_i is adjacent to a

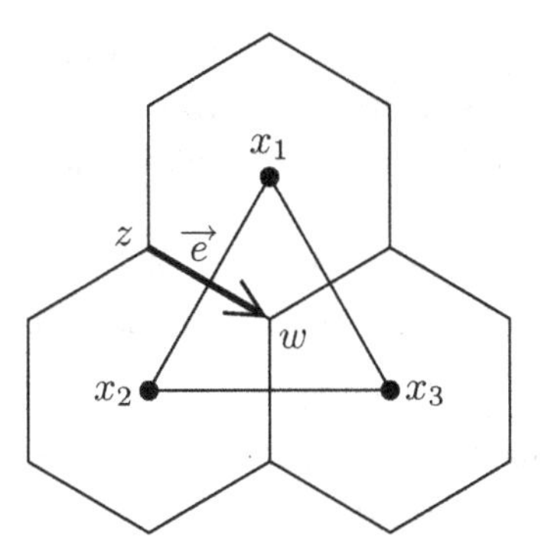

Figure 12. The centroid w of a triangle $x_1x_2x_3$ in T, and the oriented edge $\overrightarrow{e} = \overrightarrow{zw}$ of the hexagonal lattice H that separates x_1 from x_2.

site in A_i^+, there is a cycle C in the triangular lattice formed as follows: follow P_1 from x_1 to A_1. Then follow (part of) A_1^+ to its end. Then follow part of A_2^+, and, finally, follow P_2 from A_2 to x_2. The cycle C is shown by dotted lines in Figure 11. We may view C, which is a cycle in the triangular lattice, as a simple closed curve in the plane in the natural way.

As we go around the cycle C starting at x_1, we first visit black hexagons, and then white hexagons. Thus, exactly two edges of the interface graph I cross C, the edge $\overrightarrow{e}$ shown in Figure 12, and an edge yy'. These are the two edges of H shown with arrows in Figure 11. The path P starts with the edge yy', which takes it inside C. As all grey hexagons are outside C, the path P must leave C at some point, which it can only do along the edge $\overrightarrow{e}$, proving the claim. □

Let P' be the path in the interface graph I defined as follows: starting at y, continue along edges of I until either we traverse $\overrightarrow{e}$ in the positive direction, or we reach a grey hexagon. If P does traverse the edge $\overrightarrow{e}$ in the positive direction, then P' is an initial segment of P. Otherwise, P' is all of P. Let $N(P')$, the *neighbourhood* of P', be the set of sites of G corresponding to hexagons one or more of whose edges appears in P'.

Claim 8. *If P' ends with the edge $\overrightarrow{e}$, then the set $N(P') \subset G$ contains (necessarily disjoint) paths Q_1 from x_1 to A_1 and Q_2 from x_2 to A_2, with Q_1 open and Q_2 closed.*

Proof. The argument is as in the proof of Lemma 5: if P' ends with the edge $\overrightarrow{e}$, then the sites corresponding to the black hexagons on the left of

P' form a connected subgraph S of $A_1^+ \cup G$ meeting A_1^+ and containing x_1. Any such subgraph S includes a subgraph S' of G containing both x_1 and a site w of G adjacent to A_1^+. Every site $v \in S'$ is open: the corresponding hexagon is black (as $v \in S$), and $v \in G$, so v is open. Finally, any $w \in G$ adjacent to A_1^+ is in A_1, so there is an open path $Q_1 \subset S' \subset S \subset N(P')$ from x_1 to A_1. Similarly there is a closed path $Q_2 \subset N(P')$ from x_2 to A_2. □

Together, the claims above show that B_1W_2 holds if and only if P' ends with the edge $\vec{e}$. The proofs of Claims 7 and 8 also give a little more.

Claim 9. *The event $B_1W_2B_3$ holds if and only if P' ends with $\vec{e}$ and there is an open path $P_3 \subset G$ from x_3 to A_3 using no site of the neighbourhood $N(P')$ of P'. Similarly, $B_1W_2W_3$ holds if and only if P' ends with $\vec{e}$ and there is a closed path $P_3 \subset G$ from x_3 to A_3 using no site of $N(P')$.*

Proof. It suffices to prove the first statement. Suppose that $B_1W_2B_3$ holds, so there are disjoint paths P_i from x_i to A_i with P_1 open, P_2 closed, and P_3 open. Then the proof of Claim 7 shows that, apart from its initial and final edges, P' lies entirely within a cycle C formed by P_1, P_2 and parts of A_1^+ and A_2^+ (i.e., the dotted cycle in Figure 11). Thus, every site of $N(P')$ is on or inside C. But P_3 cannot cross C, so P_3 lies entirely outside C, and is disjoint from $N(P')$.

The reverse implication is immediate from Claim 8. □

It is now easy to deduce Lemma 6.

Proof of Lemma 6. As noted earlier, it suffices to show that (7) holds, i.e., that

$$\mathbb{P}(B_1W_2B_3) = \mathbb{P}(B_1W_2W_3).$$

Let Ω be the state space consisting of all $2^{|G|}$ possible assignments of states (open or closed) to the sites of G, and note that the probability measure $\mathbb{P}$ induces the normalized counting measure on Ω.

For $\omega \in \Omega$, let $P'(\omega)$ be the path P' defined as above, with respect to the configuration ω. Let ω' be obtained from ω by flipping (changing from open to closed or vice versa) the states of all sites in $G \setminus N(P'(\omega))$. The path P' may be found step-by-step, at each step examining the colour of a hexagon adjacent to the current path. Hence, the event that

P' takes a particular value is independent of the states of the sites of $G \setminus N(P')$. In particular, $P'(\omega') = P'(\omega)$, so $\omega'' = \omega$ for any $\omega \in \Omega$. Thus the map $\omega \mapsto \omega'$ is a bijection.

Suppose that $\omega \in B_1W_2B_3$. Then, by Claim 9, the configuration ω contains an open path in $G \setminus N(P'(\omega))$ from x_3 to A_3. Hence, ω' contains a closed path $G \setminus N(P'(\omega)) = G \setminus N(P'(\omega'))$ from x_3 to A_3. Thus, by Claim 9, $\omega' \in B_1W_2W_3$. Similarly, if $\omega' \in B_1W_2W_3$, then $\omega'' = \omega \in B_1W_2B_3$. As $\omega \mapsto \omega'$ is a measure-preserving bijection, $\mathbb{P}(B_1W_2B_3) = \mathbb{P}(B_1W_2W_3)$, completing the proof. □

Let us remark that, while we could have used the interface path P' to define 'innermost' open and closed paths Q_1 from x_1 to A_1 and Q_2 from x_2 to A_2, there was no need: it was simpler to work directly with properties of the interface itself.

7.2.4 Separating probabilities

The next step is to give a formal definition of the 'separating probabilities'; the limits of these probabilities will be harmonic functions on D. From this point on we shall view the discrete domains G_δ that we shall consider as subgraphs of δT rather than T: this makes no difference to the properties of G_δ as a graph, but will be convenient for taking limits later.

Let $G_\delta = (G_\delta; v_1, v_2, v_3) \subset \delta T$ be a 3-marked discrete domain with mesh δ, and let $z \in \mathbb{C}$ be the centre of a triangle in δT. As before, we may think of z as a vertex of the hexagonal lattice δH dual to δT. A key idea in Smirnov's proof of Theorem 2 is to consider the probability of the event

$$E_\delta^3(z) = \{\ G_\delta \text{ contains an open } A_1\text{–}A_2 \text{ path separating } z \text{ from } A_3^+\ \},$$

and the events $E_\delta^1(z)$ and $E_\delta^2(z)$ defined similarly, where, as before, $A_i = A_i(G_\delta)$ is the path in the inner boundary of G_δ starting at v_i and ending at v_{i+1}, and $A_i^+ = A_i^+(G_\delta) \subset \partial^+(G_\delta)$ is the corresponding path in the outer boundary of G_δ. Usually, we shall take z to be the centre of a triangle in G_δ, although the definition makes sense for points z nearer to (or even outside) the boundary of G_δ. The subscript δ in our notation is shorthand to indicate the dependence both on δ and on the discrete domain G_δ.

Needless to say, by an open A_1–A_2 path P in G_δ we mean a path P in the graph $G_\delta \subset \delta T$ all of whose sites are open, starting with a site in A_1 and ending with a site in A_2. Such a path P separates z from A_3^+ if, when we complete P to a cycle C in δT using the arcs A_1^+ and A_2^+, the point z lies in the interior of the cycle C when C is viewed as a piecewise linear closed curve; see Figure 13. Equivalently, the path P

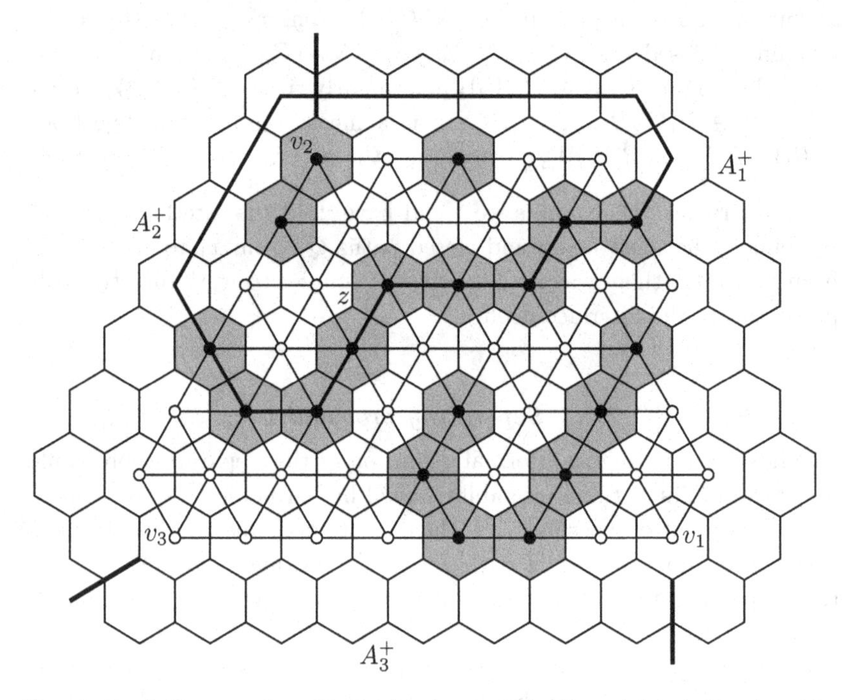

Figure 13. A discrete domain G_δ (circles and the lines joining them), with the associated coloured hexagons: open sites are shown by filled circles and shaded hexagons, closed sites by empty circles and unshaded hexagons. The outer hexagons (those not containing circles) correspond to the outer boundary $\partial^+(G_\delta)$ of G_δ, divided into three arcs A_1^+, A_2^+ and A_3^+ at the thick lines. An open A_1–A_2 path P and the associated cycle C separating z, the centre of a triangle in G, from A_3^+ are shown by thick lines.

separates z from A_3^+ if any path in δH consisting of edges dual to bonds of G_δ, starting at z and ending on (a dual site adjacent to a site on) A_3^+, crosses a bond in P.

Note that P is required to be a path, i.e., not to revisit a vertex. Thus, in the case shown on the right in Figure 14, the event $E_\delta^3(z)$ does not hold.

As before, let $w \in \mathbb{C}$ be the centre of a triangle $x_1x_2x_3$ in G_δ, with the vertices labelled in anticlockwise order, and let z be the site of δH adjacent to w that is farthest from x_3; see Figure 12. Suppose that the

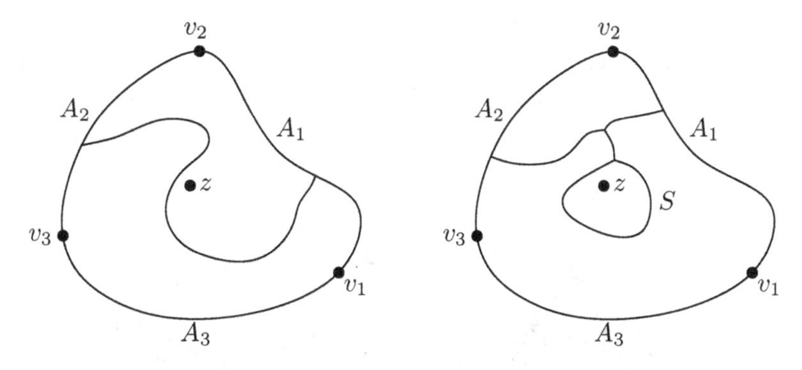

Figure 14. The figure on the left shows a schematic drawing of a 3-marked discrete domain G_δ together with an A_1–A_2 path separating z from A_3^+. If the sites on this path are open, then $E_\delta^3(z)$ holds. The figure on the right shows a connected set S meeting A_1 and A_2 and separating z from A_3^+ that does not contain an A_1–A_2 path separating z from A_3^+. If only the sites of S are open, then $E_\delta^3(z)$ does not hold.

event $E_\delta^3(z) \setminus E_\delta^3(w)$ holds, and let P be any open path in G_δ separating z from A_3^+. Completing P to a cycle C as above, the cycle C winds around z but not around w. Therefore, C must use the edge x_1x_2. If we trace C by following P from A_2 to A_1, returning anticlockwise along A_1^+ and A_2^+ outside G_δ, then C winds around z in the positive sense, so we trace the edge x_1x_2 from x_2 to x_1. Hence, P is the disjoint union of open paths P_1 from x_1 to A_1 and P_2 from x_2 to A_2. As we shall now see, there is also a closed path P_3 from x_3 to A_3.

Claim 10. *Suppose that x_1, x_2, x_3, w and z are as above (see Figures 12 and 15), and that the event $E_\delta^3(z) \setminus E_\delta^3(w)$ holds. Then $B_1B_2W_3$ holds, i.e., G_δ contains disjoint paths P_i joining x_i to $A_i = A_i(G_\delta)$, with P_1 and P_2 open and P_3 closed.*

Proof. As above, let P be an open A_1–A_2 path in G_δ separating z from A_3^+. We have already shown that P may be split into open paths P_1 from x_1 to A_1 and P_2 from x_2 to A_2. It remains to find a closed path P_3 joining x_3 to A_3: any such path is necessarily disjoint from the open paths P_1 and P_2. Note that x_3 itself is certainly closed: otherwise, the open path $P_1x_3P_2$ separates w from A_3^+.

Once again, we follow an interface. As usual, we consider the partial tiling of the plane by hexagons H_v corresponding to vertices v of $G_\delta \cup$

$\partial^+(G_\delta)$, where $\partial^+(G_\delta) = A_1^+ \cup A_2^+ \cup A_3^+$ is the outer boundary of G_δ. We colour a hexagon H_v, $v \in G_\delta$, black if v is open and white if v is closed. We colour H_v black if $v \in A_1^+ \cup A_2^+$, and white if $v \in A_3^+$; see Figure 15. Let I be the oriented interface graph whose edges are the

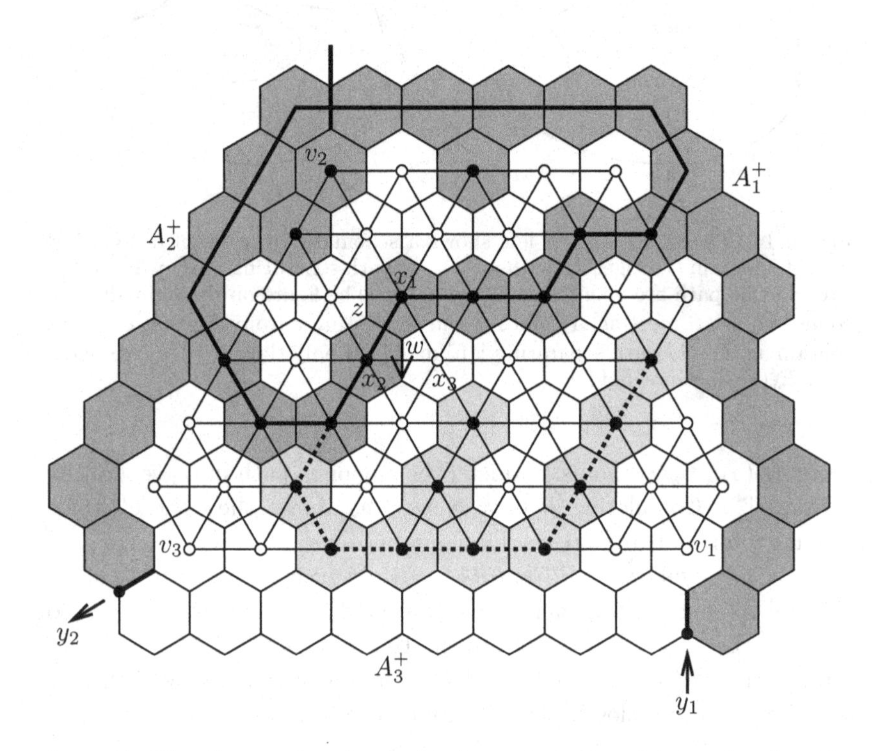

Figure 15. The discrete domain G_δ (circles and the lines joining them), with the associated coloured hexagons. The interface I between black (dark and lightly shaded) and white hexagons has exactly two vertices of degree 1, namely y_1 and y_2. An open A_1–A_2 path P and the associated cycle C separating z from A_3^+ are shown by thick lines. If x_3 is not connected to A_3 by a closed path, then the white component containing x_3 is surrounded by a connected set S of black hexagons: those not on P or $A_1^+ \cup A_2^+$ are shown lightly shaded. The union of S and P contains an A_1–A_2 path separating w from A_3^+, consisting of the dashed lines together with part of P. The oriented edge $\vec{f}$ of the hexagonal lattice crossing x_2x_3 is indicated by an arrow.

edges of the hexagonal lattice with a black hexagon on the right and a white one on the left. This graph has exactly two vertices of degree 1, the vertices y_1 and y_2 shown in Figure 15.

As x_2 is open and x_3 is closed, the oriented interface I contains the edge $\overrightarrow{f}$ of the hexagonal lattice crossing x_2x_3 with x_3 on the left; this edge is indicated with an arrow in Figure 15.

Suppose first that the component of I containing $\overrightarrow{f}$ is a path. Then it ends at a vertex of I with degree 1, which must be y_2. But then the white hexagons on the left of this path form a connected set containing x_3 and meeting A_3^+. As before, it follows that there is a closed path from x_3 to A_3, as required.

Suppose next that the component of I containing $\overrightarrow{f}$ is a cycle, as in Figure 15. Our aim is to deduce a contradiction. Let $h_1, h_2, \ldots, h_t$ be the sequence of hexagons seen on the right of this cycle as we trace it once starting from $\overrightarrow{f}$. Then each h_i is black, h_1 corresponds to x_2, h_t corresponds to x_1, and h_i is adjacent to h_{i+1} for each i. Identifying a hexagon with the corresponding vertex of the lattice, we do not have $h_ih_{i+1} = x_1x_2$ for any i: otherwise, the interface cycle would visit w twice.

Recall that we have open paths P_1 and P_2 joining x_1 to A_1 and x_2 to A_2, respectively, and that these paths can be closed to a cycle C surrounding z but not w by adding parts of A_1^+ and A_2^+. No edge of the interface I can cross C, so every vertex h_i lies on or outside C. Let r be maximal subject to $h_r \in P_2 \cup A_2^+$, and let s be minimal subject to $s \geq r$ and $h_s \in P_1 \cup A_1^+$. Note that $1 \leq r < s \leq t$, as $h_1 = x_2 \in P_2$ and $h_t = x_1 \in P_1$.

Let $S = \{h_{r+1}, \ldots, h_{s-1}\}$. Then S is a connected set of vertices of the graph $G_\delta - x_1x_2$ formed from G_δ by deleting the edge x_1x_2; the hexagons corresponding to S are lightly shaded in Figure 15. As S contains a neighbour of $P_1 \cup A_1^+$ and a neighbour of $P_2 \cup A_2^+$, it follows that $P_1 \cup P_2 \cup S$ contains a path P' in $G_\delta - x_1x_2$ joining A_1 to A_2. (The part of this path off $P_1 \cup P_2$ is shown by dashed lines in Figure 15.) The corresponding hexagons are black (drawn with dark or light shading in Figure 15), so the path P' is open. Furthermore, P' lies on or outside C, so it separates z from A_3^+. As P' does not use the edge x_1x_2, it follows that P' separates w from A_3^+. This contradicts our assumption that $E_\delta^3(w)$ does not hold. $\square$

The proof above is longer than one might like, and can probably be expressed more simply. Note, however, that one cannot give a 'purely topological' proof: in order to show that $E_\delta^3(w)$ holds, it is not enough to find a connected black set meeting A_1 and A_2 and separating w from A_3^+. Indeed, one might expect that the centre z of a triangle in δT is separated from A_3^+ by an open A_1–A_2 path if and only if z lies 'above' the unique path component of the black/white-interface in Figure 15. However, this is not the case, due to the possibility of the configuration on the right in Figure 14.

It is easy to see that the converse of Claim 10 also holds.

Claim 11. *Let $w \in \delta H$ be the centre of a triangle $x_1x_2x_3$ in G_δ, labelled in anticlockwise order. If $z \in \delta H$ is the neighbour of w opposite x_3, then $E_\delta^3(z) \setminus E_\delta^3(w)$ holds if and only if $B_1B_2W_3$ holds.*

Proof. The forward implication is exactly Claim 10. For the converse, suppose that $B_1B_2W_3$ holds, as in Figure 16. Then the disjoint open

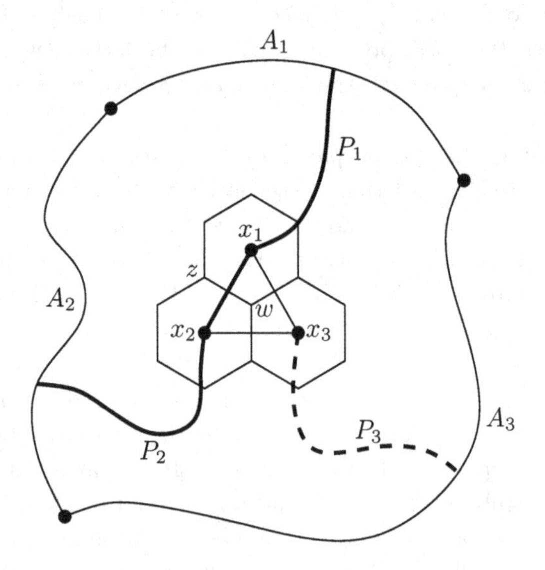

Figure 16. A schematic picture of the event $B_1B_2W_3$.

paths P_1 from x_1 to A_1 and P_2 from x_2 to A_2 together give an open path P from A_1 to A_2. As P uses the edge x_1x_2, the path P separates exactly one of z and w from A_3^+. As there is a closed path from x_3 to A_3, the point w cannot be separated from A_3^+ by an open path. Thus $E_\delta^3(z)$ holds and $E_\delta^3(w)$ does not. □

As before, given a 3-marked discrete domain G_δ and a point $z \in \delta H$, let $E_\delta^i(z)$ be the event that G_δ contains an open A_{i+1}–A_{i+2} path separating z from A_i^+, where the subscripts are taken modulo 3. In the light of Claim 11, Lemma 6 has the following immediate consequence.

Lemma 12. *Let $x_1x_2x_3$ be a triangle in a 3-marked discrete domain G_δ, with its vertices labelled in anticlockwise order. Let $w \in \delta H$ be the centre of the triangle, and let z_1, z_2, $z_3 \in \delta H$ be the neighbours of w in δH, with z_i and x_i opposite for each i. Then*

$$\mathbb{P}\big(E_\delta^1(z_1) \setminus E_\delta^1(w)\big) = \mathbb{P}\big(E_\delta^2(z_2) \setminus E_\delta^2(w)\big) = \mathbb{P}\big(E_\delta^3(z_3) \setminus E_\delta^3(w)\big).$$

Proof. Setting $z = z_3$, Claim 11 states that the events $E_\delta^3(z_3) \setminus E_\delta^3(w)$ and $B_1B_2W_3$ coincide. Permuting all subscripts cyclically, we have $E_\delta^1(z_1) \setminus E_\delta^1(w) = B_2B_3W_1$ and $E_\delta^2(z_2) \setminus E_\delta^2(w) = B_3B_1W_2$. The conclusion thus follows from Lemma 6. □

For the centre $z \in \delta H$ of a triangle in G_δ, set

$$f_\delta^i(z) = \mathbb{P}\big(E_\delta^i(z)\big), \tag{9}$$

and, if w and z are the centres of adjacent triangles in G_δ, set

$$h_\delta^i(w, z) = \mathbb{P}\big(E_\delta^i(z) \setminus E_\delta^i(w)\big).$$

Note that, trivially,

$$f_\delta^i(z) - f_\delta^i(w) = h_\delta^i(w, z) - h_\delta^i(z, w). \tag{10}$$

As noted by Smirnov, there is a surprising amount of cancellation in (10). It turns out that for z and w far from the boundary of D, the quantity $h_\delta^i(w, z)$ is of order $\delta^{2/3}$ as $\delta \to 0$; the exponent 2/3 is the '3-arm exponent' appearing in relation (54) in the next section. In contrast, $f_\delta^i(z) - f_\delta^i(w)$ is presumably of order δ. We shall not be concerned with proving these statements, as they are not needed for the proof of Smirnov's Theorem.

Roughly speaking, Lemma 12 implies that the discrete derivatives of the f_δ^i are related to each other by rotation. For a precise statement, it is easier to work with integrals around contours. We consider only *discrete triangular contours* in G_δ, i.e., equilateral triangular contours C whose corners are sites of δT and whose sides are parallel to edges of δT, such that all sites on C are vertices of G_δ, as in Figure 17. We orient C anticlockwise. Given such a contour, for $1 \le i \le 3$ we define the *discrete contour integral*

$$\oint_C^{\mathrm{D}} f_\delta^i(z)\,\mathrm{d}z$$

as the usual contour integral of the piecewise constant function $f(z)$ whose value at a point z of an edge $\overrightarrow{xy}$ of δT on C is the value of

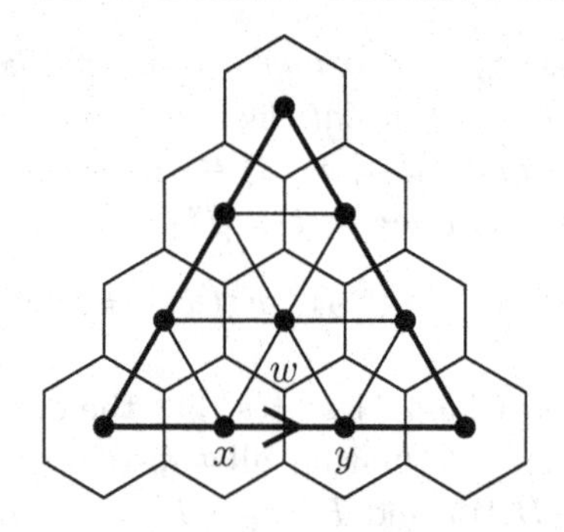

Figure 17. A discrete triangular contour C. The discrete contour integral of f_δ^i around C is defined using the values of f_δ^i at points of the hexagonal lattice δH. Along xy, for example, we integrate $f_\delta^i(w)$.

$f_\delta^i(w)$ on the nearest vertex w of δH inside C, i.e., on the vertex of δH immediately to the left of $\overrightarrow{xy}$; see Figure 17. For example, if C is a triangle with side-length δ, and w is the centre of the triangle C, then $f(z) = f_\delta^i(w)$ identically on C, so $\oint_C^{\mathrm{D}} f_\delta^i(z)\,\mathrm{d}z = 0$.

Let $\omega = (-1+\sqrt{-3})/2$ denote one of the cube roots of unity.

Lemma 13. *Let G_δ be a discrete 3-marked domain such that no point of $\mathbb{C}$ is within distance a of all three arcs of ∂G_δ, where $a > 2000\delta$. If C is a discrete triangular contour in G_δ of length L, then*

$$\left| \oint_C^{\mathrm{D}} f_\delta^{i+1}(z)\,\mathrm{d}z - \omega \oint_C^{\mathrm{D}} f_\delta^i(z)\,\mathrm{d}z \right| \le AL(\delta/a)^\alpha \tag{11}$$

for $i = 1, 2, 3$, where the superscript is taken modulo 3, α is the constant in Lemma 4, and A is an absolute constant.

Proof. We shall prove (11) for $i = 1$; the corresponding equations for $i = 2, 3$ follow by relabelling the domain.

Let $w \in \delta H$ be the centre of a triangle in G_δ, and let z be a neighbour of w in δH. From Claim 10 above, if $E_\delta^i(z) \setminus E_\delta^i(w)$ holds for some i, then there are monochromatic (entirely open or entirely closed) paths from the three sites of δT adjacent to w to the three boundary arcs of G_δ. The point w is at distance at least a from one boundary arc, A_j, say, so a monochromatic path from a point adjacent to w to A_j gives an open or closed crossing of the annulus with centre w and inner and outer radii δ and a. By Lemma 4, the probability that such a crossing exists is at most $2(1000\delta/a)^\alpha = O\big((\delta/a)^\alpha\big)$. Hence,

$$h_\delta^i(w,z) = O\big((\delta/a)^\alpha\big), \tag{12}$$

uniformly over z, w with w the centre of a triangle in G_δ and $z \sim w$: here and below, $\sim$ denotes adjacency in the graph δH.

Let

$$S_i = \sum_{w\in C^{\mathrm{o}}} \sum_{z\sim w} (z-w)h_\delta^i(w,z),$$

where C^{o} is the set of vertices of δH in the interior of C, and z runs over the three neighbours of w in δH. To evaluate $z-w$ we view w and z as complex numbers.

Fix $w \in C^{\mathrm{o}}$, and let z_1, z_2, z_3 be the neighbours of w in anticlockwise order, so

$$z_1 - w = \omega^{-1}(z_2 - w) = \omega^{-2}(z_3 - w). \tag{13}$$

From Lemma 12 we have

$$h_\delta^1(w,z_1) = h_\delta^2(w,z_2) = h_\delta^3(w,z_3). \tag{14}$$

Also, applying the same lemma with the roles of z_1, z_2 and z_3 permuted,

$$h_\delta^1(w,z_2) = h_\delta^2(w,z_3) = h_\delta^3(w,z_1)$$

and

$$h_\delta^1(w,z_3) = h_\delta^2(w,z_1) = h_\delta^3(w,z_2).$$

Setting $z_4 = z_1$, we thus have

$$\begin{aligned}\sum_{j=1}^{3}(z_j - w)h_\delta^2(w,z_j) &= \sum_{j=1}^{3}(z_{j+1} - w)h_\delta^2(w,z_{j+1})\\ &= \sum_{j=1}^{3}\omega(z_j - w)h_\delta^2(w,z_{j+1})\\ &= \omega\sum_{j=1}^{3}(z_j - w)h_\delta^1(w,z_j),\end{aligned}$$

where the first equality is simply a relabelling of the same sum, the second is from (13), and the third from (14) and the following similar equalities. Summing over $w \in C^{\mathrm{o}}$, it follows that

$$S_2 = \omega S_1. \tag{15}$$

We can write $S_i = \sum_{w\sim z,\, w\in C^{\text{o}}}(z-w)h_\delta^i(w,z)$ as $S_i'+S_i''$, where

$$S_i' = \sum_{\{w,z\}:w\sim z,\, w,z\in C^{\text{o}}} (z-w)h_\delta^i(w,z)+(w-z)h_\delta^i(z,w)$$

and

$$S_i'' = \sum_{w\sim z,\, w\in C^{\text{o}},\, z\notin C^{\text{o}}} (z-w)h_\delta^i(w,z).$$

The sum S_i'' has $O(L/\delta)$ terms each bounded by $\delta \sup_{w\sim z} h_\delta^i(w,z)$. Using (12), it follows that

$$S_i'' = O\big(L(\delta/a)^\alpha\big) \tag{16}$$

for each i.

For $w,z\in C^{\text{o}}$ with $w\sim z$, from (10) we have

$$\begin{aligned}(z-w)h_\delta^i(w,z)+(w-z)h_\delta^i(z,w) &= (z-w)(h_\delta^i(w,z)-h_\delta^i(z,w))\\ &= (z-w)(f_\delta^i(z)-f_\delta^i(w)).\end{aligned}$$

To obtain S_i' we must sum this last term over all unordered pairs $\{w,z\}$ with $w\sim z$ and $w,z\in C^{\text{o}}$. Collecting all terms of the form $xf_\delta^i(w)$, $x\in\mathbb{C}$, we thus have

$$\begin{aligned}S_i' &= \sum_{w\in C^{\text{o}}}\sum_{z\sim w,\, z\in C^{\text{o}}} (w-z)f_\delta^i(w)\\ &= \sum_{w\in C^{\text{o}}}\sum_{z\sim w}(w-z)f_\delta^i(w) - \sum_{w\in C^{\text{o}}}\sum_{z\sim w,\, z\notin C^{\text{o}}}(w-z)f_\delta^i(w).\end{aligned}$$

In the first sum in the final line, the coefficient of each $f_\delta^i(w)$ is exactly 0. There is one term in the final sum for every edge $\overrightarrow{wz}$ of δH crossing C from the inside to the outside. The edge $\overrightarrow{wz}$ of δH may be obtained from the dual edge $\overrightarrow{xy}\in C$ of δT by a clockwise rotation through $\pi/2$ followed by scaling by a factor $1/\sqrt{3}$. (For x, y and w, see Figure 17.) Thus, $z-w=(-\sqrt{-1}/\sqrt{3})(y-x)$ Hence, from our definition of the discrete contour integral,

$$S_i' = \sum_{w\in C^{\text{o}}}\sum_{z\sim w,\, z\notin C^{\text{o}}}(z-w)f_\delta^i(w) = -\frac{\sqrt{-1}}{\sqrt{3}}\oint_C^{\text{D}} f_\delta^i(z)\,\mathrm{d}z. \tag{17}$$

As $S_i = S_i'+S_i''$, the result follows from (15), (16) and (17). □

Later, we shall show that if we have a sequence of 3-marked discrete domains G_δ with mesh $\delta\to 0$ that give finer and finer approximations to a Jordan domain D_3, then the functions $f^i_\delta(z)$ converge to continuous functions f^i on $\overline{D}$. Lemma 13 will imply that the (usual) contour integrals of these functions f^i around a given contour are related by multiplication by ω.

7.2.5 Approximating a continuous domain

In this subsection we show that one can approximate a Jordan domain D by suitable discrete domains without changing the crossing probability significantly, proving the technical Lemma 14 below. This result is immediate for sufficiently 'nice' domains, such as rectangles. The reader interested only in such domains may wish to skip the proof.

Let us recall some of the definitions involved in Theorem 2. By a *crossing* of a 4-marked Jordan domain D_4 in the lattice δT, we mean a path in δT whose first and last edges cross the arcs $A_1(D_4)$ and $A_3(D_4)$, with all vertices except the first and last inside D. If the sites of a crossing are open, then it is an *open crossing* of D_4. As before, we write $P_\delta(D_4) = P_\delta(D_4, T)$ for the probability that D_4 has an open crossing.

The corresponding definitions for discrete domains are simpler: a *crossing* of a 4-marked discrete domain G_δ is simply a path in the graph G_δ joining $A_1(G_\delta)$ to $A_3(G_\delta)$. We write $P_\delta(G_\delta)$ for the probability that a discrete domain $G_\delta \subset \delta T$ has an open crossing.

Let us write $\mathrm{dist}(x, y) = |x - y|$ for the Euclidean distance between two points x, $y \in \mathbb{R}^2 \cong \mathbb{C}$, and $\mathrm{dist}(x, A)$ and $\mathrm{dist}(A, B)$ for the distance between a point and a compact set A, or between two compact sets A and B. We avoid the more usual notation $d(x, y)$ due to potential confusion with graph distance. For two compact sets A, $B \subset \mathbb{C}$, their *Hausdorff distance* is

$$\begin{aligned} d_{\mathrm{H}}(A, B) &= \sup\big(\{\mathrm{dist}(a, B) : a \in A\} \cup \{\mathrm{dist}(b, A) : b \in B\}\big) \\ &= \inf\{\varepsilon : A \subset B^{(\varepsilon)},\ B \subset A^{(\varepsilon)}\}, \end{aligned}$$

where $A^{(\varepsilon)}$ denotes the (closed) ε-neighbourhood of A. If $D_4 = (D; (P_i))$ is a 4-marked Jordan domain and $G_\delta \subset \delta T$ is a 4-marked discrete domain, then G_δ is *ε-close* to D_4 if the corresponding boundary arcs of D_4 and of G_δ are within Hausdorff distance ε, i.e., if

$$d_{\mathrm{H}}\big(A_i(D_4), A_i(G_\delta)\big) \le \varepsilon \tag{18}$$

for $1 \le i \le 4$.

To understand the definition above, recall that $A_i(D_4)$ is an open

Jordan curve in $\mathbb{C}$. In contrast, $A_i(G_\delta)$ is formally a set of vertices of G_δ, i.e., a set of points in $\delta T \subset \mathbb{C}$: condition (18) can be interpreted with this definition. Often, however, we shall view G_δ as a union of closed hexagons, so its boundary ∂G_δ is a piecewise linear curve in $\mathbb{C}$. In this case, $A_i(G_\delta)$ is naturally defined to be part of ∂G_δ. It will never make a difference which definition we use: the two interpretations of $A_i(G_\delta)$ give sets at Hausdorff distance $\delta/\sqrt{3}$ from each other, and the arguments that follow will not be sensitive to such small changes. Thus, we shall feel free to switch between these viewpoints without notice.

Lemma 14. *Let D_4 be a 4-marked Jordan domain. Then, for some $\delta_0 = \delta_0(D_4) > 0$, there are families $\{G_\delta^-, 0 < \delta < \delta_0\}$ and $\{G_\delta^+, 0 < \delta < \delta_0\}$ of discrete domains $G_\delta^\pm \subset \delta T$, with the following properties. As $\delta \to 0$, the domains $G_\delta^\pm$ are $o(1)$-close to D_4,*

$$P_\delta(G_\delta^-) - o(1) \le P_\delta(D_4) \le P_\delta(G_\delta^+) + o(1), \tag{19}$$

and, given $\gamma > 0$, there are δ_1, $\eta > 0$ such that, for all $\delta < \delta_1$, any two points w and z of $G_\delta^\pm$ with $\mathrm{dist}(w,z) < \eta$ may be joined by a path in $G_\delta^\pm$ lying in the ball $B_\gamma(w)$.

Roughly speaking, Lemma 14 shows that, to prove Theorem 2, it suffices to work with discrete domains. More specifically, using the fact that $G_\delta^\pm$ is close to D_4, we shall show that

$$P_\delta(G_\delta^\pm) \to \pi(D_4),$$

which, together with (19), implies Theorem 2. The last condition in Lemma 14 is a technicality that we shall need to ensure that certain functions g_δ we shall define are uniformly equicontinuous as δ varies.

The rest of this subsection is devoted to the proof of Lemma 14. The construction of the domains $G_\delta^\pm$ will be broken down into a series of steps, and the proof that they have the required properties into a series of claims. Before getting started, we present two simple facts about Jordan curves that we shall use.

Lemma 15. *Let D be a fixed Jordan domain with boundary Γ. Given $\varepsilon > 0$, there is an $r = r(\Gamma, \varepsilon) > 0$ such that, if x, $y \in \Gamma$ and $\mathrm{dist}(x,y) \le r$, then one of the two arcs into which x and y divide Γ lies entirely within the ball $B_\varepsilon(x)$.*

Proof. This is standard, and immediate from the fact that $\Gamma = \gamma(\mathbb{T})$, where γ is a 1-to-1 continuous map from the circle $\mathbb{T}$ into $\mathbb{C}$. Such a map γ is a homeomorphism, so both γ and its inverse are continuous functions on a compact set, and hence uniformly continuous.

Alternatively, the fact that γ is uniformly continuous implies that, given $\varepsilon > 0$, there exist $t_0 = 0 < t_1 < \cdots < t_{n-1} < t_n = 1$ such that each $\Gamma_i = \gamma([t_i, t_{i+1}])$ lies in some ball of radius $\varepsilon/4$. Any two sets Γ_i, Γ_j with $j - i \neq 0, \pm 1$ modulo n are disjoint, and so separated by a positive distance. Hence, there is an $r > 0$ such that any two points x, y of Γ within distance r lie on the same Γ_i, or an adjacent arcs Γ_i, Γ_j. In either case, there is an arc of Γ joining x to y and lying within $B_\varepsilon(x)$. □

Lemma 16. *Let Γ be a Jordan curve bounding a domain D, and let $z \in D$ be fixed. Given $\varepsilon > 0$, there is an $\eta = \eta(\Gamma, z, \varepsilon) > 0$ with the following property: for every point x of Γ, there is an $x' \in D$ with $\mathrm{dist}(x, x') < \varepsilon$ that may be joined to z by a piecewise linear path P with $\mathrm{dist}(P, \Gamma) > \eta$.*

Proof. Let $B_{\varepsilon/2}(x_i)$ be a (minimal) finite set of balls covering Γ, and pick one point $z_i \in B_{\varepsilon/2}(x_i) \cap D$ for each i, so every $x \in \Gamma$ is within distance ε of some z_i. As D is a connected open set, each z_i may be connected to z by a piecewise linear path P_i in D. The minimal distance between the disjoint compact sets $P = \bigcup_i P_i$ and Γ is strictly positive, so there is an $\eta > 0$ such that the 2η-neighbourhood of P is disjoint from Γ. □

One can show that, given a 4-marked domain D_4 and an $\varepsilon > 0$, for sufficiently small δ there are 4-marked discrete domains G_δ^-, $G_\delta^+ \subset \delta T$ that are ε-close to D_4, such that any crossing of G_δ^- in δT contains a crossing of D_4, and any crossing of D_4 in δT contains a crossing of G_δ^+. This implies that

$$P_\delta(G_\delta^-) \leq P_\delta(D_4) \leq P_\delta(G_\delta^+). \tag{20}$$

In fact, it will be cleaner to prove a somewhat weaker statement, involving only crossings that do not pass to close to the marked points P_i. This weaker statement does not imply (20). However, we shall show, using Lemma 4, that it does imply Lemma 14, which is strong enough for the proof of Theorem 2.

The basic idea of the construction is as follows. If D_4 is a rectangle with the corners marked, then we may take G_δ^- to be a slightly longer and thinner 'rectangle' in the lattice, and G_δ^+ to be a slightly shorter and fatter rectangle. The case where D is a polygon is similarly easy. The general case is not quite so easy: when we 'narrow' D in one direction we may end up with a disconnected graph, for example. Also, when we extend it in the opposite direction, we may bump into ourselves, and end up with a domain that is not simply connected. These are not 'real' problems, but, nevertheless, it takes a fair amount of work to overcome these difficulties.

For the rest of the section, let $D_4 = (D; (P_i))$ be a fixed 4-marked Jordan domain, with boundary Γ. Given $\varepsilon_1 > 0$, in the construction that follows we shall choose other small quantities

$$\varepsilon_1 > \varepsilon_2 > \varepsilon_3 > \varepsilon_4 > \varepsilon_5 = \delta,$$

where ε_{i+1} is a function of ε_i, so that ε_{i+1} is much smaller than ε_i for each i. In particular, we shall assume that $1000\varepsilon_{i+1} < \varepsilon_i$, say.

Let $\varepsilon_1 > 0$ be given. Our first step is to simplify the curve $\Gamma = \partial D$ near the marked points P_i. As before, $A_i = A_i(D)$ is the arc of the Jordan curve Γ running from P_i to P_{i+1}. The arcs A_1 and A_3 are disjoint, and so separated by a positive distance. The same applies to A_2 and A_4. Let $z_0 \in D$ be fixed throughout, and note that $\mathrm{dist}(z_0, A_i) > 0$ for each i. Reducing ε_1, if necessary, we may assume that $\mathrm{dist}(z_0, \Gamma) \geq 10\varepsilon_1$, and that

$$\mathrm{dist}(A_1, A_3),\ \mathrm{dist}(A_2, A_4) \geq 10\varepsilon_1. \tag{21}$$

In particular, $\mathrm{dist}(P_i, P_j) \geq 10\varepsilon_1$ for $1 \leq i < j \leq 4$.

Choose $\varepsilon_2 > 0$ with

$$\varepsilon_2 \leq r(\Gamma, \varepsilon_1),$$

where $r(\Gamma, \varepsilon)$ is the function in Lemma 15. Thus, if w and z are two points of Γ with $\mathrm{dist}(w, z) \leq \varepsilon_2$, then they are joined by an arc of Γ that remains inside $B_{\varepsilon_1}(w)$. Let B_i be the open ball of radius ε_2 around P_i, and let C_i be its boundary. Taking the indices modulo 4, the arc A_i of Γ joins B_i to B_{i+1}, and so contains an arc Γ_i disjoint from $B_i \cup B_{i+1}$ joining C_i to C_{i+1}; see Figure 18. Note that for $j = i+2, i+3$, we have $\mathrm{dist}(\Gamma_i, P_j) \geq \mathrm{dist}(A_i, A_{i+2}) \geq 10\varepsilon_1$. Hence, each arc Γ_i is disjoint from every B_j. For each i, the set $A_i \setminus \Gamma_i$ consists of two arcs of Γ each of which joins a point P_j to a point at distance ε_2 from P_j; this arc is thus contained in $B_{\varepsilon_1}(P_j)$. Thus, $d_{\mathrm{H}}(\Gamma_i, A_i) \leq \varepsilon_1$.

Let Γ' be the 'simplified' curve obtained by joining Γ_i to P_i and P_{i+1} by straight line segments of length ε_2, as in Figure 19, and let D' be the interior of Γ'. Let Γ'_i be the arc of Γ' starting at P_i and ending at P_{i+1},

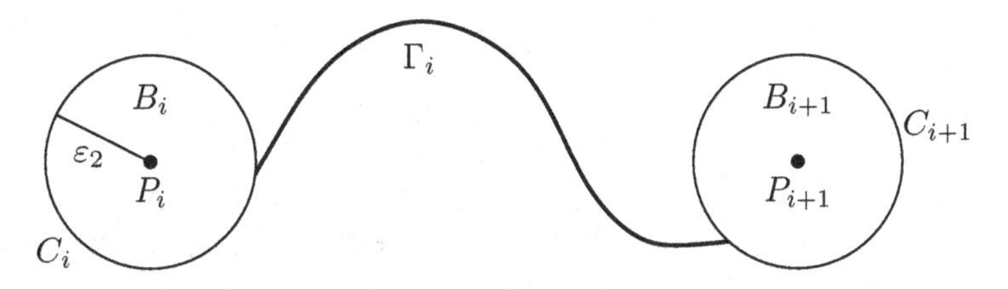

Figure 18. The arc Γ_i of Γ, joining C_i to C_{i+1}.

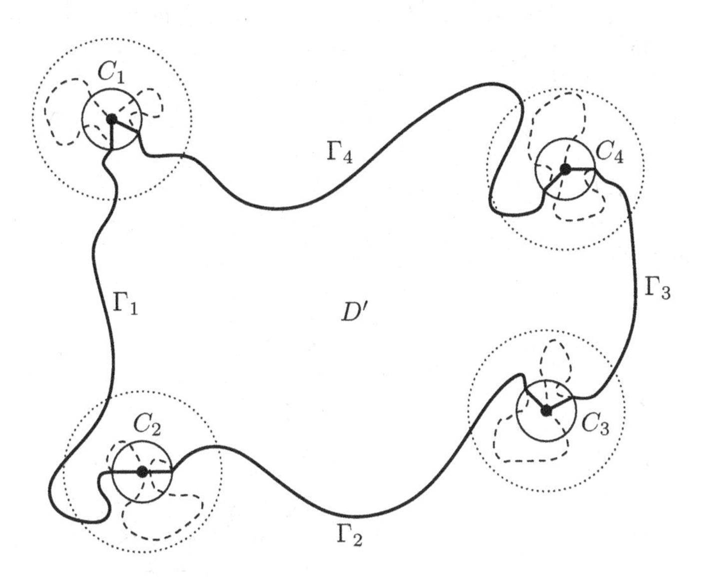

Figure 19. The 'simplified' curve Γ' (thick lines), and its interior D'. The points P_i are the centres of the circles C_i with radii ε_2 (drawn with solid lines). The dashed lines are the arcs of $\Gamma \setminus \Gamma'$, each of which remains within distance ε_1 of some P_i, as indicated by the dotted circles.

and note that

$$d_{\mathrm{H}}(\Gamma'_i, A_i) \leq \varepsilon_1 \tag{22}$$

for each i.

We shall modify Γ' in a way that is analogous to replacing a rectangle by a slightly longer and thinner one. The disjoint closed sets Γ_i, $1 \leq i \leq 4$, are separated by positive distances, so there is an $\varepsilon_3 > 0$ such that

$$\operatorname{dist}(\Gamma_i, \Gamma_j) > 10\varepsilon_3 \tag{23}$$

for $i \neq j$.

Let

$$\varepsilon_4 = r(\Gamma', \varepsilon_3/3)/2, \tag{24}$$

where, as before, r is the function in Lemma 15. Let $N_i = \Gamma_i^{(\varepsilon_4)}$ be the closed ε_4-neighbourhood of Γ_i, and let $\partial^\infty(N_i)$ denote its *external boundary*, i.e., the boundary of the infinite component of $\mathbb{C} \setminus N_i$. The Jordan curve $\partial^\infty(N_i)$ winds around Γ_i. From (23), it does not meet any other Γ_j, so it can cross Γ' only in the line segments inside B_i and B_{i+1}. In fact, it can cross only the line segments L_i and L'_i joining Γ_i to P_i and P_{i+1}, respectively. Furthermore, as Γ_i meets C_i and C_{i+1} at one point each, and lies outside $B_i \cup B_{i+1}$, the curve $\partial^\infty(N_i)$ meets each of L_i and L'_i exactly once, at the points $Q_i \in L_i$ and $Q'_i \in L'_i$ with $\mathrm{dist}(Q_i, P_i) = \mathrm{dist}(Q'_i, P_{i+1}) = \varepsilon_2 - \varepsilon_4$; see Figure 20. Hence, $\partial^\infty(N_i)$ consists of two arcs joining Q_i and Q'_i, one inside Γ' and the other outside Γ'.

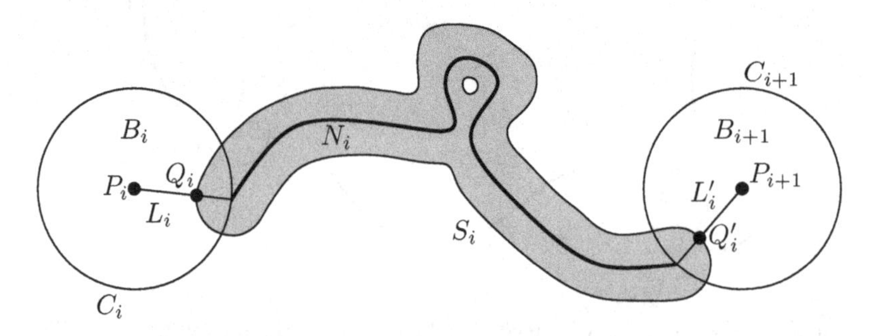

Figure 20. The curve Γ_i (thick line) together with its ε_4-neighbourhood N_i (shaded). The external boundary $\partial^\infty(N_i)$ of N_i crosses Γ' at two points, the points Q_i and Q'_i on the line segments L_i and L'_i. The curve S_i is one of the two arcs of $\partial^\infty(N_i)$ joining Q_i and Q'_i; we take S_i outside Γ' for $i = 1, 3$, and inside for $i = 2, 4$.

For $i = 1, 3$, let S_i be the arc of $\partial^\infty(N_i)$ outside Γ', and, a little dissonantly, for $i = 2, 4$, let S_i be the arc of $\partial^\infty(N_i)$ inside Γ'. (The external boundary of $N_i = \Gamma_i^{(\varepsilon_4)}$ is defined without reference to the rest of Γ', so S_i is part of the external boundary of N_i even when S_i is inside Γ'.) These choices for the arcs S_i correspond to the operation of replacing a rectangle by a longer and thinner one, by moving the first and third sides of the rectangle outwards a little, and the second and fourth sides inwards a little.

We shall write R_i for the region bounded by the closed curve formed by S_i, Γ_i, and the two line segments of length ε_4 joining the endpoints of these curves.

Claim 17. *We can parametrize the open Jordan curves S_i and Γ_i by continuous injections $s_i, g_i : [0,1] \to \mathbb{C}$ with $s_i(0) = Q_i$, $s_i(1) = Q_i'$, $g_i(0) \in C_i$, $g_i(1) \in C_{i+1}$ and $\mathrm{dist}(s_i(t), g_i(t)) < \varepsilon_3$ for all t.*

Proof. Let $\tilde{s}_i(u)$ and $\tilde{g}_i(u)$ be parametrizations of the Jordan curves S_i and Γ_i traced in the appropriate directions. Define a map $\alpha : [0,1] \to [0,1]$ as follows: for each $u \in [0,1]$, let Y_u be a point of Γ_i at distance exactly ε_4 from $X_u = \tilde{s}_i(u) \in S_i$. Such a point exists as S_i is part of the boundary of $N_i = \Gamma_i^{(\varepsilon_4)}$. Define $\alpha(u)$ by $\tilde{g}_i(\alpha(u)) = Y_u$. Note that $\alpha(0) = 0$ and $\alpha(1) = 1$, as there is a unique closest point of Γ_i to each of Q_i and Q_i'.

The straight line segment $L_u = X_uY_u$ meets $S_i \cup \Gamma_i$ only at its endpoints. Also, the interior of L_u lies in N_i and hence in R_i. (As we traverse S_i, on one side we have the unbounded component of $\mathbb{C} \setminus N_i$. On the other side we have N_i and hence R_i.) Thus, L_u separates R_i into two pieces. The boundary of one of these contains all points of S_i appearing 'before' X_u, and all points of Γ_i appearing before Y_u; the boundary of the other all points appearing after these points. As the line segments L_u, $L_{u'}$ cannot cross, it follows that $\alpha(u') \geq \alpha(u)$ for $u' \geq u$.

So far, we have found a way to trace S_i continuously so that a nearby point traces Γ_i monotonically, but not necessarily continuously. We can trace both curves continuously by 'waiting' on S_i whenever the nearby point on Γ_i jumps. More formally, writing $\alpha_-(x) = \sup\{\alpha(x') : x' < x\}$ and $\alpha_+(x) = \inf\{\alpha(x') : x' > x\}$, as $\alpha : [0,1] \to [0,1]$ is increasing, we can find continuous functions $u_1(t)$ and $u_2(t)$ from $[0,1]$ to $[0,1]$ such that each u_i is weakly increasing, and

$$\alpha_-(u_1(t)) \leq u_2(t) \leq \alpha_+(u_1(t))$$

for every t. For example, one can take $u_1(t) = \sup\{x : x + \alpha(x) \leq 2t\}$ and $u_2(t) = 2t - u_1(t)$. Let $s_i(t) = \tilde{s}_i(u_1(t))$ and $g_i(t) = \tilde{g}_i(u_2(t))$. At points where α is continuous, we have $u_2(t) = \alpha(u_1(t))$, so, by definition of α, the points $s_i(t)$ and $g_i(t)$ are at distance exactly ε_4. As both $\tilde{s}_i$ and $\tilde{g}_i$ are continuous, and each discontinuity of α may be approached from both sides by points at which α is continuous, it follows that the points $\tilde{g}_i(\alpha_\pm(u_1(t)))$ are also at distance exactly ε_4 from $s_i(t)$. Hence, these points are within distance $2\varepsilon_4$ of each other. By our choice (24) of ε_4, it follows that the arc of Γ_i joining these two points lies within a ball of radius $\varepsilon_3/3$ centred at either point, and hence within a ball of radius $\varepsilon_3/3 + \varepsilon_4 < \varepsilon_3/2$ centred at $s_i(t)$. As $g_i(t)$ is a point of this arc, we have $\mathrm{dist}(g_i(t), s_i(t)) < \varepsilon_3/2$.

So far, the functions u_j are only weakly increasing, so g_i and s_i are not injections. We can modify the u_j slightly so that they are strictly increasing, for example by adding a small multiple of u_2 to u_1 and vice versa. Using uniform continuity, this does not shift the points $g_i(t)$ or $s_i(t)$ significantly, and the result follows. □

Let Γ^- be the closed curve formed by S_1, S_2, S_3 and S_4, together with the line segments joining the endpoints of these curves to the points P_i, and let D^- be the interior of Γ^-; see Figure 21.

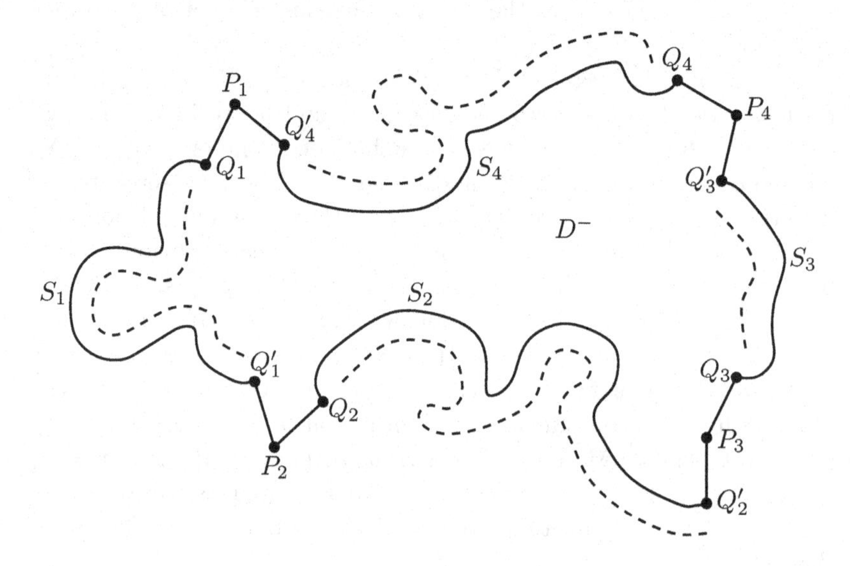

Figure 21. The curve Γ^- (solid lines) and its interior D^-. The dashed lines are the curves Γ_i. Each S_i is part of the external boundary of $N_i = \Gamma_i^{(\varepsilon_4)}$; for $i = 1, 3$, S_i is outside Γ', for $i = 2, 4$, S_i is inside Γ'.

Let Γ_i^- denote the arc of Γ^- starting at P_i and ending at P_{i+1}, so Γ_i^- consists of S_i together with two straight line segments. From Claim 17, we can find corresponding parametrizations Γ'_i and Γ_i^- that remain within distance ε_3. In particular,

$$d_{\mathrm{H}}(\Gamma'_i, \Gamma_i^-) \leq \varepsilon_3 \tag{25}$$

for $1 \leq i \leq 4$. Also, piecing together these parametrizations, we obtain parametrizations of the Jordan curves Γ^- and Γ' such that corresponding points are within distance ε_3.

Roughly speaking, we shall take G_δ^- to consist of all sites v of δT whose corresponding hexagons H_v are contained in D^-. Unfortunately, this set need not be connected, so we shall have to be more careful.

Suppose that

$$0 < 3\delta < \eta(\Gamma^-, z_0, \varepsilon_4/10), \tag{26}$$

where η is the function appearing in Lemma 16, and let D_δ^- be the set of sites $v \in \delta T$ such that $H_v \subset D^-$. Reducing δ, if necessary, we may assume that

$$\delta < \theta\varepsilon_4/100, \tag{27}$$

where θ is the smallest of the angles at the corners P_i of Γ'.

Recall that z_0 is a (fixed) point of D, with $\mathrm{dist}(z_0, \Gamma) > 10\varepsilon_1$, i.e., with $B_{10\varepsilon_1}(z_0) \subset D$. From the construction of Γ', we thus have $B_{9\varepsilon_1}(z_0) \subset D'$. As Γ' winds around z_0, it follows that Γ^- also winds around z_0, so $z_0 \in D^-$; in fact, $B_{8\varepsilon_1}(z_0) \subset D^-$. Hence, the hexagon containing z_0 is contained in D^-.

Regarding D_δ^- as an induced subgraph of δT, let G_δ^- be the component of D_δ^- containing (the site corresponding to the hexagon containing) z_0; see Figure 22. We shall take G_δ^- as one of our discrete approximations

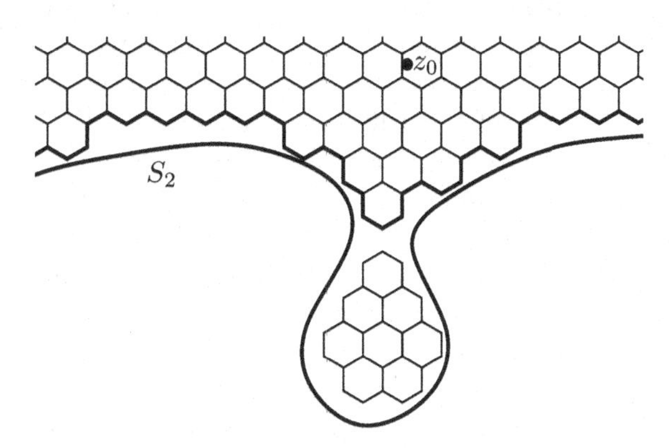

Figure 22. A part of G_δ^- (viewed as a union of hexagons), together with another (small) component of the set D_δ^- of hexagons contained in D^-. G_δ^- is the component of D_δ^- containing the point z_0.

to D. To make G_δ^- into a 4-marked discrete domain $(G_\delta^-; v_1, v_2, v_3, v_4)$, we take v_i to be the vertex of G_δ^- closest to P_i.

Claim 18. *We have* $d_{\mathrm{H}}(\partial G_\delta^-, \Gamma^-) \le \varepsilon_4/10$.

Proof. Suppose first that $x \in \partial G_\delta^-$. Then x is incident with hexagons $H_v \in G_\delta^-$ and $H_w \notin G_\delta^-$. The hexagons H_v and H_w are adjacent, and $H_v \in D_\delta^-$, so by definition of G_δ^-, we must have $H_w \notin D_\delta^-$. Thus, H_w meets Γ^-, and $\mathrm{dist}(x, \Gamma^-) < 2\delta < \varepsilon_4/10$.

Suppose next that $x \in \Gamma^-$. By (26) and Lemma 16, there is a point $x' \in D^-$ with $\mathrm{dist}(x', x) < \varepsilon_4/10$ and a path P joining x' to z_0 with $\mathrm{dist}(P, \Gamma^-) > 3\delta$. But then $P^{(3\delta)} \subset D^-$ contains a path of hexagons joining z_0 to x', so $x' \in G_\delta^-$. As $x \notin D^- \supset G_\delta^-$, the line segment xx' meets ∂G_δ^-, so $\mathrm{dist}(x, \partial G_\delta^-) < \varepsilon_4/10$. □

Recall that Γ_i^-, $1 \le i \le 4$, are the arcs into which the points P_i divide Γ^-, so Γ_i^- starts and ends with short line segments starting and ending at P_i and P_{i+1}; the rest of Γ_i^- is the curve S_i. As any point of S_i is within distance ε_4 of Γ_i, it follows from (23) that if $x \in \Gamma_i^-$ and $y \in \Gamma_j^-$ with $i \ne j$ are at distance $\mathrm{dist}(x, y) < 10\delta$, then, swapping x and y if necessary, we have $j = i + 1$, the point x is on the line segment of Γ_i^- ending at P_{i+1}, and y is on the line segment of Γ_{i+1}^- starting at P_{i+1}. As these line segments meet at an angle of at least θ, from (27) both x and y are within distance $10\delta/\theta < \varepsilon_4/10$ of P_i, i.e., well within the ball B_i.

As G_δ^- is a component of the union of the set of hexagons contained in a simply connected domain, it is simply connected. Let us trace the boundary of G_δ^- anticlockwise. Each boundary vertex v is within distance 2δ of a point of Γ^-, which must lie on some Γ_i^-. If w is the boundary vertex after v, then w is within distance 2δ of a point of some Γ_j^-. From the remark above, if $i \ne j$, then $\{i, j\} = \{k, k+1\}$, and $x, y \in B_{\varepsilon_3}(P_k)$, say. In other words, the closest boundary arc Γ_i^- can only switch when we are very close to some P_k. But in this region, we know exactly what Γ^- looks like: it meets $B_{2\varepsilon_3}(P_k)$ in two line segments. From Claim 18, the boundary of G_δ^- comes within distance $\varepsilon_4/10 < \varepsilon_3$ of P_k. Hence ∂G_δ^- meets $B_{\varepsilon_3}(P_k)$ in a single arc, and, as we trace the boundary of ∂G_δ^-, the closest curve Γ_i^- switches from Γ_k^- to Γ_{k+1}^- when we trace this arc.

Recall that v_i is the closest vertex of G_δ^- to the point $P_i \notin G_\delta^-$. Thus, v_i is a boundary vertex of G_δ^- lying in $B_{\varepsilon_4/10}(P_i)$. Let $A_i^- = A_i(G_\delta^-; v_1, v_2, v_3, v_4)$ be the boundary arcs into which the v_i divide the boundary of G_δ^-. Claim 18 and the comments above imply that

$$d_{\mathrm{H}}(A_i^-, \Gamma_i^-) \le \varepsilon_4/10 \tag{28}$$

for $1 \leq i \leq 4$. Using (25), it follows that $d_{\mathrm{H}}(A_i^-, \Gamma_i') < 2\varepsilon_3$. Hence, appealing to (22),

$$d_{\mathrm{H}}(A_i^-, A_i) < 2\varepsilon_1. \tag{29}$$

In other words, G_δ^- and D_4 are $2\varepsilon_1$-close.

Claim 19. *Let P be a crossing of G_δ^-, i.e., a path in G_δ^- from A_1^- to A_3^-. If P does not pass within distance ε_1 of any P_i, then P contains a crossing of D from A_1 to A_3, and P intersects any crossing of D from A_2 to A_4 that does not pass within distance ε_1 of any P_i.*

Proof. Let $x \in A_1^-$ and $y \in A_3^-$ be the end-vertices of P. As $x \in G_\delta^-$, the point x lies in D^-, the interior of Γ^-. From (28), $\mathrm{dist}(x, \Gamma_1^-) \leq \varepsilon_4/10$. As x is far from P_1 and P_2, it is close to a point of S_1, and outside D. Recall that R_i is the domain in the complex plane bounded by S_i and Γ_i'. We have $x \in R_1$ and $y \in R_3$, so P joins R_1 to R_3. Let P' be a minimal sub-path of P joining R_1 to R_3. Then the first edge of P' leaves R_1, which it must do across Γ_1', and hence across $\Gamma_1 \subset \Gamma$, entering D. Similarly, the last edge of P' crosses Γ_3. P can only cross Γ' on the boundary of R_1, R_3, so the rest of P' lies inside D'. Far from the corners P_i, a point is inside D, i.e., inside Γ, if and only if it is inside Γ'. Thus, all other edges of P are inside D, and P' is a crossing of D from A_1 to A_3.

Similarly, any crossing of D from A_2 to A_4 remaining far from all P_i includes a crossing of G_δ^- from A_2^- to A_4^- which, by Lemma 5, must meet P. □

Let G_δ^+ be defined as G_δ^-, but with all subscripts cycled, so we take S_2 and S_4 outside Γ', and S_1 and S_3 inside Γ'.

Claim 20. *Let D_4 and $\varepsilon > 0$ be given. If $\varepsilon_1 > 0$ is chosen small enough, and G_δ^-, G_δ^+ are constructed as above, then*

$$P_\delta(G_\delta^-) - \varepsilon \leq P_\delta(D_4) \leq P_\delta(G_\delta^+) + \varepsilon.$$

Proof. Opposite arcs of D_4 are separated by some positive distance $c > 0$. From (29), it follows that if ε_1 is small enough, then any pair of opposite arcs of D_4, G_δ^- or G_δ^+ are at distance at least $c/2$. From Lemma 4, it follows that if ε_1 is small enough, then the probability that any of these domains has an open crossing joining opposite arcs and passing within distance ε_1 of some P_i is at most ε. Assuming no such crossing exists, then by Claim 19 any crossing of G_{δ}^{-} from A_1^- to A_3^- crosses D_4 from A_1 to A_3, so

$$P_\delta(D_4) \geq P_\delta(G_\delta^-) - \varepsilon.$$

The second statement follows similarly. □

The final property that we require of our domains $G_\delta^\pm$ is that nearby points may be joined by a short path in the domain. We prove this for G_δ^-; the corresponding statement for G_δ^+ then follows by permuting the marked points P_i.

Claim 21. *Let $\gamma > 0$ be given. There is an $\eta = \eta(D_4, \gamma) > 0$ with the following property. If ε_1 is chosen small enough, then any two points w and z of G_δ^- with* $\mathrm{dist}(w,z) < \eta$ *are joined by a path in G_δ^- lying in $B_\gamma(w)$.*

Proof. We shall assume that $\eta < \gamma/2$.

We may assume without loss of generality that w and z lie on the boundary of G_δ^- (viewed as a union of closed hexagons). To see this, consider the line segment wz. If this lies in G_δ^-, we are done. Otherwise, let x and y be the first and last points of this segment on the boundary of G_δ^-, so wx, $yz \subset G_\delta^-$. Then $\mathrm{dist}(x,y) < \eta < \gamma/2$, so it suffices to join x and $y \in \partial G_\delta^-$ by a path in G_δ^- lying within $B_{\gamma/2}(y)$, say. Thus, Claim 21 for w, $z \in G_\delta^-$ follows from the same claim for w', $z' \in \partial G_\delta^-$ with γ replaced by $\gamma/2$.

We may also assume that the line segment wz does not meet ∂G_δ^- again: otherwise, listing the points x_i in which this segment meets ∂G_δ^- in order, it suffices to connect each to the next by a path in $B_{\gamma/2}(x_i) \subset B_\gamma(w)$. Then the union of these paths contains a path in $B_\gamma(w)$ joining w to z.

We shall choose $\eta(D_4, \gamma)$ so that $\eta < \gamma/100$ and

$$5\eta < r(\Gamma, \gamma/3), \tag{30}$$

where $r(\Gamma, \varepsilon)$ is the function appearing in Lemma 15. We shall choose $\varepsilon_1 < \eta$.

If the line segment wz lies inside G_δ^-, then we are done, so we may assume that it lies outside. As wz meets the simple closed curve ∂G_δ^- only at its endpoints, it divides $\mathbb{C} \setminus G_\delta^-$ into two components, one of which is bounded; see Figure 23. Let R denote the bounded component. Let B be the arc of ∂G_δ^- which, together with wz, bounds R. If $B \subset B_\gamma(w)$,

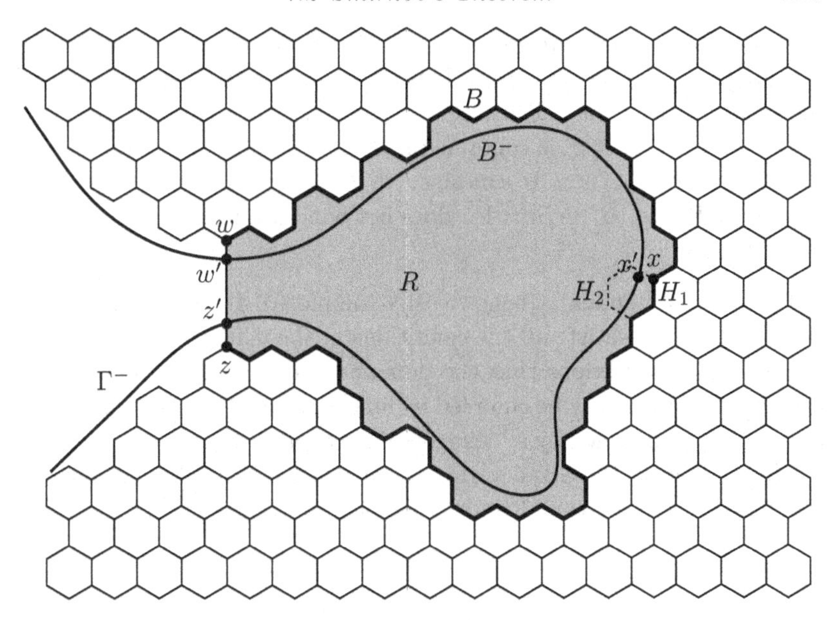

Figure 23. Part of G_δ^- (hexagons), including two points w and z on its boundary with $\text{dist}(w,z) < \eta$. The shaded region R is bounded by $B \subset \partial G_\delta^-$, a piecewise linear curve joining w to z, and the line segment wz. The curved line is part of Γ^-; the point $x' \in \Gamma^-$ is within distance 2δ of the point $x \in \partial G_\delta^-$, and $B^- \subset R$ is the arc of Γ^- from w' to z'.

then we are done, since $B \subset G_\delta^-$ joins w to z. Thus we may assume that there is a point $x \in B$ with $\text{dist}(x,w) > \gamma$.

One of the hexagons meeting at x, say H_1, lies in G_δ^-, while another, H_2, does not, and so lies in R; the hexagon H_2 is drawn with dashed lines in Figure 23. As H_1 and H_2 are adjacent, it follows from the definition of G_δ^- that Γ^- meets H_2, at a point x', say. Let w' and z' be the points at which we leave R when we trace the curve Γ^- from x' in the two possible directions, and let $B^- \subset R$ be the arc of Γ^- from w' to z' containing x'. As Γ^- cannot cross ∂G_δ^-, the points w' and z' lie on the line segment wz. In particular, $\text{dist}(w',z') < \eta$.

As noted above, it follows from Claim 17 that we may parametrize Γ^- and Γ' so that corresponding points are within distance $\varepsilon_3 < \varepsilon_1$. From the construction of Γ' (see Figure 19), it follows that we may parametrize Γ^- and Γ so that corresponding points are within distance $2\varepsilon_1 < 2\eta$. Let w'' and z'' be the points of Γ corresponding to the points w' and z' of Γ^-. Then $\text{dist}(w'',z'') < 5\eta$, so, by (30) and Lemma 15, one of the two arcs of Γ joining w'' and z'' lies in $B_{\gamma/3}(w'')$. The points of Γ^-

corresponding to this arc form an arc B' joining w' and z' remaining within $B_{\gamma/3+\varepsilon_3}(w'') \subset B_{\gamma/2}(w)$. This arc cannot be B^-, which contains the point $x' \notin B_{\gamma/2}(w)$. Thus, $B^- \cup B' = \Gamma^-$.

The curve $B^- \cup w'z'$ is contained in R, and so does not wind around x. The curve $B' \cup w'z'$ is contained in $B_\gamma(w)$, and so does not wind around x. Hence, $B^- \cup B' = \Gamma^-$ does not wind around x, contradicting $x \in G_\delta^- \subset D^-$. □

The proof of Lemma 14 is now easily completed. It remains to ensure that G_δ^- and G_δ^+ satisfy all the conditions in the definition of a discrete domain. The conditions that the domains and their outer boundaries have no cut vertex may be enforced by first replacing δ by $\delta/10$, and then replacing each hexagon by a 'hexagon of hexagons'. As noted earlier, the condition that each marked vertex has at least two neighbours outside the domain can be ensured by moving each marked vertex at most one step around the boundary.

Proof of Lemma 14. We have shown that, given a small enough $\varepsilon_1 > 0$, the construction for $G_\delta^\pm$ described above is valid for all $\delta < \varepsilon_5 = \varepsilon_5(\varepsilon_1)$. Considering a sequence of values of $\varepsilon_1 \to 0$, we may thus pick domains $G_\delta^\pm$ for all δ smaller than some δ_0 in such a way that $G_\delta^\pm$ is defined using a value $\varepsilon_1(\delta)$, with $\varepsilon_1(\delta) \to 0$ as $\delta \to 0$.

The requirement that $G_\delta^\pm$ and D_4 are $o(1)$-close as $\delta \to 0$ follows from (29). Condition (19) is given by Claim 20, and the final condition of Lemma 14 is guaranteed by Claim 21. □

Let us remark that the proof of Lemma 14 shows that, when defining $P_\delta(D_4, T)$, it does not matter exactly how we treat the boundary. Our proof of Smirnov's Theorem will be valid for any definition of a crossing of D_4 for which a crossing of the 'longer, thinner' domain G_δ^- from $A_1(G_\delta^-)$ to $A_3(G_\delta^-)$ that stays far from the corners guarantees a crossing of D_4 from A_1 to A_3, and prevents a crossing of D_4 from A_2 to A_4. For example, we could define a crossing of D_4 in δT from A_i to A_j to be open path in D starting and ending at points within distance 10δ of A_i and A_j, say.

7.2.6 Completing the Proof of Smirnov's Theorem

Let $D_4 = (D; P_1, P_2, P_3, P_4)$ be a 4-marked Jordan domain, and, as before, let $\pi(D_4)$ denote the limiting crossing probability for D_4 predicted by Cardy's formula. Thus, $\pi(D_4)$ is a conformal invariant of D_4 which, following Carleson, may be defined as follows: let φ be the unique conformal map from D_4 to the equilateral triangle that maps P_1, P_2 and P_3 to the vertices $(1, 0)$, $(1/2, \sqrt{3}/2)$ and $(0, 0)$, respectively, so φ maps P_4 into a point $(x, 0)$, $0 < x < 1$. Then $\pi(D_4) = x$.

Let $G_\delta^\pm \subset \delta T$ be the discrete domains approximating D_4 whose existence is guaranteed by Lemma 14. Note that there is a function $\varepsilon = \varepsilon(\delta)$ with $\varepsilon \to 0$ as $\delta \to 0$, such that

$$G_\delta^- \text{ is } \varepsilon\text{-close to } D_4. \tag{31}$$

Of course, the same holds for G_δ^+, but, as we shall see, it suffices to consider $G_\delta -$.

To prove Theorem 2, we shall show that

$$P_\delta(G_\delta^-) \to \pi(D_4) \tag{32}$$

as $\delta \to 0$, where $P_\delta(G_\delta^-)$ is the probability that the 4-marked discrete domain G_δ^- contains an open crossing joining its first and third boundary arcs. Indeed,

$$P_\delta(G_\delta^+) \to \pi(D_4) \tag{33}$$

can be proved in the same way as (32), and then $P_\delta(D_4) \to \pi(D_4)$ follows from (19). In fact, writing $D_4^\star = (D; P_2, P_3, P_4, P_1)$ for the 'dual' marked domain, the constructions of the approximation G_δ^+ to D_4 and the approximation G_δ^- to $D_4^\star$ given in the previous section are identical. Hence, (33) follows from (32), the relation $\pi(D_4) + \pi(D_4^\star) = 1$ (see (4)), and Lemma 5. For this reason we consider only G_δ^-.

For the moment, we shall regard the domains $D_4 = (D; P_1, P_2, P_3, P_4)$ and $G_\delta^- = (G_\delta^-; v_1, v_2, v_3, v_4)$ as 3-marked, by forgetting the fourth marked point, P_4 or v_4. We write $A_i = A_i(D_3)$, $1 \le i \le 3$, for the boundary arcs of the 3-marked domain $D_3 = (D; P_1, P_2, P_3)$. We use corresponding notation for the boundary arcs of G_δ^-.

Let $f_\delta^i(z)$ be the 'crossing-under' probabilities defined by (9), for the domain $G_\delta = G_\delta^- = (G_\delta^-; v_1, v_2, v_3)$. Thus, for z the centre of a triangle in G_δ^-, we have

$$f_\delta^i(z) = \mathbb{P}\big(E_\delta^i(z)\big),$$

where $E_\delta^i(z)$ is the event that there is an open path in G_δ^- from $A_{i+1} = A_{i+1}(G_\delta^-)$ to A_{i+2} separating z from A_i^+ (an arc of the outer boundary of G_δ^-). Here, the subscripts are taken modulo 3.

Let us extend f_δ^i to a continuous function g_δ^i on the closure $\overline{D}$ of D as follows. Recall that f_δ^i is defined at various points of δH, i.e., at the centres of various triangles (faces) of the lattice δT. At the centre $z \in \delta H$ of any face of δT meeting $\overline{D}$, set $g_\delta^i(z)$ equal to the value $f_\delta^i(w)$ of f_δ^i at the closest point w where f_δ^i is defined. (If there is more than one such w, choose one in an arbitrary way.) Note that, from (31), we have

$$\mathrm{dist}(w, z) < \varepsilon + 2\delta, \tag{34}$$

say.

We extend g_δ^i from the centres of the triangles meeting $\overline{D}$ to the entire triangles as follows: first, take the value of g_δ^i at the corner x of such a triangle to be (say) the average of the values at the centres of all triangles incident with x and meeting $\overline{D}$. Then divide each face of δT into three triangles meeting at the centre, and define g_δ^i on each of these triangles by linear interpolation. Finally, 'forget' the values of g_δ^i outside $\overline{D}$ to obtain a function with domain of definition $\overline{D}$.

We shall use two properties of the interpolating functions g_δ^i. First, each g_δ^i is a continuous function on $\overline{D}$. Second, for any i, δ and $z \in \overline{D}$, there are points w, $w' \in \delta H$ at which f_δ^i is defined, with $f_\delta^i(w) \le g_\delta^i(z) \le f_\delta^i(w')$, and $\mathrm{dist}(w, z)$, $\mathrm{dist}(w', z) < 2\varepsilon$, where $\varepsilon = \varepsilon(\delta)$ is as in (31). The first property is immediate from the definition of g_δ^i. The second follows from (34).

Claim 22. *The functions g_δ^i are uniformly equicontinuous.*

Proof. It suffices to show that the functions g_δ^1 are uniformly equicontinuous. Given $\beta > 0$, we must show that there is an $\eta > 0$ such that, for all $z, w \in \overline{D} \subset \mathbb{C}$ with $\mathrm{dist}(z, w) < \eta$ and all δ, we have $|g_\delta^1(z) - g_\delta^1(w)| < \beta$. In doing this, we may choose any $\delta_1(\beta) > 0$ and restrict our attention to the functions g_δ^1 for $\delta < \delta_1(\beta)$: whatever the values of $f_\delta^1(z)$ for $\delta \ge \delta_1(\beta)$, the linear interpolation in the definition of g_δ^1 ensures that the functions g_δ^1, $\delta \ge \delta_1(\beta)$, are uniformly equicontinuous (indeed, uniformly Lipschitz).

As no point of the 3-marked Jordan domain D_3 lies on all three arcs $A_i = A_i(D_3)$, there is a constant $c > 0$ such that

$$\max_i \mathrm{dist}(w, A_i(D_3)) > c \tag{35}$$

for any $w \in \overline{D}$.

Let us choose $\gamma > 0$ so that $\gamma < c/10$ and $(6\gamma/c)^\alpha < \beta/2$, where $\alpha > 0$ is the constant in Lemma 4. By Lemma 14, if we choose δ_1 small enough, then there is an $\eta > 0$ such that any two points x, y of G_δ^-, $\delta < \delta_1$, at distance at most η are joined by a (geometric) path in G_δ^- that stays within the ball $B_\gamma(x)$. Reducing δ_1, if necessary, we may assume that $\varepsilon(\delta) < \eta/8$ for all $\delta < \delta_1$, where $\varepsilon(\delta)$ is as in (31).

Suppose that $\delta < \delta_1$, and $w, z \in \overline{D}$ with $\mathrm{dist}(w, z) < \eta/4$. It suffices to show that $g_\delta^1(z) \leq g_\delta^1(w) + \beta$. From the second property of the interpolating functions g_δ^i listed above, there are w', $z' \in G_\delta^-$ with $\mathrm{dist}(w, w') < \eta/4$ and $\mathrm{dist}(z, z') < \eta/4$ such that $g_\delta^1(w) \geq f_\delta^1(w')$ and $g_\delta^1(z) \leq f_\delta^1(z')$. Note that $\mathrm{dist}(w', z') < \eta$. The points w', z' may be joined by a path in G_δ^- lying within $B_\gamma(w')$. As G_δ^- is 2-connected, it follows that there is a path $P' \in \delta H$ from w' to z' all of whose vertices are the centres of triangles in G_δ^-, with $P' \subset B_{2\gamma}(w')$.

Suppose that $E_\delta^1(z') \setminus E_\delta^1(w')$ holds. Then there is an edge xy of P' such that $E_\delta^1(y) \setminus E_\delta^1(x)$ holds. But then, by Claim 10, the three sites of G_δ^- immediately next to x are joined by monochromatic (i.e., all sites open, or all sites closed) paths to the three boundary arcs of G_δ^-. One of these paths must end at a distance at least $c - 2\gamma - \delta > c/2$ from w', say, where c is as in (35). Note that $\mathrm{dist}(x, w') < 2\gamma$.

We have shown that if $E_\delta^1(z') \setminus E_\delta^1(w')$ holds, then there is a monochromatic path crossing the annulus centred at w' with inner and outer radii 3γ and $c/2$. By Lemma 4 and our choice of γ, this event has probability at most

$$2\left(\frac{3\gamma}{c/2}\right)^{\alpha} \leq \beta.$$

It follows that $f_\delta^1(z') - f_\delta^1(w') \leq \beta$. Thus,

$$g_\delta^1(z) \leq f_\delta^1(z') \leq f_\delta^1(w') + \beta \leq g_\delta^1(w) + \beta,$$

as required. □

The functions g_δ^i take values in $[0, 1]$. Hence, these functions are uniformly equicontinuous and uniformly bounded. Thus, by the Arzelà–Ascoli Theorem (see, for example, Bollobás [1999, p. 90]), any subsequence of $(g_\delta^1, g_\delta^2, g_\delta^3)$ contains a subsequence that converges uniformly to a limit (g^1, g^2, g^3), with each g^i continuous.

The last two (rather substantial) pieces of the jigsaw puzzle needed to complete the proof of Smirnov's Theorem are collected in the next two claims. The first describes some crucial properties of the possible limits (g^1, g^2, g^3).

Claim 23. *Suppose, for some sequence $\delta_n \to 0$, the triples $(g_{\delta_n}^1, g_{\delta_n}^2, g_{\delta_n}^3)$*

converge uniformly to (g^1, g^2, g^3)*, with each* g^i *continuous. Then, for any contour* C *in* D*, we have*

$$\oint_C g^{i+1}(z)\,\mathrm{d}z = \omega \oint_C g^i(z)\,\mathrm{d}z, \tag{36}$$

and, for any point $z \in A_i$*, we have*

$$g^i(z) = 0 \quad and \quad g^{i+1}(z) + g^{i+2}(z) = 1, \tag{37}$$

where the superscripts are taken modulo 3.

Proof. Throughout the proof we consider only values of δ in the sequence δ_n, writing δ for δ_n to avoid cumbersome notation.

We start with (36); since the g^i are continuous functions on a compact set, they are uniformly continuous. Thus, it suffices to consider equilateral triangular contours C with sides parallel to the bonds of T: an arbitrary contour C' in D can be approximated by a sum of such contours.

Let C be an equilateral triangular contour in D with sides parallel to the bonds of T. For each δ, there is a discrete triangular contour C_δ in δT within distance δ of C. Using (31), as C is contained in the open set D, for δ sufficiently small we have $C_\delta \subset G_\delta^-$. Recall that the discrete contour integral $\oint^{\mathrm{D}}_{C_\delta} f_\delta^i(z)\,\mathrm{d}z$ is defined by integrating a function whose values on C_δ are given by values of f_δ^i at lattice points within distance δ and, at these lattice points, $f_\delta^i = g_\delta^i$. Since the functions g_δ^i are uniformly equicontinuous, it follows that

$$\oint^{\mathrm{D}}_{C_\delta} f_\delta^i(z)\,\mathrm{d}z = \oint_{C_\delta} g_\delta^i(z)\,\mathrm{d}z + o(1)$$

as $\delta \to 0$, where the second integral is the usual contour integral. Since the $g_\delta^i = g_{\delta_n}^i$ converge uniformly to g^i, we have

$$\oint_{C_\delta} g_\delta^i(z)\,\mathrm{d}z = \oint_{C_\delta} g^i(z)\,\mathrm{d}z + o(1) = \oint_C g^i(z)\,\mathrm{d}z + o(1)$$

as $\delta = \delta_n \to 0$. Combining the relations above with Lemma 13, we deduce that

$$\oint_C g^{i+1}(z)\,\mathrm{d}z = \omega \oint_C g^i(z)\,\mathrm{d}z + o(1).$$

Since neither integral depends on δ, this proves (36).

For (37), it suffices to prove the case $i = 3$, say, for a fixed $z \in A_3$.

Since the g^i are continuous, we may assume that z is not an endpoint of A_3, so

$$\operatorname{dist}(z, A_1),\ \operatorname{dist}(z, A_2) > a \tag{38}$$

for some $a > 0$.

As $\delta \to 0$, the 3-marked domain G_δ^- is ε-close to D_3 with $\varepsilon = \varepsilon(\delta) \to 0$. Hence, there are points z_δ of G_δ^- with $z_\delta \in D$ and $z_\delta \to z$. Shifting z_δ by at most 2δ, we may assume that $z_\delta \in \delta H$ is the centre of a triangle $x_\delta y_\delta u_\delta$ in G_δ^-. Thus, $f_\delta^i(z_\delta)$ is defined, and, as $g_\delta^i \to g^i$ uniformly,

$$f_\delta^i(z_\delta) = g_\delta^i(z_\delta) = g^i(z) + o(1) \tag{39}$$

as $\delta \to 0$.

From (31) we may choose a point w_δ on the boundary of G_δ^- (viewed as a union of hexagons) so that, as $\delta \to 0$, we have $w_\delta \to z$, and hence $\operatorname{dist}(w_\delta, z_\delta) \to 0$. Let $x_\delta \in G_\delta^-$ be a site of the inner boundary of G_δ^- in whose hexagon w_δ lies, so $\operatorname{dist}(x_\delta, w_\delta) < \delta$. Note that if $z = P_4$, then we may take x_δ to be v_4. The Hausdorff distance between the discrete boundary arc $A_i(G_\delta^-)$ and the continuous arc $A_i(D_3)$ tends to zero as $\delta \to 0$. Thus, from (38), for δ sufficiently small, the site x_δ lies on the arc A_3 of G_δ^- (viewed as a 3-marked domain), and the point w_δ on the corresponding arc of the boundary ∂G_δ^- of the union G_δ^- of hexagons.

Suppose that $E_\delta^3(z_\delta)$ holds. Then there is an open path from $A_1(G_\delta^-)$ to $A_2(G_\delta^-)$ separating z_δ from $A_3^+(G_\delta^-)$, and, in particular, separating z_δ from w_δ. Such a path must cross the line segment $z_\delta w_\delta$. It follows that some site of G_δ^- within distance $\operatorname{dist}(z_\delta, w_\delta) + 2\delta = o(1)$ of z_δ is joined by an open path to some site at distance at least $a - o(1)$ from z_δ. By Lemma 4, this event has probability $o(1)$, so

$$f_\delta^3(z_\delta) = \mathbb{P}\big(E_\delta^3(z_\delta)\big) = o(1).$$

Using (39), it follows that $g^3(z) = 0$.

Recall that x_δ is a boundary site of $A_3(G_\delta^-)$ within distance δ of w_δ. Let E_δ be the event that some site in $B_{\operatorname{dist}(z_\delta, w_\delta)+10\delta}(z_\delta)$ is joined by an open path to $A_1(G_\delta^-) \cup A_2(G_\delta^-)$. As above, $\mathbb{P}(E_\delta) = o(1)$. Let G'_δ be the 4-marked domain obtained from the 3-marked domain G_δ^- by taking x_δ as the fourth marked point. Note that if $z = P_4$, so $x_\delta = v_4$, then we recover the original 4-marked domain G_δ^-.

If E_δ does not hold, then no open path from $A_1(G'_\delta)$ to $A_3(G'_\delta)$ can come within distance 2δ of the line segment $z_\delta w_\delta$. Hence, $E_\delta^2(z_\delta)$ holds if and only if G'_δ has an open crossing from $A_1(G'_\delta) = A_1(G_\delta^-)$ to $A_3(G'_\delta)$; see Figure 24.

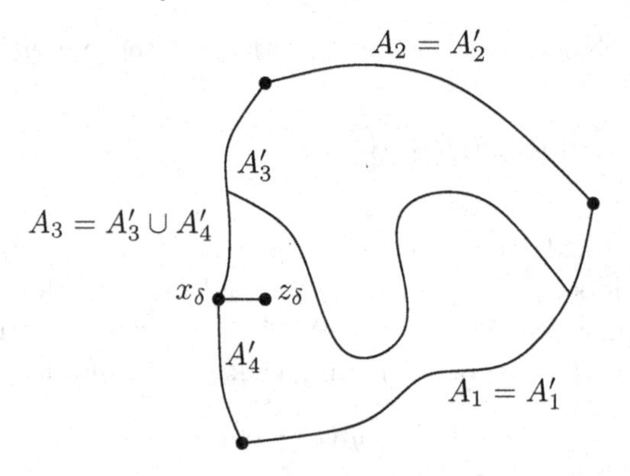

Figure 24. A 3-marked discrete domain G_δ^- with boundary arcs A_i, $1 \le i \le 3$, and the corresponding 4-marked domain G_δ' with boundary arcs A_i', $i \le i \le 4$, obtained by marking a point x_δ on A_3. If no open path joins $A_1 \cup A_2$ to the line segment $x_\delta z_\delta$, then an open path from A_1 to A_3 separates z_δ from A_2^+ if and only if it ends at a site of A_3'.

Hence,

$$\mathbb{P}\big(G_\delta' \text{ has an open crossing from } A_1' \text{ to } A_3'\big) = f_\delta^2(z_\delta) + o(1). \qquad (40)$$

Similarly, if E_δ does not hold, then $E_\delta^1(z_\delta)$ holds if and only if G_δ' has an open crossing from A_2' to A_4', so

$$\mathbb{P}\big(G_\delta' \text{ has an open crossing from } A_2' \text{ to } A_4'\big) = f_\delta^1(z_\delta) + o(1).$$

Using Lemma 5, it follows that

$$f_\delta^1(z_\delta) + f_\delta^2(z_\delta) = 1 - o(1).$$

Appealing to (39) again, $g^1(z) + g^2(z) = 1$ follows, completing the proof of the claim. □

As before, let $\omega = (-1 + \sqrt{-3})/2$ be a cube root of unity.

Claim 24. *Let Δ be the equilateral triangle with vertices $v_1 = 1$, $v_2 = \omega$ and $v_3 = \omega^2$. There is a unique triple (g^1, g^2, g^3) of continuous functions on $\overline{D}$ taking values in $[0,1]$ and satisfying (36) and (37). Furthermore, $g^i(z) = h^i(\varphi(z))$, where h^i is the linear function on Δ with $h^i(z)$ equal to two-thirds of the distance from z to the ith side of Δ, and φ is the unique conformal map from D to Δ whose continuous extension to Γ maps P_i to v_i for $1 \le i \le 3$.*

Proof. Let (g^1, g^2, g^3) be continuous functions on $\overline{D}$ satisfying (36) and

(37), and set $g = g^1 + \omega g^2 + \omega^2 g^3$. Then g is continuous on $\overline{D}$. For any contour C in D we have

$$\oint_C g = \oint_C g^1 + \omega \oint_C g^2 + \omega^2 \oint_C g^3 = \oint_C g^1 + \omega^2 \oint_C g^1 + \omega^4 \oint_C g^1 = 0.$$

Hence, by Morera's Theorem (see, e.g., Beardon [1979, p. 166]), the function g is analytic in D.

On the boundary of D, condition (37) ensures that, for $z \in A_i$, the value of $g(z)$ is a convex combination of ω^{i+1} and ω^{i+2}, i.e., that g maps z into the line segment $v_{i+1}v_{i+2}$. Since g is continuous, it follows that g maps P_i into v_i. Furthermore, as z traces the boundary of D in the anticlockwise direction, $g(z)$ remains within $\partial\Delta$ and winds exactly once around this boundary. As noted by Beffara [2005], it follows that g is a conformal map from D to Δ. Indeed, applying the argument principle (see, e.g., Beardon [1979, p. 127]), the equation $g(z) = w$ has a unique solution for each $w \in \Delta$, and no solution for w outside $\overline{\Delta}$. Since g maps P_i to v_i, the maps g and φ are identical.

Arguing similarly, the function $g^1+g^2+g^3$ is analytic in D, continuous on the boundary, and takes the constant value 1 on the boundary. Thus $g^1 + g^2 + g^3$ is identically 1. This means that the real-valued functions g^i are determined by g: for example,

$$g^1 = (2g^1 - g^2 - g^3)/3 + (g^1 + g^2 + g^3)/3 = 2\mathrm{Re}(g)/3 + 1/3.$$

As $g = \varphi$, this shows that the triple (g^1, g^2, g^3) is uniquely determined. The triple given by $g^i(z) = h^i(\varphi(z))$ satisfies conditions (36) and (37), so the claim follows. □

We now have all the pieces in place to prove Smirnov's Theorem.

Proof of Theorem 2. Let $D_4 = (D; P_1, P_2, P_3, P_4)$ be a 4-marked Jordan domain. Let G_δ^- and G_δ^+ be defined as in Lemma 14, for all $0 < \delta < \delta_0$, where $\delta_0 > 0$ is constant. It suffices to show that $P_\delta(G_\delta^-) \to \pi(D_4)$, since the same argument shows that $P_\delta(G_\delta^+) \to \pi(D_4)$, and then $P_\delta(D_4, T) \to \pi(D_4)$ follows from (19).

Let f_δ^i and g_δ^i be defined as above, using the discrete domains G_δ^-. We claim that $g_\delta^i \to h^i \circ \varphi$, where h^i and φ are defined as in Claim 24. Suppose, for a contradiction, that this is not the case, i.e., that there is an $\varepsilon > 0$ and a sequence (δ_n) such that for each n, there are $1 \le i \le 3$ and $z \in \overline{D}$ with

$$|g_{\delta_n}^i(z) - h^i(\varphi(z))| \ge \varepsilon. \tag{41}$$

By Claim 22, the functions g_δ^i are uniformly equicontinuous on the compact set $\overline{D}$. Since they are also uniformly bounded, the sequence $(g_{\delta_n}^1, g_{\delta_n}^2, g_{\delta_n}^3)$ has a uniformly convergent subsequence. By Claims 23 and 24, this subsequence converges to $(h^1 \circ \varphi, h^2 \circ \varphi, h^3 \circ \varphi)$, contradicting (41).

The proof of Claim 23 shows that there are points $z_\delta \in \delta H \cap G_\delta^-$ with $z_\delta \to P_4$ such that

$$P_\delta(G_\delta^-) = f_\delta^2(z_\delta) + o(1) = g_\delta^2(z_\delta) + o(1) = h^2(\varphi(P_4)) + o(1).$$

Indeed, the first equation is exactly (40) for $z = P_4$; as noted above, the domain G'_δ appearing in (40) is exactly the 4-marked domain G_δ^- in this case. As $h^2(\varphi(P_4)) = \pi(D_4)$ by definition, the proof is complete. □

Site percolation on the triangular lattice is the only standard percolation model for which conformal invariance is known; there are some non-standard models to which either Smirnov's Theorem, or its proof, has been adapted. Indeed, Camia, Newman and Sidoravicius [2002; 2004] have established conformal invariance for certain dependent site percolation models on the triangular lattice, obtained by running a particular deterministic cellular automaton from an initial state given by independent site percolation. They establish conformal invariance by showing that the dependent model is in a certain sense a 'small perturbation' of the independent model, and then applying Smirnov's Theorem.

Chayes and Lei [2006] defined a rather unusual class of percolation models based on dependent bond percolation on the triangular lattice, and proved the equivalent of Smirnov's Theorem for these models, by translating Smirnov's proof to this context. It remains an important challenge to prove conformal invariance for any other standard model, for example, for site or bond percolation on the square lattice, or for Gilbert's model, or random Voronoi percolation in the plane (see Chapter 8).

7.3 Critical exponents and Schramm–Loewner evolution

There is a widely held belief that the behaviour of various 'critical phenomena', including critical percolation, should be characterized by certain 'critical exponents'. This opinion originated among theoretical physicists, but by now many mathematicians have been converted. For percolation, these critical exponents should depend on the dimension, but not on the details of the particular model considered. This is a very substantial topic in its own right; a detailed discussion is beyond the scope of this book. Here, we shall briefly state the main rigorous results for percolation: the existence and values of these critical exponents for site percolation on the triangular lattice follow from Smirnov's Theorem and the work of Lawler, Schramm and Werner. Our presentation is based on that of Smirnov and Werner [2001]; we refer to the reader to their paper for further details and full references.

To describe the exponents associated to critical percolation we need a few definitions. We shall consider only site percolation on the triangular lattice, although the definitions make sense in a much broader context. We write $\mathbb{P}_p = \mathbb{P}^{\mathrm{s}}_{T,p}$ for the probability measure in which each site of the triangular lattice T is open with probability p, and the states of the sites are independent. As before, we write C_0 for the *open cluster of the origin*, i.e., the largest connected subgraph of T containing 0, all of whose sites are open. For the critical exponents, we use standard notation; see, e.g., Kesten [1987c].

Recall that the *percolation probability* $\theta(p)$ is defined as

$$\theta(p) = \mathbb{P}_p\big(C_0 \text{ is infinite}\big),$$

and that $\theta(p) = 0$ if $p \le p_{\mathrm{c}} = p^{\mathrm{s}}_{\mathrm{H}}(T) = 1/2$, while $\theta(p) > 0$ if $p > 1/2$. It is believed that, in a very general setting, the behaviour of $\theta(p)$ as p tends to p_{c} from above follows a power law, i.e., that

$$\theta(p) = (p - p_{\mathrm{c}})^{-\beta + o(1)} \quad \text{as } p \to p_{\mathrm{c}} \text{ from above}, \tag{42}$$

where $\beta > 0$ is a constant that depends on the dimension but not on the details of the percolation model.

Turning to the size of the open cluster C_0 when it is finite, as earlier, we write $\chi(p) = \mathbb{E}_p(|C_0|)$ for the expected size of (number of sites in) C_0. As $\chi(p) = \infty$ for $p > p_{\mathrm{c}}$, it is rather more informative to modify the definition slightly, and consider

$$\chi^f(p) = \mathbb{E}_p\big(|C_0|\mathbf{1}_{|C_0|<\infty}\big) = \sum_{n=0}^{\infty} n\mathbb{P}_p\big(|C_0| = n\big),$$

so that $\chi^f(p) = \chi(p)$ if $p < p_{\mathrm{c}}$. Again, power-law behaviour is expected:

$$\chi^f(p) = |p - p_{\mathrm{c}}|^{-\gamma + o(1)} \quad \text{as} \quad p \to p_{\mathrm{c}}, \tag{43}$$

where γ is a positive constant, and $p \to p_{\mathrm{c}}$ from either side.

Furthermore, at the critical probability, the tail of the distribution of

$|C_0|$ is expected to follow a power law:

$$\mathbb{P}_{p_c}\big(n \le |C_0| < \infty\big) = n^{-1/\delta + o(1)} \quad \text{as } n \to \infty, \tag{44}$$

as is the tail of the distribution of the radius $r(C_0)$:

$$\mathbb{P}_{p_c}\big(n \le r(C_0) < \infty\big) = n^{-1/\delta_r + o(1)} \quad \text{as } n \to \infty. \tag{45}$$

In the last definition, when considering a lattice in $\mathbb{R}^d$, it makes no difference whether we measure the radius $r(C_0)$ of C_0 in the graph-theoretic or the geometric sense, since these two metrics are equivalent. More precisely, we may take $r(C_0)$ to be the maximum graph distance of a site $x \in C_0$ from 0, as before, or we may take $r(C_0)$ to be the maximum Euclidean distance of a site in C_0 from the origin. For site percolation on the triangular lattice, we have $\theta(p_c) = 0$, so the condition $|C_0| < \infty$ can be omitted in (44) and (45).

The *correlation length* describes the 'typical radius' of an open cluster: writing $|y|$ for the Euclidean distance between $y \in T$ and the origin, let

$$\xi(p) = \left[\frac{1}{\chi^f(p)} \sum_{y \in T} |y|^2 \mathbb{P}_p(\{0 \to y\} \cap \{|C_0| < \infty\}) \right]^{1/2}.$$

It is conjectured that

$$\xi(p) = |p - p_c|^{-\nu + o(1)} \quad \text{for } p \ne p_c. \tag{46}$$

The reason for the term 'correlation length' is that $\xi(p)$ is expected to be closely related to the probability that two sites at a given distance ℓ are in the same finite open cluster: roughly speaking, this probability should decay exponentially with $\ell/\xi(p)$.

Finally, it is expected that, at the critical probability, we have

$$\mathbb{P}_{p_c}(0 \to y) = |y|^{2-d-\eta+o(1)} \quad \text{as } y \to \infty, \tag{47}$$

where d is the dimension (so $d = 2$ for percolation on the triangular lattice T).

The constants β, γ, δ, δ_r, ν and η defined above are called *critical exponents*, provided they exist. We have used standard notation for these exponents (though δ_r is also written as ρ); the form of (47), for example, shows that this notation is not the most natural for percolation.

In two dimensions, it is not hard to deduce from the Russo–Seymour–Welsh Theorem that if one of η and δ_r exists, then so does the other, and

$$d - 2 + \eta = 2/\delta_r. \tag{48}$$

Indeed, let x_1, x_2 be two points at some (large) distance r. If there is an open path joining x_1 and x_2, then each x_i must be joined by an open path to the boundary of the ball $B_{r/3}(x_i)$, say. But with probability bounded away from zero there are open cycles in these two balls surrounding the centres that are joined to each other; relation (48) then follows using Harris's Lemma (Lemma 3 of Chapter 2).

Kesten [1987b; 1987c] established highly non-trivial relationships between the various exponents for two-dimensional percolation. Firstly, building on his work on the 'incipient infinite cluster' (Kesten [1986]), he showed that if η exists (or, equivalently, if δ_r exists), then so does δ, with $\eta(\delta+1) = 4$. Furthermore, he showed that if the exponents ν and δ exist, then so do the other exponents, and

$$\beta = \frac{2\nu}{\delta+1}, \quad \gamma = 2\nu\frac{\delta-1}{\delta+1}, \quad \text{and} \quad \eta = \frac{4}{\delta+1}.$$

In particular, for two-dimensional percolation the 'scaling relations'

$$\gamma + 2\beta = \beta(\delta+1) \quad \text{and} \quad \gamma = \nu(2-\eta), \tag{49}$$

hold, as do the 'hyperscaling relations'

$$d\delta_r = \delta + 1 \quad \textit{and} \quad 2 - \eta = d\frac{\delta-1}{\delta+1}. \tag{50}$$

It is believed that (49) holds in all dimensions, and that (50) holds for $d \le 6$; see Grimmett [1999]. A very large number of papers have been written about the critical exponents associated to percolation and the relationships between them; see, for example, Rudd and Frisch [1970], Wu [1978], Kesten [1981], Aizenman and Newman [1984], Durrett [1985], Chayes and Chayes [1987], Tasaki [1987], Kesten and Zhang [1987], Kesten [1987a; 1988] and Hammond [2005]. Borgs, Chayes, Kesten and Spencer [1999] proved the deep result that the hyperscaling relations (50) hold as long as two assumptions are satisfied: δ_r exists, and, at $p = p_c$, the crossing probabilities for cuboids with fixed aspect ratios are bounded away from 1. (Their results are stated for bond percolation in $\mathbb{Z}^d$, but proved in a more general setting.) The latter assumption is expected to hold for $d < 6$.

Returning to two dimensions, Schramm [2000] studied a certain scaling limit of 'loop-erased random walks' in the plane, which we shall not define. He defined a family of random curves in a domain in the plane, whose distribution depends on a real parameter κ, which he called the *stochastic Loewner evolution* with parameter κ, and denoted SLE_κ. The random curve SLE_κ is often known by the name *Schramm–Loewner evolution*. Schramm showed that if, as conjectured, the scaling limit of the loop-erased random walk is conformally invariant, then it must be SLE_2. Furthermore, he showed that SLE_6 is the only possible conformally invariant 'scaling limit' of critical percolation on a lattice, in a sense that we shall not make precise.

Smirnov [2001a] proved that critical site percolation on T does indeed have a scaling limit, and that this limit is conformally invariant and thus equal to SLE_6. Considering the face percolation on the hexagonal lattice corresponding to site percolation on the triangular lattice, it follows that, for $p = 1/2$, the long-range behaviour (i.e., limiting behaviour as the lattice spacing tends to zero) of interfaces between open and closed (black and white) regions converges to SLE_6.

In a series of papers, Lawler, Schramm and Werner [2001c; 2001d; 2002b; 2002d] studied the behaviour of SLE; in particular, they determined various 'critical exponents' associated to SLE_6. Combining these results with those of Smirnov [2001a], Smirnov and Werner [2001] established the existence and values of the various critical exponents for site percolation on T.

Theorem 25. *For site percolation on the triangular lattice, the critical exponents* β, γ, ν *and* η *defined implicitly by* (42), (43), (46) *and* (47) *exist and take the values*

$$\beta = \frac{5}{36}, \quad \gamma = \frac{43}{18}, \quad \nu = \frac{4}{3} \quad \textit{and} \quad \eta = \frac{5}{24}. \qquad \square$$

These values for the critical exponents coincide with the predictions of theoretical physicists; see Kesten [1987c], Smirnov and Werner [2001] and the references therein. As noted above, $\delta_r = \eta/2 = 5/48$ follows relatively easily, and $\delta = 91/5$ follows from the results of Kesten [1987b].

The proof of Theorem 25 is based on crossings of annuli. To say a few words about this proof, denote by $A^i_{r,R}$ the event that in the site percolation on T restricted to the disc $B_R(0)$, there are (at least) i distinct open clusters each of which connects a site in $B_r(0)$ to a site near the boundary of $B_R(0)$. For fixed i, the asymptotic behaviour of $\mathbb{P}_{1/2}(A^i_{r,R})$ does not depend on r, as long as r is large enough for this probability to be positive. Writing A^i_R for $A^i_{2,R}$, say, Smirnov and Werner noted that, by the results of Kesten [1987c], Theorem 25 follows from the two relations

$$\mathbb{P}_{1/2}(A^1_R) = R^{-5/48+o(1)} \tag{51}$$

and

$$\mathbb{P}_{1/2}(A_R^2) = R^{-5/4+o(1)}. \tag{52}$$

Clearly, the first of these relations states simply that δ_r exists and takes the value $5/48$.

Lawler, Schramm and Werner [2002b] showed that relation (51) follows from Smirnov's conformal invariance results and their results on SLE_6. Smirnov and Werner [2001] then proved (52), and thus Theorem 25.

In fact, Smirnov and Werner proved a more general statement about crossings of the annulus $A(r, R)$ centred at the origin, with inner and outer radii r and R respectively. As before, we think of open paths as black and closed paths as white. Given a sequence $\mathbf{c} = (c_i)_{i=1}^j \in \{B, W\}^j$ of colours, let $H_{\mathbf{c}}(r, R)$ be the event that $A(r, R)$ contains j vertex-disjoint monochromatic paths $P_1, \ldots, P_j$, where P_i has colour c_i, each P_i starts at a site of $A(r, R)$ adjacent to the inside of the annulus and ends at a site adjacent to the outside, and the initial sites v_i of the P_i appear in the cyclic order $v_1, v_2, \ldots, v_j$ around the circle of radius r. (As the paths do not cross, their final vertices appear in the same cyclic order around the outer circle.) Let $G_{\mathbf{c}}(r, R)$ be the event corresponding to $H_{\mathbf{c}}(r, R)$, but defined in the *half-annulus*

$$A^+(r, R) = \{z \in \mathbb{C} : r < |z| < R, \operatorname{Re}(z) > 0\}.$$

It is not hard to see that, for critical site percolation on T, the probability of the event $G_{\mathbf{c}}(r, R)$ does not depend on the terms of the sequence $\mathbf{c}$, only on its length. Indeed, the proof of Lemma 5 allows us to define a 'lowest' (clockwise-most) open crossing of $A^+(r, R)$ from the inner circle to the outer circle, whenever such a crossing exists. Having found such a crossing, P_1, we may then look for a lowest open or closed crossing above P_1 in the same way. The lowest crossing P_1 may be found without examining the states of sites above P_1, so when we look for the next crossing, the probability of success does not depend on whether it is an open or a closed crossing that we seek. It follows similarly that

$$\mathbb{P}_{1/2}\big(G_{\mathbf{c}}(r, R)\big) = a_j(r, R)$$

for some $a_j(r, R)$ that depends on the length of $\mathbf{c}$ but not the actual sequence. Smirnov and Werner [2001] showed that, if $j \geq 1$ is fixed and r is large enough, then

$$a_j(r, R) = R^{-j(j+1)/6+o(1)} \tag{53}$$

as $R \to \infty$. This exponent $j(j+1)/6$ is known as the *j-arm exponent in the half-plane.*

Returning to the full annulus, one can show (see Aizenman, Duplantier and Aharony [1999]) that $\mathbb{P}_{1/2}\big(H_{\mathbf{c}}(r,R)\big)$ is independent of the colours in the sequence $\mathbf{c}$, *provided* $\mathbf{c}$ *contains at least one* B *and at least one* W. This is related to Lemma 6 above: if the sequence contains two terms of opposite colours, then it contains two consecutive such terms, and one can start by searching for an 'innermost' pair of paths of opposite colours. Then, working outwards, each remaining path may be found as the lowest crossing of a certain region, and the probability of finding the next path does not depend on the colour.

Writing $b_j(r,R)$ for $\mathbb{P}_{1/2}\big(H_{\mathbf{c}}(r,R)\big)$, where $\mathbf{c}$ is any sequence of length j containing at least one B and at least one W, Smirnov and Werner [2001] showed that if $j \geq 2$ is fixed and r is large enough, then

$$b_j(r,R) = R^{-(j^2-1)/12+o(1)} \tag{54}$$

as $R \to \infty$. This exponent $(j^2-1)/12$ is the *(multichromatic) j-arm exponent in the plane.* The values of these exponents, and of the half-plane exponents, were predicted correctly by physicists; see, for example, Saleur and Duplantier [1987], Aizenman, Duplantier and Aharony [1999] and the references therein.

If one is careful with the exact definitions at the boundary (rather than glossing over them as we do here in our brief description of the results), then the events $A^k_{r,R}$ and $H_{\mathbf{c}}(r,R)$ coincide, where $\mathbf{c}$ is the alternating sequence of length $2k$: the k black paths correspond to the k open clusters joining the inner and outer circles of $A(r,R)$, and the white paths witness the fact that these clusters are disjoint. Hence, the case $j=4$ of (54) is exactly (52).

Without going into the details, we shall say a few words about the proofs of (53) and (54), and hence of Theorem 25. As noted earlier, the key elements are the result of Smirnov [2001a] that the (suitably defined) scaling limit of an interface in critical site percolation on T exists and is equal to SLE_6, and the results of Lawler, Schramm and Werner on the behaviour of SLE_6. Putting these ingredients together, one can show that

$$a_j(r,R) = (R/r)^{-j(j+1)/6+o(1)} \tag{55}$$

as $R, r \to \infty$ with R/r fixed. (In fact, to obtain (55), one first needs an 'a priori' bound of the form $a_3(r,R) = O(R^{-1-\varepsilon})$ as $R \to \infty$, for constants r and $\varepsilon > 0$; see Smirnov and Werner [2001] and the references therein.)

To obtain (53), one also needs 'approximate multiplicativity' that, for $r_1 < r_2 < r_3$,

$$a_j(r_1, r_2)a_j(r_2, r_3) = \Theta(a_j(r_1, r_3));$$

see Kesten [1987c] and Kesten, Sidoravicius and Zhang [1998]. When considering paths of the same colour, this relation is fairly easy to derive from the Russo–Seymour–Welsh Theorem. Fortunately, for half-annuli, one can assume that all paths have the same colour. The corresponding relation for b_j is much harder; see Smirnov and Werner [2001].

Finally, let us note that, in the annulus, the restriction that not all paths be the same colour really does seem to matter. It is likely that

$$\mathbb{P}_{1/2}\big(H_{\mathbf{c}}(r, R)\big) = R^{-\gamma_j + o(1)}$$

as $r \to \infty$ with r and j fixed, where $\mathbf{c}$ is a sequence of length j with $c_i = B$ for every i. However, these 'monochromatic' exponents γ_j are very likely different from the multichromatic exponents $(j^2 - 1)/12$ above. In particular, Grassberger [1999] reported numerical evidence that $\gamma_2 = 0.3568 \pm 0.0008$. The numerical value, or even the existence, of γ_j for $j \geq 2$ is not known, although Lawler, Schramm and Werner [2002b] showed that γ_2 exists and is equal to the maximum eigenvalue of a certain differential operator.

Critical percolation is not the only discrete object known to have a conformally invariant scaling limit described by SLE: Lawler, Schramm and Werner [2004] have shown that one may define certain natural scaling limits of loop-erased random walks in the plane and of uniformly chosen spanning trees in a planar lattice; these are related to SLE_2 and to SLE_8, respectively.

The brief remarks in this section hardly scratch the surface of the the theory that has grown out of conformal invariance and the study of SLE. For a selection of related results see, for example, Schramm [2001a], Lawler, Schramm and Werner [2001a; 2001b; 2002a; 2002c; 2003], Kleban and Zagier [2003], Beffara [2004], Dubédat [2004], Zhan [2004], Morrow and Zhang [2005], and Rohde and Schramm [2005]. Informative surveys of the field have been written by Schramm [2001b], Lawler [2004; 2005], Werner [2004; 2005], Kager and Nienhuis [2004] and Cardy [2005].

8
Continuum percolation

Shortly after Broadbent and Hammersley started percolation theory and Erdős and Rényi [1960; 1961a], together with Gilbert [1959], founded the theory of random graphs, Gilbert [1961] started a closely related area that is now known as *continuum percolation.* The basic objects of study are *random geometric graphs*, both finite and infinite. Such graphs model, for example, a network of transceivers scattered at random in the plane or a planar domain, each of which can communicate with those others within a fixed distance.

Although this field has attracted considerably less attention than percolation theory, its importance is undeniable; in this single chapter, we cannot do justice to these topics. Indeed, this area has been treated in hundreds of papers and several monographs, including Hall [1988] on coverage processes, Møller [1994] on random Voronoi tessellations, Meester and Roy [1996] on continuum percolation, and Penrose [2003] on random geometric graphs. These topics are also touched upon in the books by Matheron [1975], Santaló [1976], Stoyan, Kendall and Mecke [1987; 1995], Ambartzumian [1990] and Molchanov [2005].

In the first section we present the most basic model of continuum percolation, the *Gilbert disc model* or *Boolean model*, and give some fundamental results on it, including bounds on the critical area. In the second section we take a brief look at finite random geometric graphs, with emphasis on their connectedness. The most important part of the chapter is the third section, in which we shall sketch a proof of the analogue of the Harris–Kesten result for continuum percolation: the critical probability for random Voronoi percolation in the plane is $1/2$.

We shall frequently encounter sequences of events (A_n) with $\mathbb{P}(A_n) \to 1$ as $n \to \infty$; using standard shorthand, we say that A_n holds *whp* or *with high probability* in this case. As usual, an event holds *almost surely*, or a.s., if it has probability 1.

8.1 The Gilbert disc model

For $r > 0$, the vertex set of the *standard Gilbert disc model*, or the *Boolean model*, G_r, is a set of points distributed 'uniformly' in the plane, with density 1. To obtain G_r, join two points by an edge if the distance between them is at most r. The trouble with this 'definition' is that it is not clear how we can choose points uniformly, with a certain density. In fact, it is easy to turn this hopelessly loose idea into a definition of a Poisson process in the plane, the 'proper' way of selecting points uniformly.

Let λ be a positive real number, and let $\mathcal{P}_\lambda \subset \mathbb{R}^2$ be a random countably infinite set of points in the plane. Let us write $\mu_\lambda(U)$ for the number of points of $\mathcal{P}_\lambda$ in a bounded Borel set U; note that $\mu_\lambda(U)$ is a random variable. We call $\mathcal{P}_\lambda$ a *homogeneous Poisson process of intensity (density)* λ if the following two conditions hold.

(i) If $U_1, \ldots, U_n$ are pairwise disjoint bounded Borel sets, then the random variables $\mu_\lambda(U_1), \ldots, \mu_\lambda(U_n)$ are independent;
(ii) For every bounded Borel set U, the random variable $\mu_\lambda(U)$ is a Poisson random variable with mean $\lambda|U|$, where $|U|$ is the standard (Lebesgue) measure of U.

It is easily seen that there is at most one random point process (a random countably infinite subset of the plane) satisfying these conditions; in fact, as pointed out by Rényi in the 1950s, condition (ii) alone defines the Poisson process $\mathcal{P}_\lambda$. In the other direction, it is not hard to show that *there is* a point process satisfying conditions (i) and (ii); for example, one can use the following concrete construction.

For $\lambda > 0$, let $\{X_{i,j} : (i,j) \in \mathbb{Z}^2\}$ be independent Poisson random variables, each with mean λ. Thus,

$$\mathbb{P}(X_{i,j} = k) = e^{-\lambda}\frac{\lambda^k}{k!}$$

for $k = 0, 1, \ldots$. Let $Q_{i,j}$ be the unit square with bottom left vertex $(i,j) \in \mathbb{Z}^2$:

$$Q_{i,j} = \{(x,y) : i \le x \le i+1,\ j \le y \le j+1\}.$$

For every $(i,j) \in \mathbb{Z}^2$, select $X_{i,j}$ points independently and uniformly from $Q_{i,j}$: then the union of all these sets has properties (i) and (ii), so we may take this as the definition of $\mathcal{P}_\lambda$.

Although we shall not study it here, let us remark in passing that a Poisson process $\mathcal{P}_f$ of intensity f satisfies (i) and (ii), except that the mean of $\mu_f(U)$ is given by $\int_U f$, where f is a non-negative Lebesgue integrable function.

Returning to $\mathcal{P}_\lambda$, note that if Z is a Borel set of measure 0, then the probability that $\mathcal{P}_\lambda$ has *any* point in Z is 0; hence, in what follows, we shall assume that $\mathcal{P}_\lambda \cap Z = \emptyset$ for all measure 0 sets Z that we consider. For example, we shall assume that no point of $\mathcal{P}_\lambda$ is on a given (fixed) polygon; a little more generally, given a plane lattice, every point of $\mathcal{P}_\lambda$ will be assumed to be an interior point of a face.

For $\lambda > 0$ and $r > 0$, let $G_{r,\lambda}$ be the random geometric graph whose vertex set is $\mathcal{P}_\lambda$, with an edge joining two points of $\mathcal{P}_\lambda$ if they are at distance at most r; see Figure 1. We call $G_{r,\lambda}$ the *Gilbert model with parameters r and λ*, or the *Boolean model with parameters r and λ*. With this notation, the standard Gilbert (or Boolean) model is $G_r = G_{r,1}$.

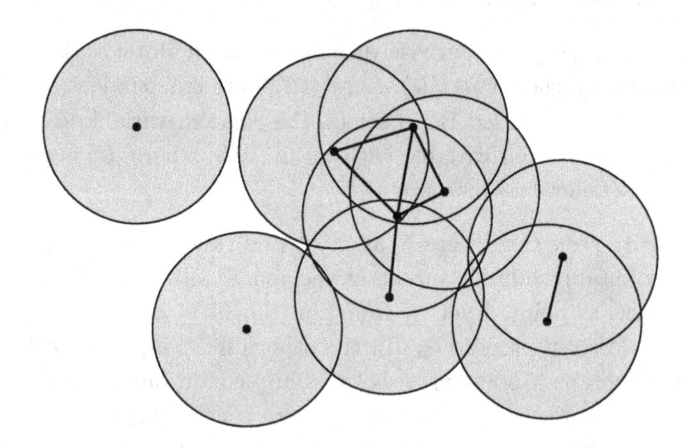

Figure 1. Part of the graph $G_{r,\lambda}$ (dots and lines). Two vertices are joined if they are within distance r, i.e., if each lies in the shaded circle centred on the other.

The model $G_{r,\lambda}$ has numerous variants and extensions. First, let us rescale r by writing $H_{r,\lambda}$ for $G_{2r,\lambda}$. This rescaling is not entirely pointless, since it allows us to define a random subset of $\mathbb{R}^2$ in a natural way, by writing $D_{r,\lambda} = D_{r,\lambda}(\mathcal{P}_\lambda)$ for the union of the discs of radius r about the points of $\mathcal{P}_\lambda$; see Figure 2. Note that the probability that there are two points of our Poisson process at distance exactly r is 0, so it makes no difference whether we take open discs or closed discs. There is a one-to-one correspondence between the components of $G_{2r,\lambda} = G_{2r,\lambda}(\mathcal{P}_\lambda)$ and those of $D_{r,\lambda} = D_{r,\lambda}(\mathcal{P}_\lambda)$; in particular, as $\mathcal{P}_\lambda$ has (a.s.)

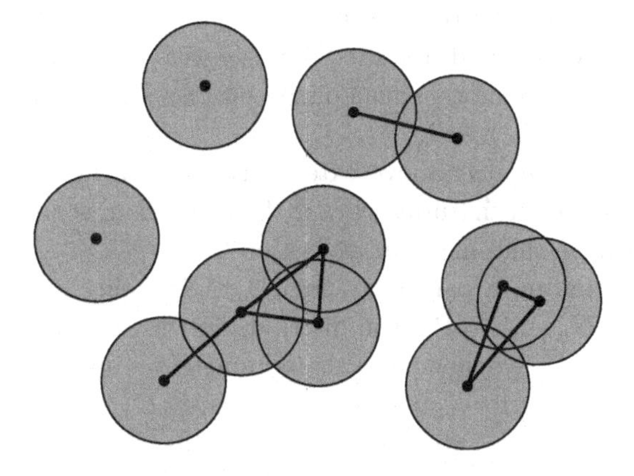

Figure 2. Part of the graph $G_{2r,\lambda}$ (dots and lines). The shaded region is $D_{r,\lambda}$. Two vertices of $G_{2r,\lambda}$ are adjacent if and only if the corresponding shaded discs meet.

no accumulation points, $G_{2r,\lambda}$ has an infinite component if and only if $D_{r,\lambda} \subset \mathbb{R}^2$ has an unbounded component. Note that although $D_{r,\lambda}$ is very close to $H_{r,\lambda}$, the two models are not isomorphic, since the disc of radius r about a point may be covered by other discs making up $D_{r,\lambda}$. The advantage of considering $D_{r,\lambda}$ rather than $G_{2r,\lambda}$ is that the complement of $D_{r,\lambda}$, the 'empty (or vacant) space' $E_{r,\lambda} = \mathbb{R}^2 \setminus D_{r,\lambda}$ not covered by the discs, is just as interesting as the original set $D_{r,\lambda}$. In particular, we can study the the component structure of $E_{r,\lambda}$ as well: we can look for an unbounded vacant component, i.e., an unbounded component of $E_{r,\lambda}$.

We may define variants of the graphs $G_{r,\lambda}$ and $H_{r,\lambda}$, and sets $D_{r,\lambda}$ and $E_{r,\lambda}$, by replacing the circular disc by an arbitrary centrally symmetric subset of $\mathbb{R}^2$. Thus, given an open symmetric set $A \subset \mathbb{R}^2$, for $W \subset \mathbb{R}^2$ we write $G_A(W)$ for the graph with vertex set W in which two vertices x and y are joined if $x - y \in A$. Equivalently, for $x \in \mathbb{R}^2$, set $B_x = B + x$, where $B = \frac{1}{2}A$, and join two points x and y of W if $B_x \cap B_y \neq \emptyset$. Again, the union $\bigcup_{x \in W} B_x$ reflects the structure of $G_A(W)$. If A is the disc of radius r, then we may write $G_r(W)$ for $G_A(W)$. In two of the most natural variants A is taken to be a square and an annulus. Of course, the model also generalizes to d dimensions in a natural way.

Taking for W the point set of a homogeneous Poisson process $\mathcal{P}_\lambda$ of intensity λ, and letting A be the disc of radius r centred at the origin, we see that $G_A(\mathcal{P}_\lambda) = G_{r,\lambda}$. More generally, in $G_A(W)$ both A and W

may be chosen to be random: one of the simplest cases is when we assign independent, identically distributed non-negative random variables $r(x)$ to the points x of a homogeneous Poisson process $\mathcal{P}_\lambda$, and take the union of the discs $B_{r(x)}(x)$, $x \in \mathcal{P}_\lambda$.

Clearly, all the models above extend trivially to higher dimensions. There are many other 'natural' ways to define random geometric graphs, some of which we shall mention in Section 2.

If we condition on a particular point $x \in \mathbb{R}^2$ being in $\mathcal{P}_\lambda$, then the degree of x in $G_{r,\lambda}$ has a Poisson distribution with mean $\pi r^2 \lambda$. In fact, the structure of $G_{r,\lambda}$ depends on the parameters r and λ only through the expected degree: for $\lambda_0 r_0^2 = \lambda_1 r_1^2$, the graphs G_{r_0,λ_0} and G_{r_1,λ_1} have the same distribution as abstract random graphs. Thus, when studying, for example, the various critical phenomena concerning these graphs, we are free to change either r or λ, provided we keep $a = \pi r^2 \lambda$ constant. In view of this, we shall write $G(a)$ for any of the random graphs $G_{r,\lambda}$ with $a = \pi r^2 \lambda$, the canonical representative being G_r; we call a the *degree* of $G_{r,\lambda}$. The quantity a is also known as the *connection area*, or simply *area*: G_r is defined by joining each point $x \in \mathcal{P}_1$ to all other points of $\mathcal{P}_1$ in a disc with area a.

In an obvious sense, the random graph $G_{r,\lambda}$ models an infinite communication network in which two transceivers can communicate if their distance is at most r.

As in the discrete case (when studying percolation on lattices or lattice-like infinite graphs), we say that $G_{r,\lambda}$ *percolates* if it has an infinite component. We write $\theta(r, \lambda) = \theta(a)$ for the probability that the component of the origin is infinite. To make sense of this definition, we shall condition $G_{r,\lambda}$ on the origin being one of the points; equivalently, we shall assume that the origin is in $\mathcal{P}_\lambda$. This assumption does not change the distribution of the remaining points of $G_{r,\lambda}$.

The first question concerning this 'continuum percolation' is when the 'percolation probability' $\theta(r, \lambda) = \theta(a)$ is strictly positive. It is trivial that $\theta(a) = 0$ if a is sufficiently small and $\theta(a) \to 1$ as $a \to \infty$. Since $\theta(a)$ is monotone increasing, there is a *critical degree* or *critical area* a_c: if $a < a_c$, then the probability that $G(a)$ percolates is 0, and if $a > a_c$, then this probability is strictly positive. Also, as in the discrete case, Kolmogorov's 0-1 law (Theorem 1 of Chapter 2) implies that if $a < a_c$ then every component of $G(a)$ is finite a.s., and if $a > a_c$ then $G(a)$ has an infinite component a.s. Note that the critical degree a_c corresponds to the critical probability p_H for percolation. As we shall see later, in Theorem 3, the natural analogue a_T of p_T is equal to a_c.

Our main aim in the rest of this section is to give bounds on the critical degree a_c. An easy way of bounding a_c is by comparing $G_{r,\lambda}$ to a discrete percolation model with good bounds on its critical probability. Perhaps the most natural model is the *face percolation* on the hexagonal lattice in which each *face* is open with the same probability p, independently of the states of the other faces, and closed otherwise; two faces are neighbours if they share an edge. (As every vertex has degree 3, somewhat misleadingly, this is equivalent to sharing a vertex.) This model is more usually described as site percolation on the triangular lattice. Theorem 8 of Chapter 5 tells us that if $p < 1/2$, then a.s. there is no face percolation, and if $p > 1/2$, then a.s. we have face percolation.

To obtain a face percolation model from $\mathcal{P}_\lambda$, let Λ be a hexagonal lattice, with each face a regular hexagon of side-length s. Define a face percolation model on Λ by setting a face to be *open* if it contains at least one point of $\mathcal{P}_\lambda$, and *closed* otherwise, so each face is closed with probability $e^{-\lambda A}$ and open with probability $1 - e^{-\lambda A}$, where $A = 3\sqrt{3}s^2/2$ is the area of a hexagonal face (see Figure 3). From Theorem 8

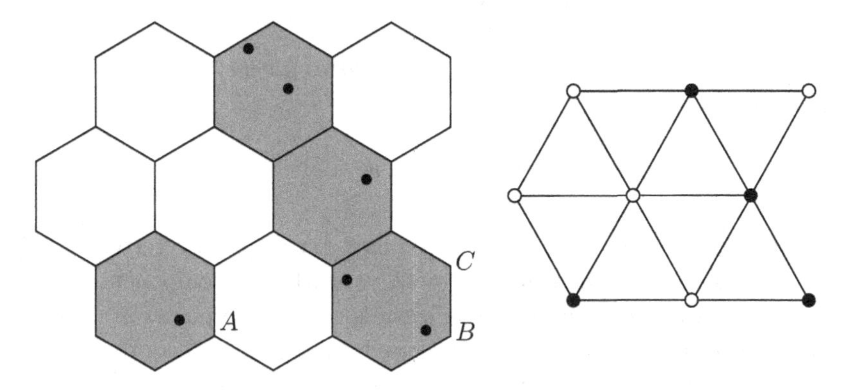

Figure 3. Comparison between Gilbert's model and face percolation on the hexagonal lattice, i.e., site percolation on the triangular lattice: a hexagon is shaded if it contains one or more points of the Poisson process. The corresponding site in the triangular lattice is then taken to be open. If the hexagons have side-length s, then $AB = 2\sqrt{3}s$, so $AC = \sqrt{13}s$.

of Chapter 5, if $1 - e^{-\lambda A} < 1/2$ then we do not have face percolation, and if $1 - e^{-\lambda A} > 1/2$ then we do.

All that remains is to draw the appropriate conclusions about the disc model $G_{r,\lambda}$. Note that any two points in neighbouring faces are at distance at most $\left((2\sqrt{3})^2 + 1^2\right)^{1/2} s = s\sqrt{13}$, and any two points in non-neighbouring faces are at distance at least s. Consequently, if

$1 - e^{-\lambda A} < 1/2$ and $r < s$ then we do not have percolation in $G_{r,\lambda}$, and if $1 - e^{-\lambda A} > 1/2$ and $r > s\sqrt{13}$ then we do. Taking $\lambda = 1$ and substituting $A = (3\sqrt{3}/2)s^2$, we find that if $(3\sqrt{3}/2)s^2 > \log 2$ then $a_c \geq \pi s^2$, and if $(3\sqrt{3}/2)s^2 < \log 2$ then $a_c \leq 13\pi s^2$, i.e.,

$$\frac{2\pi \log 2}{3\sqrt{3}} \leq a_c \leq \frac{26\pi \log 2}{3\sqrt{3}}.$$

We have given these inequalities to show what one can read out of the approximation of a Poisson process by face percolation on the hexagonal lattice, although the lower bound above is worse than the trivial bound 1 implied by the simplest branching process argument. On the other hand, the upper bound, given by Gilbert [1961] under the assumption (unproved at the time, but proved now) that the critical probability for site percolation on the triangular lattice was 1/2, has not been improved much in over four decades. The slightly better upper bound in the theorem below is due to Hall [1985b], while the lower bound is the bound from the seminal paper of Gilbert [1961]. (In fact, the numerical value for the formula given there was 1.75...!)

Theorem 1. *Let a_c be the critical degree (area) for the Gilbert (Boolean) disc model G_r. Then*

$$\frac{6\pi}{2\pi + 3\sqrt{3}} = 1.642\ldots \leq a_c \leq 10.588\ldots.$$

Proof. Let us start with the lower bound. Let C_0 be the vertex set of the component of the origin in $G_r = G_{r,1}$, briefly, the component of the origin. We shall use a very simple process to find the points in C_0 one by one, but in order to achieve a concise formulation of this process, we describe it in a rather formal way.

We shall construct a sequence of pairs of disjoint (finite) sets of points of the plane, $(D_0, L_0), (D_1, L_1), \ldots$, say. The points in D_t are the points of the component C_0 that are *dead* at time t: they belong to the component C_0, and so do all their neighbours; the points in L_t are *live* at time t: they belong to C_0, but we have made no attempt to find their neighbours. To start the sequence, we set $D_0 = \emptyset$ and $L_0 = \{X_0\}$, where $X_0 = 0$ is the origin. Next, let N_0 be the set of neighbours of X_0, and set $D_1 = \{X_0\}$ and $L_1 = N_0$. Having found (D_t, L_t), if $L_t = \emptyset$ then we terminate the sequence; otherwise, we pick a point X_t from L_t, and define $D_{t+1} = D_t \cup \{X_t\} = \{X_0, X_1, \ldots, X_t\}$ and $L_{t+1} = N_t \cup L_t \setminus \{X_t\}$, where N_t is the set of neighbours of X_t that are not neighbours of any of the points in D_t. Since G_r is locally finite (no disc of radius r contains infinitely many points of $\mathcal{P}_1$), the sets D_t and L_t are disjoint finite sets.

By construction, $D_t \cup L_t \subset C_0$; furthermore, if $L_\ell = \emptyset$ then $C_0 = D_\ell$. Since

$$D_t \setminus \{X_0\} = \{X_1, X_2, \dots, X_{t-1}\} \subset \bigcup_{i=0}^{t-2} N_i,$$

we have

$$|D_t| - 1 = t - 1 \le \sum_{i=0}^{t-2} |N_i|. \qquad (1)$$

Let V_t be the disc of radius r with centre X_t, and set $U_t = \bigcup_{s=0}^t V_s$. Conditioning on the points $X_0, \dots, X_t$, we find that $|N_t|$ is a Poisson random variable with mean $|V_t \setminus U_{t-1}|$. Since the centre X_t of V_t is in a disc V_s, $s \le t-1$, Figure 4 tells us that

$$|V_t \setminus U_{t-1}| \le \left(\frac{\pi}{3} + \frac{\sqrt{3}}{2}\right) r^2 = b.$$

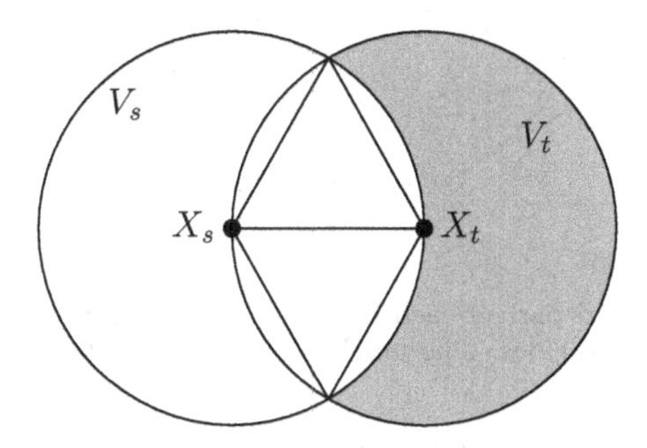

Figure 4. Two discs V_s and V_t of radius r with centres X_s and X_t, such that $X_t \in V_s$. The area of $V_t \setminus V_s$ (shaded) is maximized if $\mathrm{dist}(X_s, X_t) = r$, as shown. In this case the area is $b = (\frac{\pi}{3} + \frac{\sqrt{3}}{2})r^2$.

Now, to bound $|C_0|$, let $Z_0, Z_1, \dots$ be independent Poisson random variables, with $\mathbb{E}(Z_0) = \pi r^2$, the area of the disc V_0, and $\mathbb{E}(Z_i) = b$ for $i \ge 1$. Then inequality (1) implies that

$$\mathbb{P}\big(|C_0| \ge k\big) \le \mathbb{P}\left(\sum_{i=0}^{k-2} Z_i \ge k-1\right).$$

If $b < 1$, it follows easily that $\mathbb{P}\big(|C_0| \ge k\big) \to 0$ as $k \to \infty$, i.e., $\theta(r, 1) = 0$.

Since for

$$a = \pi r^2 < \frac{\pi}{\frac{\pi}{3} + \frac{\sqrt{3}}{2}} = \frac{6\pi}{2\pi + 3\sqrt{3}}$$

we have $b < 1$, the critical area is indeed at least as large as claimed.

For the upper bound on a_c, Hall [1985b] tweaked Gilbert's argument that gave the trivial bound $\frac{26\pi \log 2}{3\sqrt{3}} = 10.89\ldots$, by replacing the cells of a hexagonal tessellation by 'rounded hexagons'. To give the details of this argument, consider the lattice of hexagons, each with side-length 1. For each hexagon H, let H^* be the *rounded hexagon* in H, the intersection of the six discs with centres at the mid-points of the sides and touching the opposite sides, as in Figure 5. By construction, each of these discs

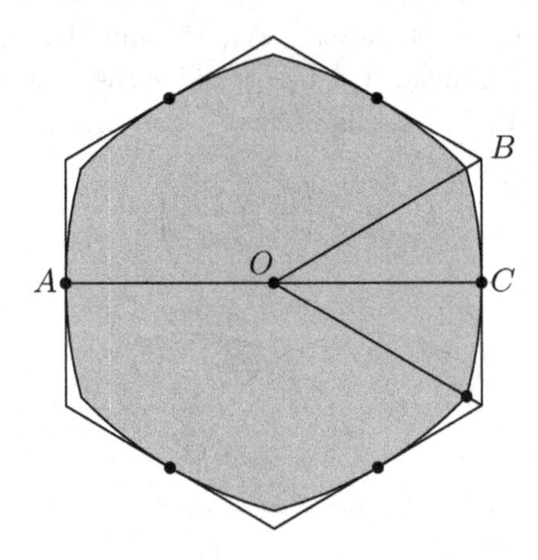

Figure 5. The shaded part of each hexagon is the region within distance 1 of the midpoints of all 6 sides. The labels correspond to those in Figure 6.

has radius $\sqrt{3}$, so any two points of two neighbouring rounded hexagons are at distance at most $2\sqrt{3}$. Hence, if the probability that a rounded hexagon contains at least one point of the Poisson process $\mathcal{P}_\lambda$ is greater than 1/2 then, appealing to Theorem 8 of Chapter 5 as before, we find that $G_{2\sqrt{3},\lambda}$ percolates with strictly positive probability. Therefore, if the area of a rounded hexagon is a, then $e^{-\lambda a} < 1/2$ implies that $a_c \leq 12\pi\lambda$. Hence,

$$a_c \leq \frac{12\pi \log 2}{a}.$$

Finally, the area a of a rounded hexagon H^* can be read out of Figure 6. First, the area of the sector ACD is $3\psi/2$. To calculate ψ, note that $\sin\varphi = \frac{AO}{AD}\sin(5\pi/6) = 1/4$, so $\cos\varphi = \sqrt{15}/4$. Also,

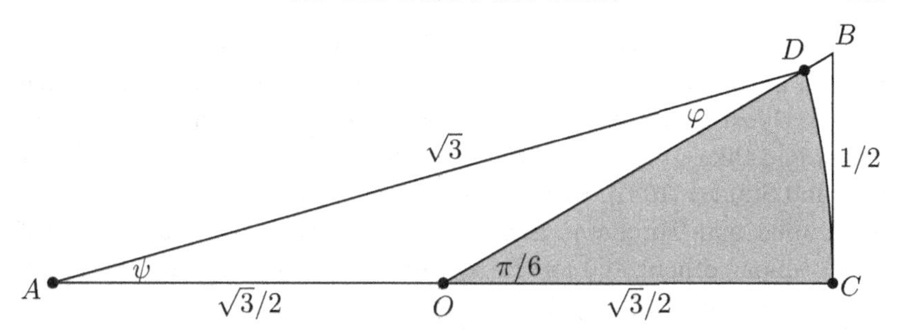

Figure 6. The points A and C are midpoints of two opposite sides of a hexagon H; O is the centre of H and BC is half of a side. Thus, $AO = OC = \sqrt{3}/2$, $OB = 1$, $BC = 1/2$ and $\angle COB = \pi/6$. The shaded region is part of a circle with centre A, so $AD = \sqrt{3}$. We write φ for $\angle ADO$, and ψ for $\angle DAO$, so that $\varphi + \psi = \pi/6$. The shaded domain DOC is one twelfth of the area of a rounded hexagon H^*.

$\sin\psi = \sin(\frac{\pi}{6} - \varphi) = \frac{\sqrt{3}}{8}(\sqrt{5}-1)$, while $\psi = \frac{\pi}{6} - \arcsin(\frac{1}{4}) = 0.2709\ldots$. Furthermore,

$$OD = \frac{\sin\psi}{\sin(5\pi/6)} AD = \frac{3}{4}(\sqrt{5}-1),$$

and so the area of the triangle AOD is

$$\frac{1}{2} AO \left(BC \frac{OD}{OB} \right) = \frac{3\sqrt{3}(\sqrt{5}-1)}{32}.$$

The area a of H^* is twelve times the area of DOC, so

$$a = 12 \left(\frac{3}{2}\psi - \frac{3\sqrt{3}(\sqrt{5}-1)}{32} \right) = 3\pi - 18\arcsin(1/4) - \frac{9\sqrt{3}(\sqrt{5}-1)}{8},$$

so that $a = 2.467\ldots$ and $a_c \leq 10.588\ldots$, as claimed. □

The simplest form of the branching process argument above implies that if A is any bounded open symmetric set in $\mathbb{R}^d$, then the critical degree for $G_A(\mathcal{P}_\lambda)$ is at least 1.

Hall [1985b] improved Gilbert's lower bound as well: this improvement involves a more substantial modification of Gilbert's argument than the tweaking of the upper bound presented above. Indeed, to obtain his improvement, Hall compares disc percolation to a multi-type branching process, with the 'type' of a child defined as its distance from the father.

Theorem 2. *The critical degree for the Gilbert (Boolean) disc model is greater than* 2.184. □

There has been much numerical work on the critical degree for the Boolean model, and on its square variant. Simulation methods have been used for the disc by Roberts [1967], Domb [1972], Pike and Seager [1974], Seager and Pike [1974], Fremlin [1976], Haan and Zwanzig [1977], Gawlinski and Stanley [1981], Rosso [1989], Lorenz, Orgzall and Heuer [1993], Quintanilla and Torquato [1999], and Quintanilla, Torquato and Ziff [2000], among others, and for the square by Dubson and Garland [1985], Alon, Drory and Balberg [1990], Garboczi, Thorpe, DeVries and Day [1991], and Baker, Paul, Sreenivasan and Stanley [2002]. (Not surprisingly, several of the bounds obtained happen to contradict each other.) For the critical degree of disc percolation, Quintanilla, Torquato and Ziff [2000] gave lower and upper bounds of 4.51218 and 4.51228; for square percolation, Baker, Paul, Sreenivasan and Stanley [2002] suggested 4.388 and 4.396.

In studying the critical degrees in these two models, Balister, Bollobás and Walters [2005] gave rigorous reductions of the problems to complicated numerical integrals, which they calculated by Monte Carlo methods. Indeed, their method was the basis of the discussion of rigorous 99% confidence intervals for site and bond percolation critical probabilities in Chapter 6. The basic idea is to use results about k-independent percolation to prove that a certain bound on a_c holds, as long as a certain event E defined in terms of the restriction of G_r to a finite region has at least some probability p_0. (For example, they consider the event E that the largest components of the subgraphs of G_r induced by the squares $[0,\ell]^2$ and $[\ell, 2\ell] \times [0,\ell]$ are part of a single component in the subgraph induced by $[0, 2\ell] \times [0,\ell]$. Considering events isomorphic to E, there is a natural way to define a 1-independent bond percolation measure on $\mathbb{Z}^2$ such that every bond is open with probability $\mathbb{P}(E)$. Another possible choice for E is described in Chapter 6.) Unfortunately, one cannot evaluate $\mathbb{P}(E)$ exactly; however, there are Monte Carlo methods for evaluating $\mathbb{P}(E)$ that come with rigorous bounds on the probability of errors of certain magnitudes; see Chapter 6 for more details of the basic method. Using this technique, Balister, Bollobás and Walters [2005] proved that, with confidence 99.99%, the critical degree for the disc percolation is between 4.508 and 4.515, and that for the square percolation is between 4.392 and 4.398.

The results and methods we discussed in earlier chapters for discrete percolation models are easily applied to the Gilbert Boolean disc model to prove the uniqueness of the infinite open cluster above the critical degree a_{c} and exponential decay below a_{c}. We start with a a result of Roy [1990] giving exponential decay below a_{c}, which implies the analogue of $p_{\mathrm{H}} = p_{\mathrm{T}}$ for Gilbert's model. This result holds in any dimension; for notational simplicity we state and prove it only for dimension two.

Theorem 3. *Let $\pi r^2\lambda = a < a_{\mathrm{c}}$, where a_{c} is the critical degree for the Gilbert model, and let $|C_0(G_{r,\lambda})|$ denote the number of points in the component of the origin in $G_{r,\lambda}$. Then*

$$\mathbb{P}\big(|C_0(G_{r,\lambda})| \geq n\big) \leq \exp(-c_a n),$$

where $c_a > 0$ does not depend on n. In particular, $a_{\mathrm{T}} = a_{\mathrm{c}}$, where

$$a_{\mathrm{T}} = \inf\big\{\pi r^2\lambda : \mathbb{E}\big(|C_0(G_{r,\lambda})|\big) = \infty\big\}.$$

Proof. The result is more or less immediate from Menshikov's Theorem (see Theorems 7 and 9 of Chapter 4), using the natural approximation of continuum percolation by discrete percolation. Roughly speaking, we shift the points of the Poisson process $\mathcal{P}_\lambda$ slightly, by rounding their coordinates to multiples of a small constant δ. If the shifted points are at distance at most $r - 2\sqrt{2}\delta$, then the original points must have been connected in $G_{r,\lambda}$. On the other hand, if the shifted points are at distance more than $r + 2\sqrt{2}\delta$, then the original points cannot have been connected in $G_{r,\lambda}$.

Let us fill in the details of the argument. Pick $\lambda_1 > \lambda$ and $r_1 > r$ so that $a < \pi r_1^2\lambda_1 < a_{\mathrm{c}}$, and $\delta > 0$ such that $r + 2\sqrt{2}\delta < r_1$. Let $\delta\mathbb{Z}^2 = (\delta\mathbb{Z})^2$ be the square lattice with faces

$$F^\delta_{i,j} = \{(x,y) : \ i\delta \leq x \leq (i+1)\delta,\ j\delta \leq y \leq (j+1)\delta\ \}.$$

Let $\Lambda(\delta, r_1)$ be the graph whose vertices are the faces $F^\delta_{i,j}$, in which two faces are joined if the maximum distance between two points of their union is at most r_1. We couple G_{r_1,λ_1} with independent site percolation on $\Lambda(\delta, r_1)$ in the obvious way: declare a site $F^\delta_{i,j}$ of $\Lambda(\delta, r_1)$ to be open if and only if it contains at least one point of $\mathcal{P}_{\lambda_1}$. Note that the states of the sites (i.e., faces) are independent, and that each is open with probability $p_1 = 1 - e^{-\lambda_1\delta^2}$.

Suppose that there is an infinite open cluster in $\Lambda(\delta, r_1)$. If x and y are adjacent open sites of $\Lambda(\delta, r_1)$, then there is at least one point of $\mathcal{P}_{\lambda_1}$ in the faces of $\delta\mathbb{Z}^2$ corresponding to x and y, and any two such points are joined in G_{r_1,λ_1}. Hence, G_{r_1,λ_1} has an unbounded open cluster. But $a_1 < a_{\mathrm{c}}$, so this event has probability zero. Thus, $p_1 \leq p^{\mathrm{s}}_{\mathrm{H}}(\Lambda(\delta, r_1))$.

The graph $\Lambda(\delta, r_1)$ satisfies the conditions of Menshikov's Theorem, Theorem 7 of Chapter 4. (As usual, we regard $\Lambda(\delta, r_1)$ as an oriented graph by replacing each edge by two oppositely oriented edges.) Set $p = 1 - e^{-\lambda\delta^2} < p_1$. If each site of $\Lambda(\delta, r_1)$ is now taken to be open with probability p, independently of the other sites, then by Theorems 7 and 9 of Chapter 4, we have exponential decay of the number of sites in the open cluster of the origin in $\Lambda(\delta, r_1)$.

Returning to Gilbert's model, this time we take a face of $\delta\mathbb{Z}^2$ to be open if it contains at least one point of $\mathcal{P}_\lambda$, so the faces are open independently with probability p. We condition on the origin lying in $\mathcal{P}_\lambda$. If two points of $\mathcal{P}_\lambda$ are within distance r, then any two points of the corresponding faces are within distance $r + 2\sqrt{2}\delta < r_1$, so the faces are joined in $\Lambda(\delta, r_1)$. Thus, for every vertex of $G_{r,\lambda}$ in the component C_0 of the origin, the corresponding face is in C_0', the open cluster of the origin in $\Lambda(\delta, r_1)$. Given C_0', the number of points of $\mathcal{P}_\lambda$ in each face of C_0' has a Poisson distribution conditioned on being at least 1. An easy calculation shows that the total number of points in such faces is very unlikely to be much larger than $|C_0'|$, and the result follows. □

Meester and Roy [1994] proved uniqueness for both the occupied and vacant clusters. Again, this result holds in any dimension.

Theorem 4. *In the Gilbert model $G_{r,\lambda}$, a.s. there is at most one infinite component, and at most one unbounded component of $E_{r,\lambda}$.* □

Unlike Theorem 3, this does not follow easily from the corresponding discrete results: the event that there is at most one unbounded (occupied or vacant) cluster is neither increasing nor decreasing, so there is no straightforward way to bound its probability by that of an event in a discrete approximation to Gilbert's model. However, the Burton–Keane proof of the Aizenman–Kesten–Newman uniqueness result for lattices (Theorem 4 of Chapter 5) can easily be adapted to Gilbert's model.

Once we know that (with probability 1) $G_{r,\lambda}$ has a unique infinite component, it is reasonable to expect that inside a 'big box' there is only one 'big' component. Penrose and Pisztora [1996] showed that the heuristics, if properly stated, are indeed true, but they have to work rather hard to prove them.

Other results for percolation on lattices have analogues for continuum percolation. For example, Alexander [1996] proved an analogue of the Russo–Seymour–Welsh Theorem for occupied clusters in $G_{r,\lambda}$, i.e., for $D_{r,\lambda}$. A corresponding result for vacant clusters was proved by Roy [1990].

Even in the plane, the Gilbert model has numerous natural extensions. For example, instead of a disc, we may use a convex domain $K \subset \mathbb{R}^2$ to define a random set

$$D_{K,\lambda} = \bigcup_{i=1}^{\infty}(x_i + K),$$

where $\mathcal{P}_\lambda = \{x_1, x_2, \dots\}$ is a Poisson process of intensity λ in the plane. This process *percolates* if $D_{K,\lambda}$ has an unbounded component. Note that K is not assumed to be symmetric; we generalize the interpretation of the Gilbert model shown in Figure 2. The corresponding graph is exactly $G_A(\mathcal{P}_\lambda)$ as defined earlier, with $A = K - K$. In particular, the expected degree of a vertex is λ times the area of $K - K$. Writing $\lambda_c(K)$ for the critical intensity of this percolation process, Jonasson [2001] proved the beautiful result that, among convex domains K of area 1, $\lambda_c(K)$ is minimal for a triangle. Roy and Tanemura [2002] showed that an analogous result holds in higher dimensions, with a simplex replacing the triangle.

To conclude this section, we note that there are natural models with critical degrees close to the minimum. First, as shown by Penrose [1993], the critical degree for 'long range percolation' in $\mathbb{Z}^2$ (or $\mathbb{Z}^d$) tends to 1 as the 'range' tends to infinity. (See Bollobás and Kohayakawa [1995] for a combinatorial proof.) More surprisingly, let A_ε be the annulus with radii 1 and $1+\varepsilon$ (and so area $\pi(2\varepsilon+\varepsilon^2)$); then the critical degree for $G_{A_\varepsilon}(\mathcal{P}_\lambda)$ tends to 1 as $\varepsilon \to 0$. This was proved, independently, by Franceschetti, Booth, Cook, Meester and Bruck [2005], and Balister, Bollobás and Walters [2004]. In the latter paper, it was shown that the corresponding assertion for 'square annuli' is *false*: let S_ε be the square annulus with the inner square having side-length 1, and the outer $1+\varepsilon$, $0 < \varepsilon < 1$, say. Then the critical degree for $G_{S_\varepsilon}(\mathcal{P}_\lambda)$ is at least $c > 1$, with c independent of ε.

Turning to ball percolation in higher dimensions (corresponding to the Boolean disc model in dimension two), let $a_c^{(d)}$ be the critical degree (volume) in $\mathbb{R}^d$, so that $a_c = a_c^{(2)}$. Penrose [1996] showed that $a_c^{(d)} \to 1$ as $d \to \infty$. Balister, Bollobás and Walters [2004] proved a general result about models with critical degrees tending to 1, the minimum possible value; this implies trivially the results for annuli and high-dimensional balls.

Rather than asking for an unbounded component, one may ask for all or almost all of our space to be covered by a collection of random sets. Usually, the random sets are not translates of the same set, but are chosen with a certain probability distribution. For example, we may take independent identically distributed compact sets $K_1, K_2, \dots$ in $\mathbb{R}^d$ and take the random set

$$\Sigma = \bigcup_{i=1}^{\infty}(x_i + K_i),$$

where $\{x_1, x_2, \dots\}$ is a random sequence of points in $\mathbb{R}^d$. Conditions implying that Σ is the entire space $\mathbb{R}^d$ were given by Stoyan, Kendall and Mecke [1987], Hall [1988], Meester and Roy [1996], and Molchanov and Scherbakov [2003]; Athreya, Roy and Sarkar [2004] gave conditions for the sets to cover all but a bounded part of $\mathbb{R}^d$.

8.2 Finite random geometric graphs

As random geometric graphs frequently model 'real-life' networks, e.g., a network of transceivers distributed in a bounded domain, it is natural to study *finite* random geometric graphs, with the quintessential example being the restriction of the Boolean model to a finite set. To define this, let V_n be the restriction of a Poisson process $\mathcal{P}_n$ of intensity n to the square $[0,1]^2$. For $r > 0$, let $G_r(V_n)$ be the random geometric graph in which two points of V_n are joined if their distance is at most r. It is easy to check that $G_r(V_n)$ is close to the model $G_r(U_n)$, in which U_n consists of n points chosen uniformly from the unit square, and two points are joined as before. Clearly, scaling makes no difference to the model: multiplying r by ℓ and taking the restriction of the Poisson process $\mathcal{P}_{n/\ell^2}$ to $[0,\ell]^2$, we get a model isomorphic to $G_r(V_n)$. For example, it is natural to replace the unit square by $[0,\sqrt{n}]^2$ and then take the restriction of the Poisson process of intensity 1 to this square. What *does* matter is the relationship between the expected degree and the expected number of points.

In fact, to obtain a mathematically more elegant model, we shall choose our points from the torus $\mathbb{T}_\ell^2$ obtained from $[0,\ell]^2$ by identifying 0 with ℓ. To be precise, let $G_r(\mathbb{T}_\ell^2, \lambda)$ be the random graph whose vertex set is a Poisson process $\mathcal{P}_\lambda$ on $\mathbb{T}_\ell^2$ with intensity λ, in which two points of $\mathcal{P}_\lambda$ are joined if their distance is at most r. The expected number of vertices of this random graph is $n = \lambda\ell^2$, and for $r \le \ell/2$, given that $x \in \mathcal{P}_\lambda$, the expectation of the degree of x in $G_r(\mathbb{T}_\ell^2, \lambda)$ is $d = \pi r^2 \lambda$. Once again, the scaling is irrelevant: n and d determine $G_r(\mathbb{T}_\ell^2, \lambda)$ so, with a slight abuse of notation, we may write $G_{n,d}$ for this model. The main advantage of using the torus, and so $G_{n,d}$, is the homogeneity of the model: however, to all intents and purposes, for $d = d(n) = o(n)$ the models $G_{n,d}$, $G_r(V_n)$ and $G_r(U_n)$ are interchangeable provided $d = \pi r^2 n$.

The first question we should like to answer about the finite graphs $G_{n,d}$ is the following. For what values of the parameters n and d is the random geometric graph $G_{n,d}$ likely to have isolated vertices? This was answered by Steele and Tierney [1986]. Later, Penrose [1997; 1999] extended this result by giving detailed information about the distribution of the minimal degree $\delta(G_{n,d})$ of $G_{n,d}$. It is fascinating that this result is the exact analogue of the classical result of Erdős and Rényi [1961b] on random graphs (see also Bollobás [2001, Theorem 3.5]). Here we present only a weaker form of Penrose's theorem.

Theorem 5. *Let $d = d(n) = \log n + k \log\log n + \alpha(n)$, where k is a fixed non-negative integer. If $\alpha(n) \to -\infty$ then*

$$\mathbb{P}\big(\delta(G_{n,d}) \le k\big) \to 1,$$

and if $\alpha(n) \to +\infty$ then

$$\mathbb{P}\big(\delta(G_{n,d}) \ge k+1\big) \to 1.$$

Proof. To simplify the calculations, we sketch a proof for $k = 0$. Thus, we set $d = \log n + \alpha(n)$, and let ℓ, λ and r satisfy $n = \lambda \ell^2$ and $d = \pi r^2 \lambda$, so that $G_{n,d} \equiv G_r(\mathbb{T}_\ell^2, \lambda)$. We write X_0 for the number of isolated vertices in $G_r(\mathbb{T}_\ell^2, \lambda)$. The probability that a fixed disc of radius r in $\mathbb{T}_\ell^2$ contains no points of our Poisson process $\mathcal{P}_\lambda$ is $e^{-\pi r^2 \lambda}$. It follows from basic properties of Poisson processes that

$$\mathbb{E}(X_0) = ne^{-\pi r^2 \lambda} = ne^{-d} = e^{-\alpha}. \tag{2}$$

Indeed, dividing $\mathbb{T}_\ell^2$ into $(\ell/\varepsilon)^2$ small squares S_i of side $\varepsilon < r/2$, by linearity of expectation, $\mathbb{E}(X_0)$ is $(\ell/\varepsilon)^2$ times the probability that S_1 contains an isolated vertex of $G_r(\mathbb{T}_\ell^2, \lambda)$. (No S_i can contain two such vertices.) As $\varepsilon \to 0$, this probability is asymptotically

$$\lambda \varepsilon^2 e^{-\lambda \varepsilon^2} e^{-\pi r^2 \lambda},$$

so

$$\mathbb{E}(X_0) = \lim_{\varepsilon \to 0} (\ell/\varepsilon)^2 \lambda \varepsilon^2 e^{-\lambda \varepsilon^2} e^{-\pi r^2 \lambda} = \ell^2 \lambda e^{-\pi r^2 \lambda} = ne^{-d},$$

implying (2).

If $\alpha \to \infty$, then (2) implies that

$$\mathbb{P}\big(\delta(G_{n,d}) = 0\big) = \mathbb{P}(X_0 \geq 1) \leq \mathbb{E}(X_0) = o(1).$$

Now, suppose that $\alpha(n) \to -\infty$. To prove that in this case $G_{n,d} \equiv G_r(\mathbb{T}_\ell^2, \lambda)$ is very likely to have isolated vertices, we need another (rather trivial) step. We know that $m_1 = \mathbb{E}(X_0) \to \infty$. It is easily checked that the second moment $\mathbb{E}(X_0^2)$ of X_0 is $m_2 = (1 + o(1))m_1^2$; hence, by Chebychev's inequality,

$$\mathbb{P}\big(\delta(G_{n,d}) \geq 1\big) = \mathbb{P}(X_0 = 0) \leq \frac{m_2 - m_1^2}{m_1^2} = o(1).$$

The general case can be proved along the same lines, with a little more calculation. □

In fact, it is easy to prove that, for $d = d(n)$ in the appropriate range, the number X_k of vertices of degree k has asymptotically Poisson distribution; this implies very good bounds on the probability $\mathbb{P}\big(\delta(G_{n,d}) \geq k\big)$. Although this is not immediately obvious, it is not hard to show that Theorem 5 does indeed carry over to the model $G_r(V_n)$ defined on the square rather than the torus, i.e., that the 'boundary effects' do not matter.

Our next aim is to state a considerably weightier result of Penrose establishing a close connection between the properties of s-connectedness and of having minimal degree at least s. We start by defining the hitting radius for an arbitrary property and an arbitrary set. Given a point set P and a monotone increasing property Q of graphs (i.e., a property Q such that, if G has Q and G' is obtained by adding an edge to G, then G' also has Q), the *hitting radius* $\rho_Q(P)$ of Q on P is defined as

$$\rho_Q(P) = \min\{r : \ G_r(P) \in Q\}.$$

Note that if $Q \subset Q'$ then $\rho_Q(P) \geq \rho_{Q'}(P)$ for *every* set P. In particular, the hitting radius of s-connectedness is at least as large as that of having minimal degree at least s:

$$\rho_{s-\mathrm{conn}}(P) \geq \rho_{\delta \geq s}(P)$$

for *every* set P.

A basic result in the theory of random graphs is the result of Bollobás and Thomason [1985] that, for almost every random graph process, the hitting time of s-connectedness equals the hitting time of having minimal degree at least s (see Bollobás [2001], Theorem 7.4). Penrose [1999] (see also Penrose [2003, pp. 302–305]) showed that this result carries over to the Boolean model on the torus.

Theorem 6. *Let $\mathcal{P}_\lambda$ be a Poisson process on the torus $\mathbb{T}_1^2$ with intensity λ. Then*

$$\lim_{\lambda\to\infty} \mathbb{P}\big(\rho_{s-\mathrm{conn}}(\mathcal{P}_\lambda) = \rho_{\delta\geq s}(\mathcal{P}_\lambda)\big) = 1. \qquad \square$$

Thus, for λ large enough, if we start with a set $\mathcal{P}_\lambda$ of isolated points, and add edges one by one, always choosing the shortest possible edge to add, then, with high probability, the very moment this graph has minimal degree s, it is also s-connected.

Combining this result with Theorem 5, we can identify the critical degree for s-connectedness.

Theorem 7. *Let s be a fixed non-negative integer, and let $d = d(n) = \log n + (s-1)\log\log n + \alpha(n)$. If $\alpha(n) \to -\infty$ then*

$$\mathbb{P}\big(G_{n,d} \text{ is } s\text{-connected}\big) \to 0,$$

and if $\alpha(n) \to +\infty$ then

$$\mathbb{P}\big(G_{n,d} \text{ is } s\text{-connected}\big) \to 1. \qquad \square$$

As shown by Penrose [1999], all these results carry over to random graphs on the square and, *mutandis mutatis*, to the m-dimensional cube $[0,1]^m$ and torus. The analogous problems for the simpler graph defined using the ℓ_∞-distance, rather than the Euclidean distance ℓ_2, were studied by Appel and Russo [1997; 2002].

The random graph $G_{n,d}$ above is connected if and only if the minimal weight spanning tree in the complete graph on the same vertex set, with each edge weighted by the distance between its end-vertices, has no edge longer than d. Such spanning trees, generated by a (possibly non-uniform) Poisson process in the m-dimensional cube $[0,1]^m$, have been studied by a number of people, including Henze [1983], Steele and Shepp [1987], and Kesten and Lee [1996].

Our next aim is to study the connectedness of one of the many models of random geometric graphs related to the Boolean model. Let V_n be the restriction of a Poisson process of intensity 1 to a square S_n of area n, and join each point $x \in V_n$ to the k points nearest to it. Let $H_{n,k}$ be the random geometric graph obtained in this way. Note that n is the *expected* number of vertices of $H_{n,k}$; also, if $H_{n,k}$ has $m \geq k+1$ vertices, then it has at least $km/2$ and at most km edges.

The random geometric graph $H_{n,k}$ is again a model of an ad hoc network of transceivers and, as such, it has been studied by many people, including Kleinrock and Silvester [1978], Silvester [1980], Hajek [1983], Takagi and Kleinrock [1984], Hou and Li [1986], Ni and Chandler [1994], Gonzáles-Barrios and Quiroz [2003], and Xue and Kumar [2004].

We should like to know for which functions $k = k(n)$ the graph $H_{n,k}$ is likely to be connected as $n \to \infty$. More precisely, we should like to find a function $k_0(n)$ such that, if $\varepsilon > 0$ is constant, then

$$\lim_{n\to\infty} \mathbb{P}\big(H_{n,k} \text{ is connected}\big) = \begin{cases} 0 & \text{if} \quad k \leq (1-\varepsilon)k_0(n), \\ 1 & \text{if} \quad k \geq (1+\varepsilon)k_0(n). \end{cases}$$

Such a function $k_0(n)$ is considerably harder to determine than the critical degree d for connectedness in the Boolean model given by Theorem 7, since the trivial obstruction to connectedness, the existence of an isolated vertex, is ruled out by the definition of $H_{n,k}$.

As we shall now see, simple back-of-an-envelope calculations give us the order of $k_0(n)$. Nevertheless, we are very far from determining the asymptotic value of $k_0(n)$. In the arguments that follow, the inequalities are claimed to hold only if n is sufficiently large.

Let us see first that if $c > e$ then $k = k(n) = \lfloor c \log n \rfloor$ is an upper bound for $k_0(n)$. If every disc of area $a = \pi r^2$ centred at a point $x \in V_n$ contains at most k other points of V_n, then $H_{n,k}$ contains $G_{n,a}$ as a subgraph, where $G_{n,a}$ is the graph on V_n obtained by joining two points if they are within distance r. The variant of Theorem 7 for the square (rather than the torus) tells us that the graph $G_{n,a}$ is connected whp (with high probability, i.e., with probability tending to 1 as $n \to \infty$) if $a = \log n + \log\log n$, say. (In the application of the theorem, $s = 1$, and $\alpha(n) = \log\log n$.) For this a, the probability that a fixed disc of area a contains at least $k+1$ points is at most

$$e^{-a} \sum_{\ell=k+1}^{\infty} \frac{a^\ell}{\ell!} < \frac{1}{n} \frac{a^{k+1}}{(k+1)!} < \frac{1}{n}(e/c')^k < n^{-1-\varepsilon},$$

where $e < c' < c$ and $\varepsilon > 0$. From basic properties of Poisson processes, it follows that the probability that the disc of area a about *some* point of V_n contains more than k other points of V_n is at most $n^{-\varepsilon}$. Consequently, $H_{n,k}$ is connected whp, as claimed.

Next, we show that if $\varepsilon > 0$, then $k = \lfloor(1-\varepsilon)\log n/8\rfloor - 1$ is a lower bound for $k_0(n)$. To this end, define $r > 0$ by $\pi r^2 = k+1$, and consider a family $\mathcal{D}$ of three concentric discs D_1, D_3 and D_5 contained in S_n, where D_i has centre x and radius ir; see Figure 7. We say that the family $\mathcal{D}$ is

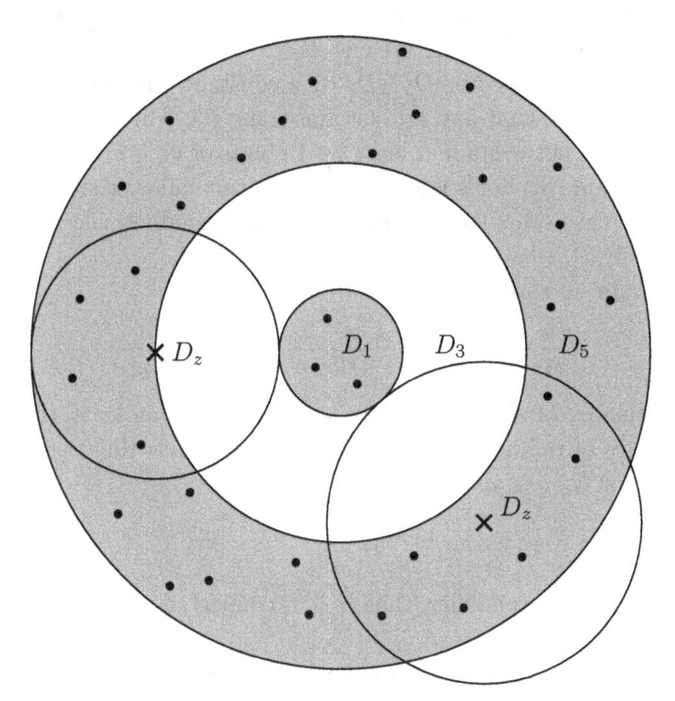

Figure 7. A family $\mathcal{D} = \{D_1, D_3, D_5\}$ of three concentric discs, and two possible discs D_z touching D_1. For $k = 2$, the family $\mathcal{D}$ is bad, so $H_{n,2}$ contains no edges from the vertices in D_1 to the rest of the graph.

bad for the graph $H_{n,k}$ (or set V_n) if the following three conditions hold:

(i) D_1 contains at least $k+1$ points of V_n,

(ii) $D_3 \setminus D_1$ contains no point of V_n, and

(iii) for every $z \in D_5 \setminus D_3$, the region $D_z \cap (D_5 \setminus D_3)$ contains at least $k+1$ points of V_n, where D_z is the disc with centre z and radius $\operatorname{dist}(x,z) - 1$.

Clearly, if some family $\mathcal{D}$ is bad for $H_{n,k}$, then the graph $H_{n,k}$ is disconnected. A family is good if it is not bad.

Let us show that the probability that a fixed family $\mathcal{D}$ is bad for $H_{n,k}$ is not too small. First, the probability that D_1 contains at least $k+1$

points is approximately 1/2; say, at least 1/3. Next, condition (ii) holds with probability

$$e^{-8\pi r^2} = e^{-8(k+1)} \geq n^{-1+\varepsilon}.$$

Finally, condition (iii) holds if it holds for the points z with $\mathrm{dist}(z,x) = 3r$. For such a point z, the area of $D_z \cap (D_5 \setminus D_3)$ is $c\pi r^2$ for some $c > 2$. Thus, one can cover $D_5 \setminus D_3$ by a constant number C of regions R_i of area $2\pi r^2$ so that any D_z contains some R_i. The probability that a given R_i does not contain at least $k+1$ points of V_n is $o(1)$. Hence, the probability that (iii) holds is $1-o(1)$, and in particular at least 1/2 for n large. Since the events (i), (ii) and (iii) are independent, the probability that a family $\mathcal{D}$ is bad for $H_{n,k}$ is at least $n^{-1+\varepsilon}/6$.

The square S_n contains at least

$$\frac{n}{101r^2} \geq \frac{n}{101((1-\varepsilon)\log n)/(8\pi)} > \frac{n}{5\log n}$$

disjoint squares of side-length $10r$, so it contains at least this many disjoint discs of radius $5r$. Therefore, the probability that every family $\mathcal{D}$ is good for $H_{n,k}$ is at most

$$\left(1 - n^{-1+\varepsilon}/6\right)^{n/5\log n} < e^{-n^{\varepsilon/2}}.$$

Consequently, the probability that $H_{n,k}$ is connected is at most $e^{-n^{\varepsilon/2}} = o(1)$.

The argument above can be considerably simplified if all we want is that connectedness happens around $k = \Theta(\log n)$: there is no need to use Penrose's Theorem, Theorem 7, which is quite a big gun for such a small sparrow. Be that as it may, the remarks above tell us that if $c < 1/8$ then $H_{n,\lfloor c\log n\rfloor}$ is disconnected whp, and if $c > e$ then $H_{n,\lfloor c\log n\rfloor}$ is connected whp. Xue and Kumar [2004] were the first to publish bounds on $k_0(n)$: they proved that $0.074\log n$ is a lower bound and $5.1774\log n$ is an upper bound, with the upper bound following from Penrose's Theorem. (In fact, the upper bound $3.8597\log n$ is implicit in Gonzáles-Barrios and Quiroz [2003].)

Balister, Bollobás, Sarkar and Walters [2005] considerably improved the constants 1/8 and e in the trivial bounds above; in particular, they disproved the natural conjecture that $k_0(n)$ is asymptotically $\log n$, which is the analogue of Theorem 6 for $H_{n,k}$.

Theorem 8. *If $c \leq 0.3043$ then $H_{n,\lfloor c\log n\rfloor}$ is disconnected whp, and if $c > 1/\log 7 \approx 0.5139$ then $H_{n,\lfloor c\log n\rfloor}$ is connected whp.*

For the rather involved proof and a number of related results we refer the reader to the original paper.

Instead of asking for the connectedness of certain random geometric graphs, we may ask for the ground domain to be covered by the random sets we used to define the graph. These *coverage* questions have a long history, and are prominent in three books: Stoyan, Kendall and Mecke [1987], Hall [1988] and Meester and Roy [1996]. Here we shall say a few words about some of the many results.

Just before the dawn of percolation, Dvoretzky [1956] raised a beautiful question concerning covering a circle by random arcs. Let $0 < \ell_1, \ell_2, \dots < 1$. Drop arcs of lengths $\ell_1, \ell_2, \dots$ independently at random onto a circle with perimeter 1. For what sequences (ℓ_i) do our arcs cover the entire circle whp? Independently of Dvoretzky, Steutel [1967] raised a similar question and proved a formula that has turned out to be very useful. Shepp [1972] gave a delicate argument to prove that a necessary and sufficient condition is that

$$\sum_{n=1}^{\infty} n^{-2} e^{\ell_1 + \dots + \ell_n} = \infty.$$

Flatto [1973] considered random arcs of the same length α, $0 < \alpha < 1$. (Surprisingly, this problem had been considered by Stevens [1939] several years before Dvoretzky posed his question, in a journal on *eugenics*.) To state Flatto's result, let m be a fixed natural number, and drop the arcs on the circle (of perimeter 1) at random, one by one, stopping as soon as every point of the circle is covered at least m times. Write $N_{\alpha,m}$ for the number of arcs used. (Thus, $N_{\alpha,m}$ is the 'hitting time' for having an m-fold cover.) Flatto showed that

$$\lim_{\alpha \to 0} \mathbb{P}\Big(N_{\alpha,m} \le \alpha^{-1}\big(\log(1/\alpha) + m \log\log(1/\alpha) + x\big)\Big) = e^{-e^{-x}/(m-1)!},$$

which is once again reminiscent of the classical Erdős–Rényi result. Siegel [1979] proved results about the distribution of the length of the uncovered part of the circle; later, Hall [1985a] extended this result to higher dimensions. The results of Stevens were generalized by Siegel and Holst [1982].

The number, total length and sizes of the gaps left by the covering arcs were studied by Holst and Hüsler [1984] and Huillet [2003], among others. In particular, Huillet made use of Steutel's identity to prove exact and asymptotic results about these quantities.

A related problem, due to Rényi [1958], concerns choosing *disjoint* short subintervals of an interval at random; this problem and variants of it gave rise to much research (see, e.g., Ney [1962], Dvoretzky and Robbins [1964], Solomon and Weiner [1986], and Coffman, Flatto and Jelenković [2000]); these problems are also studied under the name of random sequential adsorption models.

Turning to random covers in higher dimensions, Maehara [1988] studied the threshold for random caps to cover the unit sphere $S^2 \subset \mathbb{R}^3$. Suppose we put N spherical caps of area $4\pi p(N)$ independently and at random on S^2, and that

$$\lim_{N\to\infty} \frac{p(N)N}{\log N} = c.$$

Maehara proved that if $c < 1$, then whp the sphere is not covered completely, while if $c > 1$ then whp it is. Further results of Maehara [1990; 2004] concern the intersection graphs of random arcs and random caps: these graphs are the analogues of the Gilbert graphs on the torus and square we studied earlier in this section.

In a different vein, Aldous [1989] used Stein's method to prove sharp results about covering a square by random small squares.

Generalizing several earlier results, Janson [1986] used ingenious and long arguments to prove that, under rather weak conditions and after appropriate normalization, the (random) number of random small sets needed to cover a larger set tends to the extreme-value distribution $\exp(-e^{-u})$ as the measure of the small sets tends to 0.

To conclude this section, we shall say a few words about the random convex hull problem, yet another old problem about random points in a convex domain. Pick n points at random from a convex domain $K \subset \mathbb{R}^2$. What can one say about the (random) number X_n of sides of the convex hull H_n of these n points? Rényi and Sulanke [1963] proved that the distribution of X_n depends very strongly on the smoothness of the boundary: if K has a smooth boundary then $\mathbb{E}(X_n) = O(n^{1/3})$, while if K is a convex k-gon then $\mathbb{E}(X_n) = \frac{2k}{3}(\log n + C) + c(K) + o(1)$, where C is Euler's constant and $c(K) = o(k)$ depends on K and is maximal for regular polygons and their affine equivalents. In a follow-up paper, Rényi and Sulanke [1964] studied the area and perimeter length of the convex hull H_n. These results of Rényi and Sulanke have spurred much research, including papers by Groeneboom [1988], Cabo and Groeneboom [1994], Hsing [1994], Hueter [1994; 1999a; 1999b], Bräker, Hsing and Bingham [1998], Bräker and Hsing [1998], and Finch and Hueter [2004]. For example, Groeneboom proved that, as $n \to \infty$, the normalized random variable $(X_n - 2\pi c_1 n^{1/3})/(n^{1/6}\sqrt{2\pi c_2})$ tends to a standard normal random variable, where $c_1 = (3\pi/2)^{-1/3}\Gamma(5/3) \approx 0.538\ldots$ and c_2 is expressed in terms of complicated double integrals. Finch and Hueter gave an explicit expression for c_2.

8.3 Random Voronoi percolation

In this section we shall consider yet another continuum percolation process associated to a Poisson process in $\mathbb{R}^d$ for $d \geq 2$. Up to now, our Poisson process was used to define a random geometric graph, and the question was whether this graph has an infinite component or not. This time we go further: we consider a graph defined by the Poisson process, and then consider site percolation on this graph.

To be precise, let $\mathcal{P}$ be a Poisson process (of intensity 1, say) in $\mathbb{R}^d$. For every Poisson point x (i.e., $x \in \mathcal{P}$), the *open Voronoi cell* U_x of x with respect to $\mathcal{P}$ is the set of all points closer to x than to any other Poisson point. The closure V_x of U_x is the *closed Voronoi cell of* x, or simply the *Voronoi cell of* x, with respect to $\mathcal{P}$. Thus,

$$V_x = V_x(\mathcal{P}) = \{y \in \mathbb{R}^d : \mathrm{dist}(y, x) \leq \mathrm{dist}(y, x') \text{ for all } x' \in \mathcal{P}\}.$$

It is easily seen that, for a Poisson process $\mathcal{P}$, with probability 1 each cell $V_x = V_x(\mathcal{P})$ is a d-dimensional convex polytope with finitely many $(d-1)$-dimensional faces, and any two Voronoi cells are either disjoint or meet in a full $(d-1)$-dimensional face. Also, for any k-dimensional face of V_x, there are exactly $d+1-k$ Voronoi cells V_z containing it. For convenience, we shall assume that these conditions *always* hold. We call $V(\mathcal{P}) = \{V_x : x \in \mathcal{P}\}$ the *Voronoi tessellation* associated to $\mathcal{P}$. The random tessellation $V = V(\mathcal{P})$ is a *random Voronoi tessellation* of $\mathbb{R}^d$; see Figure 8.

The Voronoi tessellation defines a graph $G_{\mathcal{P}}$ on $\mathcal{P}$: two Voronoi cells are *adjacent* if they share a $(d-1)$-dimensional face, and we join two points $x, y \in \mathcal{P}$ if their Voronoi cells are adjacent; see Figure 8.

The terminology is in honour of Voronoi [1908] who, at the beginning of the last century, used these cells and tessellations to study quadratic forms. In fact, concerning tessellations in two and three dimensions, Dirichlet [1850] had anticipated Voronoi by over fifty years, so one may also talk of *Dirichlet domains* and *Dirichlet tessellations*. In 1911, these tessellations were again rediscovered (see Thiessen and Alter [1911]), so in some circles they go under the name of *Thiessen polygonalizations*.

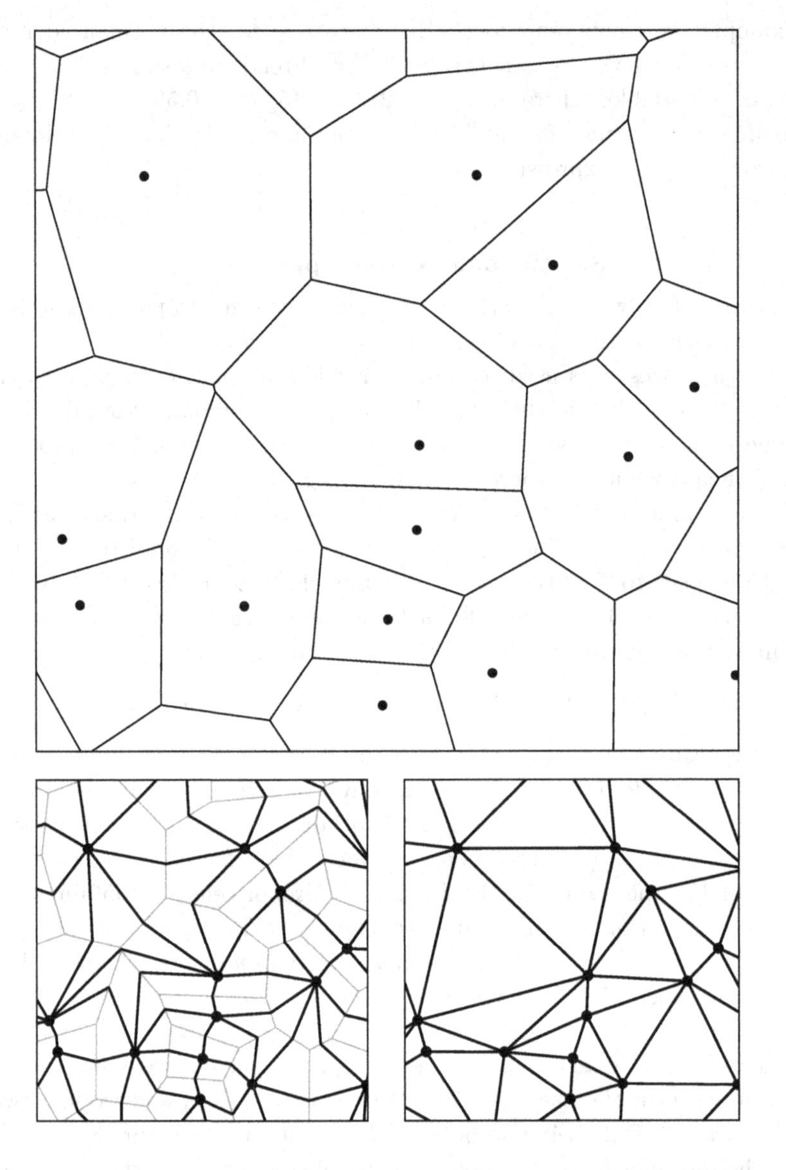

Figure 8. The upper figure shows part of the Voronoi tessellation $V(\mathcal{P})$ associated to a Poisson process $\mathcal{P} \subset \mathbb{R}^2$; the points of $\mathcal{P}$ are also shown. If the Voronoi cells associated to $x, y \in \mathcal{P}$ meet, then their common edge is part of the perpendicular bisector of xy. The lower figures show the graph $G_{\mathcal{P}}$ associated to $V(\mathcal{P})$: two points of $\mathcal{P}$ are joined if the corresponding Voronoi cells meet. On the left, each bond of $G_{\mathcal{P}}$ is drawn using two straight line segments, on the right, using one. The first embedding of $G_{\mathcal{P}}$ is clearly planar. It is easy to show that the second is also.

Other self-explanatory terms are *Voronoi diagram* and *Dirichlet diagram* for the tessellation, *Voronoi polygon* and *Dirichlet polygon* for a cell, and *Poisson–Voronoi tessellation* for our random tessellation. In what follows, we shall use the terms Voronoi tessellation and random Voronoi tessellation.

The study of Voronoi (Dirichlet) tessellations has a very long history, especially in discrete geometry, in connection with sphere packings and other problems; see, for example, the books by Fejes Tóth [1953; 1972], Rogers [1964] and Böröczky [2004]. Although deterministic problems are outside the scope of this book, let us remark that the sphere packing problem asks for the maximum density of a packing of congruent spheres in $\mathbb{R}^d$. In 1929, Blichfeldt gave the upper bound $2^{-d/2-1}(d+2)$ for this density. Using an argument suggested by H. E. Daniels, Rogers [1958] improved this bound to a certain constant σ_d, with $\sigma_d \sim 2^{-d/2}d/e$ as $d \to \infty$. Twenty years later, Kabatjanskiĭ and Levenšteĭn [1978] gave a bound which is better for $d \geq 43$; recently Bezdek [2002] proved a lower bound for the surface area of Voronoi cells which enabled him to improve Rogers's bound for all $d \geq 8$.

There has been much research on *random* Voronoi tessellations as well, first in crystallography and then in mathematics. For solids composed of different kinds of crystals, Delesse [1848] estimated the fraction of the volume occupied by the ith crystal; a century later, Chayes [1956] gave statistics for these estimates. Johnson and Mehl [1939] gave a model for crystal growth; the cells of this model depend not only on a Poisson process, but also on the 'arrival times' of the nuclei. These cells need not even be convex, although they are star-domains from their nuclei. Meijering [1953] introduced Voronoi tessellations into crystallography, without being aware of the considerably earlier papers of Dirichlet and Voronoi. Since then, this model (attributed to Meijering, rather than Dirichlet or Voronoi) and the related Johnson–Mehl model have been much studied.

Gilbert [1962] proved results about the expectations of the surface area, the number of faces, the total edge length, and other parameters of a cell (polyhedron) in a random Voronoi tessellation. In particular, he noted in passing that Euler's formula implies that in the plane the expected number of vertices of a polygon is 6.

Since the 1960s, a considerable body of results has been proved about the 'typical' cell of a random Voronoi tessellation; see Møller [1994] and Stoyan, Kendall and Mecke [1995] for many results. Here, we shall note only some of the more recent results concerning planar tessellations. Hayen and Quine [2000] gave a formula for the probability that the cell containing the origin is a triangle, and Calka [2003] gave an explicit formula for the distribution of the number of sides. Calka and Schreiber [2005] proved results about the asymptotic number of vertices and the area of a cell conditioned to contain a disc of radius r, as $r \to \infty$. Continuing the work of Foss and Zuyev [1996], Calka [2002a; 2002b], made use of the result of Stevens [1939] mentioned in the previous section to study the radius of the smallest circle containing the cell of the origin, and the radius of the largest circle in the cell.

Somewhat surprisingly, although Gilbert introduced his disc percolation model, and studied the random Voronoi tessellation as a model of crystal growth, he did not pose the problem of face percolation on a random Voronoi tessellation. However, a little later, in one of the early papers devoted to percolation theory, Frisch and Hammersley [1963] called for attempts to pioneer 'branches of mathematics that might be called stochastic geometry or statistical topology'. Eventually, this challenge was taken up by physicists; for example, Pike and Seager [1974] and Seager and Pike [1974] performed computer analyses of the percolation and the conductance of the Gilbert model and some of its variants. Detailed computer studies of these models were carried out by many people, including Haan and Zwanzig [1977], Vicsek and Kertész [1981], Gawlinski and Stanley [1981], and Gawlinski and Redner [1983], among others.

Percolation on random Voronoi tessellations was studied by Hatfield [1978], Winterfeld, Scriven and Davis [1981], Jerauld, Hatfield, Scriven and Davis [1984], and Jerauld, Scriven and Davis [1984]. In particular, based on computer simulations and the cluster moment method of Dean [1963], Winterfeld, Scriven and Davis estimated that 0.500 ± 0.010 is the critical probability for random Voronoi percolation in the plane (to be defined below). Nevertheless, no proof was offered even for the simple fact that the critical probability is strictly between 0 and 1. As we shall see later, this is not entirely trivial, unlike for lattices such as the example $\mathbb{Z}^d$ in Chapter 1.

In a series of papers, Vahidi-Asl and Wierman [1990; 1992; 1993] studied first-passage percolation on random Voronoi tessellations. More recently, Gravner and Griffeath [1997] proved substantial results about random Voronoi tessellations; in the problem they consider, the faces are coloured at random with k colours, and then the colours change according to the deterministic, discrete-time rules of a cellular automaton. The question is what happens in the long run.

Here, we consider *cell* or *face percolation* on the random Voronoi tessellation Vassociated to a Poisson process $\mathcal{P}$ on $\mathbb{R}^d$ of intensity 1: each d-dimensional cell, or face, V_x, $x \in \mathcal{P}$, of V is open with probability p, independently of the other faces. See Figure 9 for part of a random

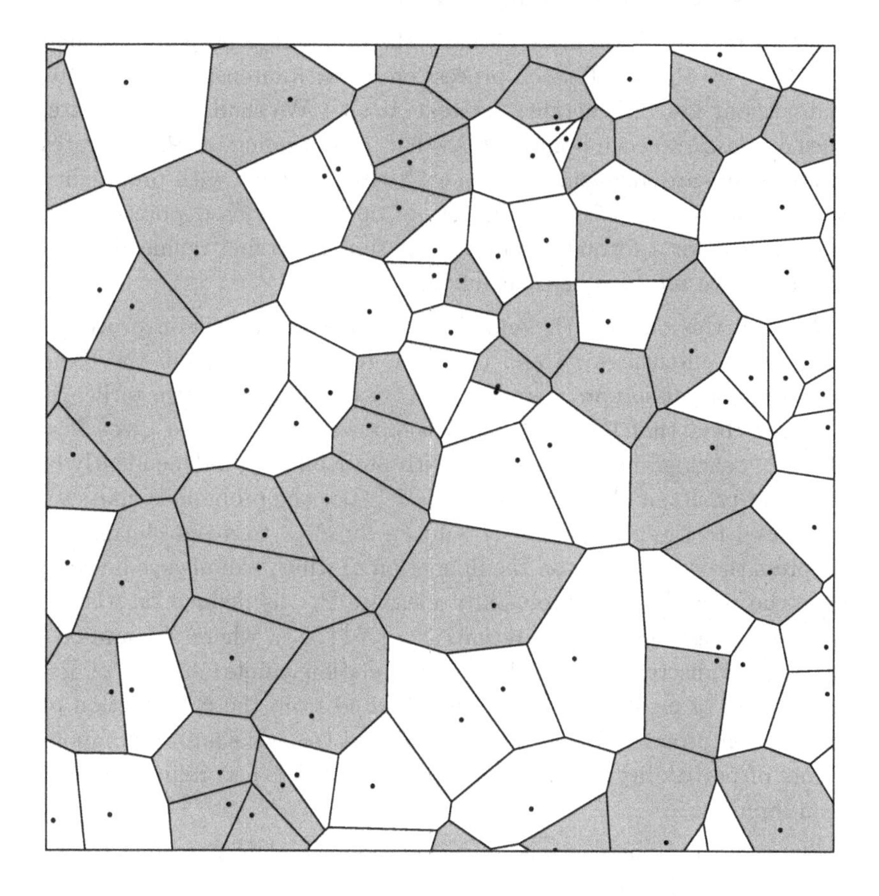

Figure 9. Random Voronoi percolation in the plane at $p = 1/2$.

Voronoi tessellation of the plane; the open faces are shaded. Equivalently, we shall study site percolation on the random graph $G_{\mathcal{P}}$, in which two vertices $x, y \in \mathcal{P}$ are adjacent if their Voronoi cells V_x and V_y meet in a $(d-1)$-dimensional face. In this viewpoint, given $\mathcal{P}$, each site of $G_{\mathcal{P}}$ is open with probability p, independently of the other sites.

Since $\mathcal{P}$ has (a.s.) no accumulation points, the graph $G_{\mathcal{P}}$ is a.s. locally finite, infinite and connected.

This setting is rather different from that of previous chapters: we are now considering percolation in a *random environment*, i.e., studying a random subgraph of a graph that is itself random. The fundamental concepts of percolation on fixed graphs carry over very naturally to this setting; we shall go over the basics in detail.

Let us define a little more formally the probability measure we shall use. Let $\mathcal{P} = \mathcal{P}_\lambda$ be a Poisson process on $\mathbb{R}^d$ with intensity λ. (Usually, and without loss of generality, we take $\lambda = 1$.) We shall assign a state, *open* or *closed*, to each point of $\mathcal{P}$ so that, conditioning on $\mathcal{P}$, the states of the points are independent, and each $x \in \mathcal{P}$ is open with probability p. Writing $\mathcal{P}^+$ and $\mathcal{P}^-$ for the sets of open and closed points in $\mathcal{P}$, respectively, for a formal definition it is simpler to first define $\mathcal{P}^+$ and $\mathcal{P}^-$, and then set $\mathcal{P}$ to be their union.

To spell this out, let $\mathcal{P}^+$ and $\mathcal{P}^-$ be independent Poisson processes on $\mathbb{R}^d$ with intensities $p\lambda$ and $(1-p)\lambda$, respectively. From the basic properties of Poisson processes, or from the construction given earlier, it is easy to check that $\mathcal{P}^+$ and $\mathcal{P}^-$ are disjoint a.s. Furthermore, given $\mathcal{P} = \mathcal{P}^+ \cup \mathcal{P}^-$, each point of $\mathcal{P}$ is in $\mathcal{P}^+$ with probability p, independently of the other points in $\mathcal{P}$. We write $\mathbb{P}_{\lambda,p} = \mathbb{P}^d_{\lambda,p}$ for the probability measure associated to the pair $(\mathcal{P}^+, \mathcal{P}^-)$, and $\mathbb{P}_p$ for $\mathbb{P}_{1,p}$; to avoid clutter, we suppress the dependence on the dimension d, which will always be clear from the context. The probability measure $\mathbb{P}_{\lambda,p}$ is defined on a state space Ω consisting of *configurations* $\omega = (X^+, X^-)$, where X^+ and X^- are disjoint discrete (i.e., without accumulation points) subsets of $\mathbb{R}^2$. The σ-field of measurable events is inherited from the construction of the Poisson process. When talking of $\mathcal{P}$ and $G_{\mathcal{P}}$, we shall often ignore events of probability 0; for example, we say that $G_{\mathcal{P}}$ is infinite, rather than infinite a.s.

Many of the basic concepts of percolation translate naturally to the random Voronoi context, although in some cases there are two descriptions, one graph-theoretic, and one geometric. Let us colour a Voronoi cell V_z *black* if the point $z \in \mathcal{P}$ is open, i.e., if $z \in \mathcal{P}^+$, and *white* otherwise. We say that $x \in \mathbb{R}^d$ is *black* if x lies in a black cell, i.e., if $x \in V_z$ for some $z \in \mathcal{P}^+$. Similarly, $x \in \mathbb{R}^d$ is *white* if it lies in a white cell; see Figure 9. Thus, a point $x \in \mathbb{R}^d$ is black (white) if and only if a point of $\mathcal{P}$ at minimal distance from x is open (closed). Note that points that lie in the faces of the Voronoi cells may be both black and white. In what follows, we shall mostly write z or w for a point of $\mathcal{P}$, and x or y for a point of $\mathbb{R}^d$.

As usual, an *open cluster* is a maximal connected subgraph of the graph $G_{\mathcal{P}}$ all of whose sites are open. Alternatively, a *black cluster* is a maximal connected set of black points in $\mathbb{R}^d$, i.e., a component of the set $\{x \in \mathbb{R}^d : x \text{ is black}\}$. Our assumptions on $\mathcal{P}$ ensure that two Voronoi cells V_z and V_w meet if and only if they share a $(d-1)$-dimensional face, i.e., if and only if $zw \in E(G_{\mathcal{P}})$. Hence, open clusters and black clusters are in one-to-one correspondence.

Let E_∞ be the event that $G_{\mathcal{P}}$ contains an infinite open cluster. Note that $\mathbb{P}_p(E_\infty)$ is an increasing function of p. Also, since $\mathcal{P}$ has no accumulation points, E_∞ is exactly the event that there is an unbounded black cluster.

For any L, the event E_∞ is independent of the values of $\mathcal{P}^+ \cap [-L, L]^d$ and $\mathcal{P}^- \cap [-L, L]^d$. Hence, as $\mathcal{P}^\pm$ is the union of the independent random sets $\mathcal{P}^\pm \cap \prod_{i=1}^d [a_i, a_i + 1)$, $a_1, \ldots, a_d \in \mathbb{Z}$, the event E_∞ may be viewed as a tail event in a product probability space. Thus, by Kolmogorov's 0-1 law (Theorem 1 of Chapter 2), $\mathbb{P}_p(E_\infty)$ is 0 or 1 for any p. Since $\mathbb{P}_p(E_\infty)$ is increasing, it follows that there is a $p_{\mathrm{H}} = p_{\mathrm{H}}(d) \in [0, 1]$ such that

$$\mathbb{P}_p(E_\infty) = \begin{cases} 0 & \text{if } p < p_{\mathrm{H}}, \\ 1 & \text{if } p > p_{\mathrm{H}}. \end{cases} \tag{3}$$

Since $G_{\mathcal{P}}$ is locally finite, infinite, and connected, as noted in Chapter 1, there is a critical probability $p_{\mathrm{H}}(G_{\mathcal{P}}) = p_{\mathrm{H}}^{\mathrm{s}}(G_{\mathcal{P}})$ associated to site percolation on $G_{\mathcal{P}}$. Note that $p_{\mathrm{H}}(G_{\mathcal{P}})$ is a random variable, as it depends on the Poisson process $\mathcal{P}$. From (3), if $p < p_{\mathrm{H}}$, we have $\mathbb{P}_p(E_\infty \mid G_{\mathcal{P}}) = 0$ a.s., and, for $p > p_{\mathrm{H}}$ we have $\mathbb{P}_p(E_\infty \mid G_{\mathcal{P}}) = 1$ a.s. Thus,

$$p_{\mathrm{H}}(G_{\mathcal{P}}) = p_{\mathrm{H}} \quad a.s., \tag{4}$$

i.e., the random variable $p_{\mathrm{H}}(G_{\mathcal{P}})$ is essentially deterministic. We shall call p_{H} *the critical probability for random Voronoi percolation in* $\mathbb{R}^d$, noting that p_{H} is also the critical probability for site percolation on almost every Voronoi tessellation associated to a Poisson process in $\mathbb{R}^d$.

We next turn to the percolation probability $\theta(p)$. With probability 1, the origin lies in a unique Voronoi cell V_{z_0}, $z_0 \in \mathcal{P}$. If z_0 is open, let C_0^G be the open cluster in $G_{\mathcal{P}}$ containing the site z_0. Also, let C_0 be the *black cluster of the origin*, i.e., the component of black points in $\mathbb{R}^d$ containing the origin. If z_0 is closed, then both C_0^G and C_0 are taken to be empty. Note that C_0 is the union of the cells V_z, $z \in C_0^G$. Set

$$\theta(p) = \mathbb{P}_p\big(|C_0^G| = \infty\big) = \mathbb{P}_p\big(C_0 \text{ is unbounded}\big).$$

Since $\theta(G_{\mathcal{P}}; p) = 0$ for $p < p_{\mathrm{H}}(G_{\mathcal{P}})$ and $\theta(G_{\mathcal{P}}; p) > 0$ for $p > p_{\mathrm{H}}(G_{\mathcal{P}})$, from (4) we have $\theta(p) = 0$ if $p < p_{\mathrm{H}}$ and $\theta(p) > 0$ if $p > p_{\mathrm{H}}$, just as for site or bond percolation on a fixed graph.

Let us note that, unlike $p_{\mathrm{H}}(G_{\mathcal{P}})$, for a fixed $p > p_{\mathrm{H}}(d)$ the percolation probability $\theta(G_{\mathcal{P}}; p)$ associated to $G_{\mathcal{P}}$ is not a constant but a random variable. Let us consider $d = 2$ for simplicity; we shall see later that $p_{\mathrm{H}}(2) = 1/2$, so, for $p > 1/2$, we have $\theta(p) = \mathbb{E}\big(\theta(G_{\mathcal{P}}; p)\big) > 0$. It is easy to see that, for any k, there is a positive probability that $G_{\mathcal{P}}$ contains k disjoint triangles $T_1, \ldots, T_k$ each of which surrounds the origin. When this happens, we have $\theta(G_{\mathcal{P}}; p) \le (1-(1-p)^3)^k$: if all three sites in any T_i are closed, then there is a closed cycle surrounding the origin, and C_0 is bounded. It follows that, even for p close to 1, the random variable $\theta(G_{\mathcal{P}}; p)$ takes values arbitrarily close to 0 with positive probability. In particular, $\theta(G_{\mathcal{P}}; p) < \theta(p)$ with positive probability, so $\theta(G_{\mathcal{P}}; p)$ is not almost surely constant. In the other direction, it may happen that there is one enormous Voronoi cell centred on the origin, with hundreds of neighbours around it; in this case, for any fixed $p > 1/2$, the random variable $\theta(G_{\mathcal{P}}; p)$ can presumably take values arbitrarily close to 1.

As we saw in Chapter 1, for percolation on lattices, it is trivial that the critical probability is bounded away from 0 and 1. As we remarked earlier, the corresponding result for random Voronoi percolation is not entirely trivial. The approach that first springs to mind is comparison with independent percolation on a lattice; we applied this to Gilbert's model in the previous section. Unfortunately, the tessellation $V(\mathcal{P})$ can and will contain arbitrarily large cells. Thus, given any two points $x, y \in \mathbb{R}^d$, there is a positive probability that x and y lie in the same Voronoi cell, so the events 'x is black' and 'y is black' are *not* independent. One can get around this problem by comparing random Voronoi percolation to a suitably chosen 1-independent percolation model, and considering the event that the restriction of $(\mathcal{P}^+, \mathcal{P}^-)$ to a certain local region R forces certain points in $\mathbb{R}^d$ to be black, whatever $\mathcal{P}^+$ and $\mathcal{P}^-$ look like outside R.

Lemma 9. *For every $d \ge 2$, we have $p_{\mathrm{H}}(d) < 1$.*

Proof. We shall rescale, considering a Poisson process $\mathcal{P} = \mathcal{P}_\lambda$ on $\mathbb{R}^d$ with density λ much larger than 1.

For each bond e of the graph $\mathbb{Z}^d$, let L_e be the corresponding line segment of length 1 in $\mathbb{R}^d$, and write

$$R_e = L_e^{(1/2)} = \{\, x \in \mathbb{R}^d \,:\, \exists y \in L_e,\ \mathrm{dist}(x, y) \leq 1/2 \,\}$$

for the (closed) 1/2-neighbourhood of L_e.

Fixing the bond e for the moment, we can cover L_e by three closed balls $\overline{B}_{e,i}$, $i = 1, 2, 3$, of radius 1/4 whose interiors are disjoint; see Figure 10. Let $\overline{B}^2_{e,i}$ be the corresponding balls with the same centres but radius 1/2.

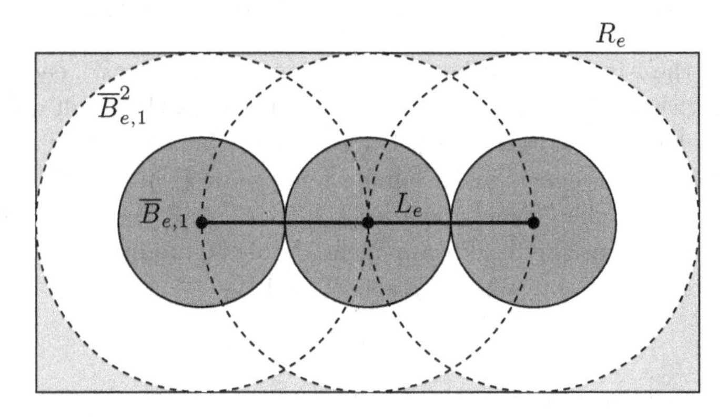

Figure 10. Three (closed) balls $\overline{B}_{e,i}$, $i = 1, 2, 3$, of radius 1/4 covering the unit line segment L_e, and the corresponding balls $\overline{B}^2_{e,i} \subset R_e$ with radius 1/2 (labelled only for $i = 1$). If each $\overline{B}_{e,i}$ contains a point of $\mathcal{P}^+$ and no $\overline{B}^2_{e,i}$ contains any point of $\mathcal{P}^-$, then every point of L_e lies in a black (open) Voronoi cell.

Let U_e be the event that each $\overline{B}_{e,i}$ contains at least one point of $\mathcal{P}^+$, and no $\overline{B}^2_{e,i}$ contains a point of $\mathcal{P}^-$. For $x \in \overline{B}_{e,i}$, every point of $\overline{B}_{e,i}$ is strictly closer to x than any point outside $\overline{B}^2_{e,i}$ is. Hence, whenever U_e holds, every point $x \in L_e \subset \bigcup_i \overline{B}_{e,i}$ is black.

The event U_e depends on the restriction of $(\mathcal{P}^+, \mathcal{P}^-)$ to the region R_e. If e and $f \in E(\mathbb{Z}^d)$ do not share a vertex, then R_e and R_f are disjoint, or meet only at boundary points of both. We may ignore the probability zero event that the boundary of any R_e contains a point of $\mathcal{P} = \mathcal{P}^+ \cup \mathcal{P}^-$. Thus, U_e and U_f are independent. In fact, if S, $T \subset E(\mathbb{Z}^d)$ are sets of edges at graph distance at least 1 from each other, then the sets $\bigcup_{e\in S} R_e$ and $\bigcup_{e\in T} R_e$ meet only in their boundaries, so the families $\{U_e : e \in S\}$

and $\{U_e : e \in T\}$ are independent. Thus, taking a bond e of $\mathbb{Z}^d$ to be open if and only if U_e holds, we have defined a 1-independent bond percolation model M on $\mathbb{Z}^d$.

The probability of the event U_e is

$$\mathbb{P}_{\lambda,p}(U_e) = \Big(1 - \exp\big(-p\lambda v_d\big)\Big)^3 \exp\big(-(1-p)\lambda u_d\big),$$

where v_d is the volume of a d-dimensional ball with radius $1/4$, and $u_d \leq 3 \times 2^d v_d$ is the volume of the union of the balls $\overline{B}^2_{e,i}$.

Choosing λ large enough that $0.9\lambda v_d > 10$, say, and then $0.9 < p < 1$ close enough to 1 that $(1-p)\lambda u_d < 1/10$, we have

$$\mathbb{P}_{\lambda,p}(U_e) \geq \Big(1 - e^{-10}\Big)^3 e^{-1/10} = 0.904\cdots > 0.8639.$$

Thus, in the 1-independent bond percolation model M on $\mathbb{Z}^d$, each bond is open with probability at least 0.8639. Hence, from the result of Balister, Bollobás and Walters [2005], Lemma 15 of Chapter 3, with positive probability the origin is in an infinite open path P in M. When this event happens, the geometric path $\bigcup_{e\in E(P)} L_e \subset \mathbb{R}^d$ consists entirely of black points, so the black component C_0 of the origin is unbounded. Hence, for the chosen values of p and λ, we have

$$\theta(p) = \mathbb{P}_{\lambda,p}(C_0 \text{ is unbounded}) > 0,$$

showing that $p_{\mathrm{H}} = p_{\mathrm{H}}(d) \leq p < 1$. □

In the other direction, we shall compare random Voronoi percolation with a 2-independent site percolation measure on $\mathbb{Z}^d$; the argument is due to Balister, Bollobás and Quas [2005].

Lemma 10. *For each $d \geq 2$ we have $p_{\mathrm{H}}(d) > 0$.*

Proof. For $v = (v_1, \ldots, v_d) \in \mathbb{Z}^d$ let H_v be the hypercube

$$H_v = \prod_{i=1}^{d} [v_i, v_i + 1],$$

and, for $r > 0$, let $H_v^{(r)}$ denote the r-neighbourhood of H_v. Let S_v be the event that some ball of radius $1/6$ with centre $x \in H_v$ contains no point of $\mathcal{P} = \mathcal{P}^+ \cup \mathcal{P}^-$, and let U_v be the event that some black component (union of open Voronoi cells) meets both H_v and a point of the boundary $\partial\big(H_v^{(1/6)}\big)$ of $H_v^{(1/6)}$. Note that if C_0 is unbounded, then U_v holds for every v such that C_0 meets H_v.

Clearly, the event S_v depends only on the restriction of $(\mathcal{P}^+, \mathcal{P}^-)$ to $H_v^{(1/6)}$. Also, U_v is the event that there is a black path in $H_v^{(1/6)}$ joining H_v to $\partial(H_v^{(1/6)})$. Hence, U_v depends only on the colours of the points x (of $\mathbb{R}^d$) with $x \in H_v^{(1/6)}$. If S_v does not hold, then the closest point of $\mathcal{P}$ to any $x \in H_v^{(1/6)}$ is within distance $1/3$, so the colour of x is determined by the restriction of $(\mathcal{P}^+, \mathcal{P}^-)$ to $H_v^{(1/2)}$. Thus, $U_v \setminus S_v$, and hence $U_v \cup S_v = S_v \cup (U_v \setminus S_v)$, is determined by the restriction of $(\mathcal{P}^+, \mathcal{P}^-)$ to $H_v^{(1/2)}$.

If $v, w \in \mathbb{Z}^d$ are such that $|v_i - w_i| \geq 2$ for some i, then the interiors of the sets $H_v^{(1/2)}$ and $H_w^{(1/2)}$ are disjoint. Let G be the graph with vertex set $\mathbb{Z}^d$ in which v and w are adjacent if $|v_i - w_i| \leq 1$ for every i, so G is a $(3^d - 1)$-regular graph. Taking v to be open if $U_v \cup S_v$ holds, we obtain a 2-independent site percolation model $\widetilde{M}$ on G. As $\lambda \to \infty$ and $p \to 0$ with $\lambda p \to 0$, we have $\mathbb{P}_{\lambda,p}(U_v \cup S_v) \to 0$. By Lemma 11 of Chapter 3, it follows that there are $\lambda > 0$ and $0 < p < 1$ such that, with probability 1, there is no infinite open cluster in $\widetilde{M}$. But then C_0 is finite with probability 1, so $p_{\mathrm{H}}(d) \geq p > 0$. □

The arguments above are very crude; we have included them only to give an indication of the basic techniques needed to handle the long-range dependence in random Voronoi percolation. For large d, however, it turns out that the method of Lemma 10 gives a bound that is not that far from the truth: the lower bound in the result of Balister, Bollobás and Quas [2005] below was obtained in this way. The upper bound is much more difficult.

Theorem 11. *If d is sufficiently large, then the critical probability $p_{\mathrm{H}}(d)$ for random Voronoi percolation on $\mathbb{R}^d$ satisfies*

$$2^{-d}(9d \log d)^{-1} \leq p_{\mathrm{H}}(d) \leq C 2^{-d} \sqrt{d} \log d,$$

where C is an absolute constant. □

We now turn to the main topic of this section, the critical probability for random Voronoi percolation in the plane. If $\mathcal{P}$ is a Poisson process in the plane, then with probability 1 the associated Voronoi tessellation has three cells meeting at every vertex; we shall assume this always holds from now on. Thus the graph $G_{\mathcal{P}}$ is a triangulation of the plane, and Euler's formula implies that the average degree of a vertex of $G_{\mathcal{P}}$ is 6. Thus, 'on average', $G_{\mathcal{P}}$ looks like the triangular lattice. This suggests that random Voronoi percolation in the plane may be similar to site percolation on the triangular lattice. Of course, critical probability is not just a function of average degree. For example, it is easy to see that for any (small) $\varepsilon > 0$ and any (large) d, $k \geq 3$, there is a k-connected lattice Λ with all degrees at least d such that $p_c^s(\Lambda) > 1 - \varepsilon$.

The numerical experiments of Winterfeld, Scriven and Davis [1981] mentioned earlier suggested that $p_H(2) = 1/2$. There is also a compelling mathematical reason for believing this: random Voronoi percolation in the plane has a self-duality property that implies analogues of Lemma 1 and Corollary 3 of Chapter 3, the basic starting point for the proof of the Harris–Kesten Theorem, and of Kesten's result that $p_H^s(T) = p_T^s(T) = 1/2$, where T is the triangular lattice (Theorem 8 of Chapter 5).

Let $R = [a, b] \times [c, d]$ be a rectangle in the plane. By a *black horizontal crossing* of R we mean a piecewise linear path $P \subset R$ starting on the left-hand side of R and ending on the the right-hand side, such that every point $x \in P$ is black, i.e., lies in the Voronoi cell of some $z \in \mathcal{P}$ with z open; see Figure 11. We write $H(R) = H_b(R)$ for the event that

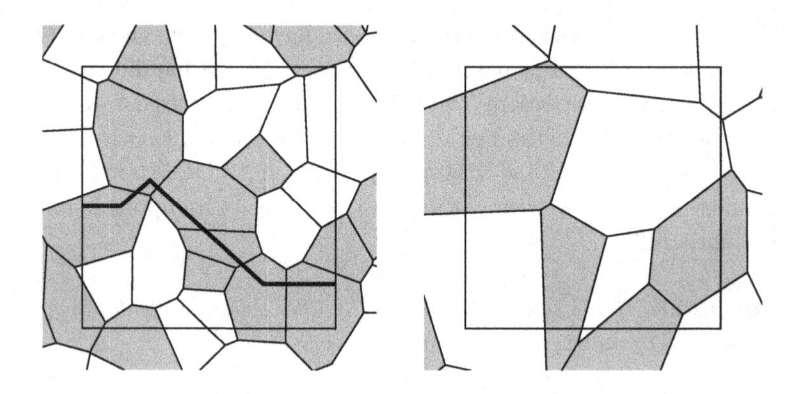

Figure 11. In the figure on the left, $H(R)$ holds: the thick line is a black horizontal crossing of R. In the figure on the right, $H(R)$ does not hold.

R has a black horizontal crossing. Equivalently, $H(R)$ is the event that the set of black points in R has a component meeting both the left- and right-hand sides of R.

Ignoring probability zero events, the event $H(R)$ holds if and only if there is an open path $P = v_1 v_2 \ldots v_t$ in $G_{\mathcal{P}}$ such that V_{v_1} meets the left-hand side of R, V_{v_t} meets the right-hand side, and, for $1 \leq i \leq t-1$, the cells V_{v_i} and $V_{v_{i+1}}$ meet inside R.

We write $H_w(R)$ for the event that R has a white horizontal crossing, and $V_b(R)$, $V_w(R)$ for the events that R has a black or white vertical crossing, respectively.

Lemma 12. *Let R be a rectangle in $\mathbb{R}^2$. Then precisely one of the events $H_b(R)$ and $V_w(R)$ holds.*

Proof. The proof is essentially the same as that of the corresponding result for bond percolation on $\mathbb{Z}^2$ (or site percolation on the triangular lattice). Defining the colours of the points in R from the Poisson processes $(\mathcal{P}^+, \mathcal{P}^-)$, colour the points outside R as in Figure 12.

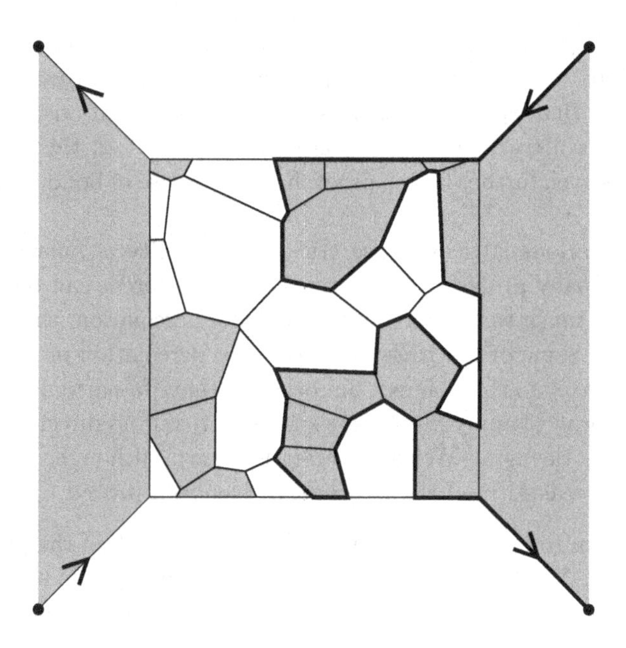

Figure 12. The shading inside the (square) rectangle R is from the Poisson process. $H_b(R)$ holds if and only if the outer black regions are joined by a black path, and $V_w(R)$ holds if and only if the outer white regions are joined by a white path. Tracing the interface between black and white regions shows that one of these events must hold.

Let I be the *interface graph*, i.e., the set of all points that are both black and white (lie in the boundaries of both black and white regions). With probability 1, the set I is a drawing in the plane of a graph in which every vertex has degree exactly 3, apart from four vertices of degree 1. As in previous proofs of this type, considering a path component of the interface shows that one of $H_{\mathrm{b}}(R)$ and $V_{\mathrm{w}}(R)$ must hold. If both hold, K_5 can be drawn in the plane. □

Corollary 13. *If S is any square $[x, x+a] \times [y, y+a]$, $a > 0$, then* $\mathbb{P}_{1/2}(H(S)) = 1/2$.

Proof. By Lemma 12, for $0 < p < 1$ we have $\mathbb{P}_p(H_{\mathrm{b}}(S)) + \mathbb{P}_p(V_{\mathrm{w}}(S)) = 1$. Flipping the states of all sites from open to closed or vice versa, we see that $\mathbb{P}_p(V_{\mathrm{w}}(S)) = \mathbb{P}_{1-p}(V_{\mathrm{b}}(S))$. From the symmetry of the Poisson process, $\mathbb{P}_{1-p}(V_{\mathrm{b}}(S)) = \mathbb{P}_{1-p}(H_{\mathrm{b}}(S))$. Thus, $\mathbb{P}_p(H_{\mathrm{b}}(S)) + \mathbb{P}_{1-p}(H_{\mathrm{b}}(S)) = 1$. Taking $p = 1/2$, the result follows. □

Freedman [1997] proved an analogue of Corollary 13 concerning 'essential loops' in random Voronoi percolation on the projective plane. He remarks that $p_{\mathrm{H}}(2) = 1/2$ by symmetry considerations. However, although Corollary 13 is the 'reason why' $p_{\mathrm{H}}(2) = 1/2$, this trivial observation is even further from a proof than in the case of bond percolation on $\mathbb{Z}^2$.

For the various lattices whose critical probability is known exactly, any of the many proofs of the Harris–Kesten Theorem can be adapted without too much work. For random Voronoi percolation, the situation is different. Some of the tools used to study percolation on lattices do carry over easily to this context, but others do not. In particular, there is no obvious way to apply Menshikov's Theorem, and no direct equivalent of the Russo–Seymour–Welsh Theorem is known, although, as we shall see later, a weak form of the latter result has been proved.

Let us turn to our main aim in this section, a sketch of the proof that $p_{\mathrm{H}}(2) = 1/2$. We start with Harris's Lemma, the first of the basic results for ordinary percolation that do extend to the Voronoi setting. To state its analogue for Voronoi percolation, we need to define 'increasing events' in this context.

We call an event E defined in terms of the Poisson processes $(\mathcal{P}^+, \mathcal{P}^-)$ *black-increasing*, or simply *increasing*, if, for every configuration $\omega_1 = (X_1^+, X_1^-)$ in E and every configuration $\omega_2 = (X_2^+, X_2^-)$ with $X_1^+ \subset X_2^+$ and $X_1^- \supset X_2^-$, we have $\omega_2 \in E$. In other words, E is preserved by the addition of open points, or the deletion of closed points. The definition of an *increasing function* $f(\mathcal{P}^+, \mathcal{P}^-)$ is analogous: f is increasing if adding open points or deleting closed points cannot decrease the value of f.

Since a point of $\mathbb{R}^2$ is black if the (a) nearest point of $\mathcal{P}$ is open, adding open points and deleting closed points can change the colour of a point

$x \in \mathbb{R}^2$ from white to black, but not vice versa. Thus, any event defined by the existence of certain black sets is increasing; an example is the event $H(R) = H_b(R)$.

It is easy to see from Harris's Lemma that, with respect to $\mathbb{P}_p$, increasing events are positively correlated, at least if they are sufficiently well behaved. In fact, all increasing events are positively correlated. To see this, let us first consider the case of a single Poisson process Y of intensity 1 on $\mathbb{R}^2$. Let us write $(\Omega_1, \mathbb{P}_1)$ for the corresponding probability space, so Ω_1 is the set of all discrete subsets of $\mathbb{R}^2$, and $\mathbb{P}_1$ is the probability measure defining the Poisson process. We shall write $\mathbb{E}_1$ for the associated expectation. Let g_i, $i = 1, 2$, be two bounded measurable functions on $(\Omega_1, \mathbb{P}_1)$, and suppose that each g_i is increasing, in the sense that $\omega \subset \omega'$ implies $g_i(\omega) \le g_i(\omega')$.

Following Roy [1991], let Σ_k be the σ-field generated by the following information: set $n = 2^k$, divide $[-n, n]^2$ into $4n^4$ squares of side-length $1/n$, and decide for each whether or not it contains points of $Y \in \Omega_1$. Thus Σ_k partitions Ω_1 into 2^{4n^4} parts, with each part consisting of all $Y \in \Omega_1$ that have at least one point in each of certain small squares, and no points in certain other small squares. Any two discrete sets lie in different parts of Σ_k for large enough k, so the sequence Σ_k is a filtration of $(\Omega_1, \mathbb{P}_1)$.

As $\mathbb{E}_1(g_i \mid \Sigma_k)$ may be viewed as an increasing function on a discrete product space, from Lemma 4 of Chapter 2 (a simple corollary of Harris's Lemma) we have

$$\begin{aligned} \mathbb{E}_1\big(\mathbb{E}_1(g_1 \mid \Sigma_k)\mathbb{E}_1(g_2 \mid \Sigma_k)\big) &\ge \mathbb{E}_1\big(\mathbb{E}_1(g_1 \mid \Sigma_k)\big)\mathbb{E}_1\big(\mathbb{E}_1(g_2 \mid \Sigma_k)\big) \quad (5) \\ &= \mathbb{E}_1(g_1)\mathbb{E}_1(g_2). \end{aligned}$$

Since g_i is bounded and Σ_k is a filtration, by the Martingale Convergence Theorem (see, for example, Williams [1991])

$$\mathbb{E}_1(g_i \mid \Sigma_k) \to g_i$$

almost everywhere. Hence, by dominated convergence, the left-hand side of (5) converges to $\mathbb{E}_1(g_1 g_2)$, proving that

$$\mathbb{E}_1(g_1 g_2) \ge \mathbb{E}_1(g_1)\mathbb{E}_1(g_2) \qquad (6)$$

whenever g_1 and g_2 are measurable, bounded and increasing.

The corresponding result for the coloured process follows immediately. We state this as a lemma, giving only the characteristic function case – although the proof for increasing functions is the same, we shall not need the result. Here, then, is the equivalent of Harris's Lemma for random Voronoi percolation in the plane. Of course, a corresponding result holds for random Voronoi percolation in $\mathbb{R}^d$.

Lemma 14. *Let $E_i = E_i(\mathcal{P}^+, \mathcal{P}^-)$, $i = 1, 2$, be two black-increasing events. Then for any $0 \le p \le 1$ we have*

$$\mathbb{P}_p(E_1 \cap E_2) \ge \mathbb{P}_p(E_1)\mathbb{P}_p(E_2),$$

where $\mathbb{P}_p$ is the probability measure associated to random Voronoi percolation in the plane.

Proof. We shall write $\mathbb{E}$ for $\mathbb{E}_p$, i.e., expectation with respect to $\mathbb{P}_p$. Let f_i be the characteristic function of E_i, so f_i is black-increasing. Fixing $\mathcal{P}^-$, for each i, $f_i(\mathcal{P}^+, \mathcal{P}^-)$ is increasing in $\mathcal{P}^+$, so by (6),

$$\mathbb{E}(f_1 f_2 \mid \mathcal{P}^-) \ge \mathbb{E}(f_1 \mid \mathcal{P}^-)\mathbb{E}(f_2 \mid \mathcal{P}^-),$$

for every possible $\mathcal{P}^-$. Taking the expectation of both sides (over $\mathcal{P}^-$),

$$\mathbb{E}(f_1 f_2) \ge \mathbb{E}\big(\mathbb{E}(f_1 \mid \mathcal{P}^-)\mathbb{E}(f_2 \mid \mathcal{P}^-)\big).$$

But as f_1, f_2 are decreasing in $\mathcal{P}^-$, so are the functions $g_i = \mathbb{E}(f_i \mid \mathcal{P}^-)$. Applying (6) to $1 - g_1$ and $1 - g_2$, which are increasing functions of $\mathcal{P}^-$ only, we see that $\mathbb{E}(g_1 g_2) \ge \mathbb{E}(g_1)\mathbb{E}(g_2)$, i.e., that

$$\begin{aligned} \mathbb{E}\big(\mathbb{E}(f_1 \mid \mathcal{P}^-)\mathbb{E}(f_2 \mid \mathcal{P}^-)\big) &\ge \mathbb{E}\big(\mathbb{E}(f_1 \mid \mathcal{P}^-)\big)\mathbb{E}\big(\mathbb{E}(f_2 \mid \mathcal{P}^-)\big) \\ &= \mathbb{E}(f_1)\mathbb{E}(f_2). \end{aligned}$$

Combining the last two inequalities, we see that $\mathbb{E}(f_1 f_2) \ge \mathbb{E}(f_1)\mathbb{E}(f_2)$, as claimed. □

The uniqueness theorem of Aizenman, Kesten and Newman [1987], Theorem 4 of Chapter 5, also carries over to the Voronoi setting (in any dimension). Just as for the Gilbert model, the proof given by Burton and Keane [1989] goes through, although a little care is needed with the details; we shall not spell these out.

Theorem 15. *For any $\lambda > 0$ and any $0 < p < 1$, there is almost surely at most one infinite open cluster in $G_{\mathcal{P}}$.* □

In Chapter 5, we presented a proof of Harris's Theorem due to Zhang. This proof relied on the basic crossing lemma (Lemma 1 of Chapter 3),

Harris's Lemma, and the symmetry of the square lattice. In his unpublished M.Sc. thesis, Zvavitch [1996] pointed out that Zhang's proof carries over to the random Voronoi setting, giving the following result.

Theorem 16. *For random Voronoi percolation in the plane, $\theta(1/2) = 0$. Hence, $p_{\mathrm{H}}(2) \geq 1/2$.* □

Unfortunately, none of the many proofs of Kesten's Theorem seems to adapt to the random Voronoi setting: there seems to be no easy way of proving the analogue of Kesten's Theorem. Nevertheless, using an argument which is considerably more involved that any of the proofs of Kesten's Theorem, Bollobás and Riordan [2006a] did prove this analogue.

Theorem 17. *For random Voronoi percolation in the plane, $p_{\mathrm{H}} = 1/2$.*

As we shall see at the end of this section, the proof of this result also gives an analogue of Kesten's exponential decay result, Theorem 12 of Chapter 3, and so establishes that the natural analogue of p_{T} is also equal to 1/2.

As remarked above, the proof of Theorem 17 is rather involved, so we shall describe only its key ideas; for full details we refer the reader to the original paper.

Perhaps surprisingly, the lack of independence in the random Voronoi model turns out not to be the main problem (or even one of the main problems). Indeed, it is easy to see that for rectangles R and R', say, separated by a moderate distance, any events E, E' depending only on the colours of the points in R and R' respectively are almost independent. In particular, there are independent events $\widetilde{E}$, $\widetilde{E}'$ that almost coincide with E and E': this property turns out to be just as good as independence. We make this idea more precise in the lemma below. Here, we think of s as the *scale* of the rectangles R and R'; we shall take $s \to \infty$ with all other parameters fixed. From now on, we fix the irrelevant scaling of the Poisson process $\mathcal{P} = \mathcal{P}_\lambda$ by setting $\lambda = 1$.

Given $\rho > 1$, $s > 1$ and a ρs by s rectangle $R_s \subset \mathbb{R}^2$ with any orientation, we write $F(R_s) = F_r(R_s)$ for the event that every ball $B_r(x)$, $x \in R_s$, contains at least one point of $\mathcal{P} = \mathcal{P}^+ \cup \mathcal{P}^-$, where $r = 2\sqrt{\log s}$. (It does not matter whether we consider open or closed balls; up to a probability zero event, which we shall ignore, this makes no difference.)

Lemma 18. *Let $\rho \geq 1$ be constant, let $R_s \subset \mathbb{R}^2$ be a ρs by s rectangle, $s > 1$, and set $r = 2\sqrt{\log s}$. Then $F(R_s) = F_r(R_s)$ holds with probability $1 - o(1)$ as $s \to \infty$. Also, if $E(R_s)$ is any event defined only in terms of the colours of the points in R_s, then $E(R_s) \cap F(R_s)$ depends only on the restriction of $(\mathcal{P}^+, \mathcal{P}^-)$ to the r-neighbourhood of R_s.*

Proof. The first statement is immediate from the properties of a Poisson process. Indeed, we may cover R_s with $O(s^2/r^2) = o(s^2)$ disjoint (half-open) squares S_i of side-length $r/\sqrt{2}$. The area of each S_i is $r^2/2 = 2\log s$, so the number of points of $\mathcal{P}$ in S_i has a Poisson distribution with mean $2\log s$. Hence, the probability that a particular S_i contains no points of $\mathcal{P}$ is exactly s^{-2}. Thus, with probability $1 - o(1)$ every S_i contains a point of $\mathcal{P}$, and $F(R_s)$ holds.

The second statement is immediate from the definition of $F(R_s)$; the argument is as in the proof of Lemma 10 above. Indeed, $F(R_s)$ clearly depends only on the restriction of $(\mathcal{P}^+, \mathcal{P}^-)$ to the r-neighbourhood $R_s^{(r)}$ of R_s. If $F(R_s)$ holds, then the closest point of $\mathcal{P}$ to any $x \in R_s$ is a point of $B_r(x)$, and hence a point of $\mathcal{P} \cap R_s^{(r)}$. Thus the colour of x is determined by the closest point of the restriction of $(\mathcal{P}^+, \mathcal{P}^-)$ to $R_s^{(r)}$. □

Of course, there is nothing special about rectangles in Lemma 18; choosing the shorter side of the rectangle as our 'scale parameter' s will be convenient later. Also, Lemma 18 has an analogue in any dimension, concerning cuboids, say.

Turning now to the specific study of dimension two, the fundamental quantity that we work with is the probability of a black crossing of a rectangle. Let

$$f_p(\rho, s) = \mathbb{P}_p\big(H([0, \rho s] \times [0, s])\big)$$

be the $\mathbb{P}_p$-probability that a ρs by s rectangle has a black horizontal crossing. We start with two easy observations about the behaviour of $f_p(\rho, s)$. First,

$$f_p(\rho_1, s) \leq f_p(\rho_2, s)$$

whenever $\rho_1 \leq \rho_2$, as the corresponding events are nested.

The second observation is that

$$f_p(\rho_1 + \rho_2 - 1, s) \geq f_p(\rho_1, s) f_p(\rho_2, s) f_p(1, s) \tag{7}$$

for all $\rho_1, \rho_2 > 1$. The argument is exactly as in Chapter 3: let R_1 and R_2 be $\rho_1 s$ by s and $\rho_2 s$ by s rectangles intersecting in an s by s square

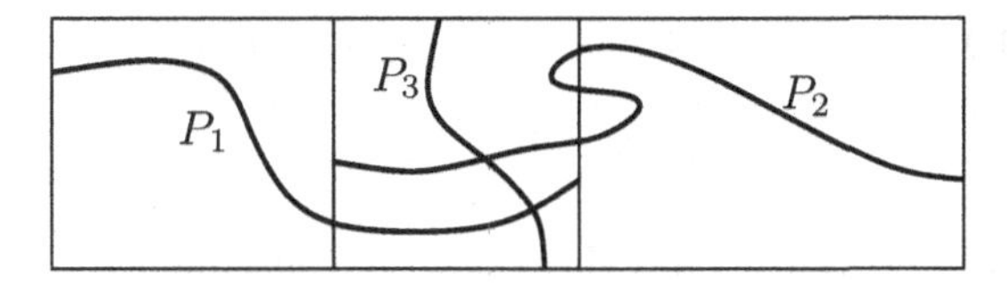

Figure 13. Two rectangles R_1 and R_2 intersecting in a square S. If P_1 and P_2 are black horizontal crossings of R_1 and R_2, respectively, and P_3 is a black vertical crossing of S, then $P_1 \cup P_2 \cup P_3$ contains a black horizontal crossing of $R_1 \cup R_2$.

S, as in Figure 13. The events $H(R_1)$, $H(R_2)$ and $V(S)$ are increasing. Thus, by Harris's Lemma, in the form of Lemma 14,

$$\mathbb{P}_p\big(H(R_1) \cap H(R_2) \cap V(S)\big) \geq \mathbb{P}_p(H(R_1))\mathbb{P}_p(H(R_2))\mathbb{P}_p(V(S)).$$

If $H(R_1) \cap H(R_2) \cap V(S)$ holds, then so does $H(R_1 \cup R_2)$ (see Figure 13). As $\mathbb{P}_p(V(S)) = \mathbb{P}_p(H(S)) = f_p(1, s)$, the relation (7) follows.

The natural analogue of the Russo–Seymour–Welsh (RSW) Theorem would state that if $f_p(1, s) > \varepsilon > 0$ then $f_p(\rho, s) > g(\rho, \varepsilon) > 0$, for some function $g(\rho, \varepsilon)$ not depending on s (so, in particular, $f_{1/2}(2, s)$ is bounded away from zero). Such a statement has not been proved for random Voronoi percolation.

It might seem that the proof of the RSW-type theorem we presented in Chapter 3 would carry over to the random Voronoi setting. The start is indeed promising: there is no problem defining the 'left-most black vertical crossing' $LV(R)$ of R whenever $V(R)$ holds; we simply follow a black-white interface as before. The problem is the next step: the event $LV(R) = P$ is independent of the *states* of the points of $\mathcal{P}$ to the right of P, but not of their *positions*: we can find $LV(R)$ without looking at the colours of the cells to the right of $LV(R)$, but knowing $LV(R)$ tells us where the centres of these cells are, and that there are no points of $\mathcal{P}$ in certain discs, parts of which are to the right of $LV(R)$. Thus, there is no simple way of showing that the probability that $LV(R)$ is joined by a black path to the right side of R is at least $\mathbb{P}(H(R))$. Of course, a general problem with random Voronoi percolation is that no two regions are independent, although well-separated regions are asymptotically independent.

The first step towards the proof of Theorem 17 is the following result; although this is considerably weaker than a direct analogue of the RSW Theorem, it turns out to be sufficient for the determination of $p_{\mathrm{H}}(2)$.

Theorem 19. *Let* $0 < p < 1$ *and* $\rho > 1$ *be fixed. If* $\liminf_{s\to\infty} f_p(1,s) > 0$, *then* $\limsup_{s\to\infty} f_p(\rho,s) > 0$.

Lemma 12 states that the hypothesis of this theorem is satisfied for $p = 1/2$; thus Theorem 19 has the following consequence.

Corollary 20. *Let* $\rho > 1$ *be fixed. There is a constant* $c_0 = c_0(\rho) > 0$ *such that for every* s_0 *there is an* $s > s_0$ *with* $f_{1/2}(\rho,s) \geq c_0$. □

The proof of Theorem 19 is rather lengthy, so we shall give only a sketch.

Proof. We proceed in two stages, throughout assuming for a contradiction that the result does not hold. Thus, there is a constant $c_1 > 0$ such that

$$f_p(1,s) \geq c_1 \tag{8}$$

for all large enough s, but, for some fixed $\rho > 1$,

$$f_p(\rho,s) \to 0. \tag{9}$$

Here, and throughout, the limit is taken as $s \to \infty$ with all other parameters, for example ρ, fixed. In this context, an event holds *with high probability*, or *whp*, if it holds with probability $1 - o(1)$ as $s \to \infty$.

The assumptions above imply that, for any fixed $\varepsilon > 0$,

$$f_p(1+\varepsilon,s) \to 0 \tag{10}$$

as $s \to \infty$. Indeed, taking $k \geq (\rho-1)/\varepsilon$ and using (7) k times,

$$f_p(\rho,s) \geq f_p(1+k\varepsilon,s) \geq f_p(1+\varepsilon,s)f_p(1,s)f_p(1+(k-1)\varepsilon,s) \geq \cdots \geq \big(f_p(1+\varepsilon,s)f_p(1,s)\big)^k f_p(1,s) \not\to 0,$$

contradicting (9).

Condition (10) imposes very severe restrictions on the possible black paths crossing a square, for example: roughly speaking, no segment of such a path can cross horizontally a rectangle that is even slightly longer than high. Thus, for any $\varepsilon > 0$, whp all black horizontal crossings of a given s by s square S pass within distance εs of the top and bottom of S: otherwise, with positive probability we could find a black horizontal crossing of one of two s by $(1-\varepsilon)s$ rectangles, contradicting (10).

The next observation is that whp any black horizontal crossing of an s by s square S starts and ends near the midpoints of the vertical sides. Indeed, if there is a positive probability that some black crossing P starts at least $\varepsilon s/2$ above the midpoint of the left-hand side, then, reflecting S in a horizontal line and shifting it vertically by εs to obtain a square S', there is also a positive probability that some black horizontal crossing P' of S' starts at least $\varepsilon s/2$ below the middle height of S', and hence below P. By Harris's Lemma, crossings P and P' with the stated properties then exist simultaneously with positive probability. But, whp, P must pass within distance $\varepsilon s/3$, say, of the bottom of S, and hence, below S'. This forces P and P' to meet; see Figure 14. As P' passes

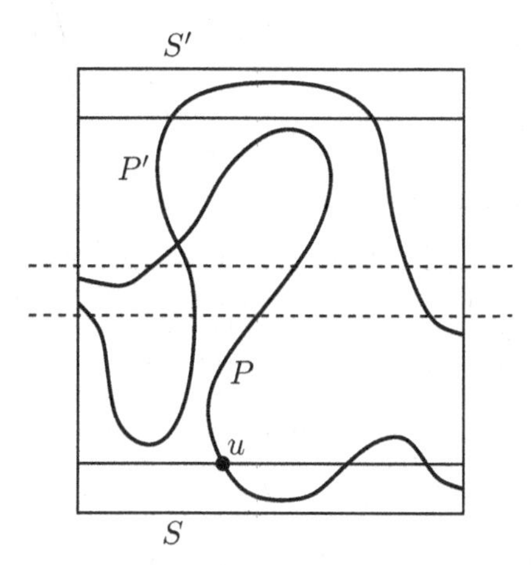

Figure 14. Two squares S and S'; their horizontal axes of symmetry are shown by dotted lines. Each of the horizontal crossings P, P' of S and S' passes near the top and bottom of the square it crosses. If P starts above P', then the paths cross: P' must leave the region bounded by part of the boundary of S' and the initial segment of P up to the point u.

within distance $\varepsilon s/3$ of the top of S', we find a black vertical crossing of a rectangle taller than it is wide. Using rotational symmetry of the model, this contradicts (10).

In a moment, we shall state precisely a consequence of (10). For now, we continue with our sketch of the argument. Suppose that P is a black horizontal crossing of the square $S = [0, s] \times [0, s]$. Let P_1 be the initial segment of P, starting on the line $x = 0$ and stopping the first time we reach $x = 0.99s$. Similarly, let P_2 be the final segment of P, obtained by tracing P backwards from the line $x = s$ until the first time we reach $x = 0.01s$; see Figure 15. Arguments similar to those above show that each of P_1 and P_2 starts and ends very close to the line $y = s/2$. Furthermore, one can show that, whp, since each P_i crosses a

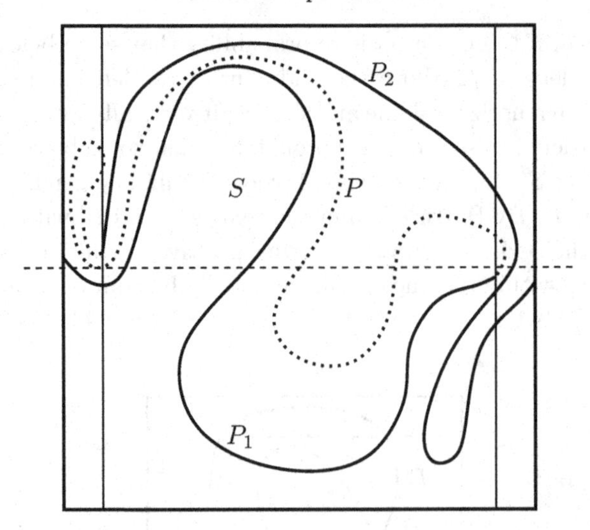

Figure 15. A square S (outer lines) and a black path P crossing it (solid and dotted curves). The path P_1 is the initial segment of P stopping at the inner vertical line. The final segment P_2 of P is defined similarly. Roughly speaking, each P_i crosses a square smaller than S, and so cannot approach too close to the top and bottom of S. Thus, $P_1 \cup P_2 \neq P$: otherwise a slightly flattened rectangle in S would have a black horizontal crossing. The dotted part of P in fact passes very close to the top and bottom of S.

rectangle of width $0.99s$, it remains within height $\pm(0.495+\varepsilon)s$ of its starting points. It follows that P_1 and P_2 are whp disjoint: otherwise, $P = P_1 \cup P_2$ is a black crossing of S that does not come within height $0.004s$, say, of the top and bottom of S; the probability that such a crossing exists is $o(1)$ by (10). But then each of P_1 and P_2 similarly contains two disjoint crossings of rectangles of width 0.98, and so on.

It is not hard to make the vague argument just outlined precise, although it turns out to be better to work in a rectangle, $[0, s] \times [-Cs, Cs]$, say, of bounded aspect ratio $2C > 1$. In Bollobás and Riordan [2006a], the following consequence of (10) is proved.

Claim 21. *Let $C > 0$ be fixed, and let $R = R_s$ be the s by $2Cs$ rectangle $[0, s] \times [-Cs, Cs]$. For $0 \leq j \leq 4$, set $R_j = [js/100, (j+96)s/100] \times [-Cs, Cs]$. Assuming that (10) holds, whp every black path P crossing R horizontally contains 16 disjoint black paths P_i, $1 \leq i \leq 16$, where each P_i crosses some R_j horizontally.* □

In other words, the path P contains 16 disjoint paths P_i winding backwards and forwards in a manner similar to the paths in Figure 15. Of course, in proving Claim 21, one must be a little more careful than in the vague outline above. Nevertheless, the proof is fairly straightforward; it uses only Harris's Lemma, certain symmetries of the model, and (10). The last condition is applied to show that whp none of a fixed number $N = N(C, \varepsilon)$ of rectangles with aspect ratio $1 + \varepsilon$ is crossed the long way by a black path.

The conclusion of Claim 21 is clearly absurd: R has a black horizontal crossing with probability at least $f_p(1, s)$, which is bounded away from zero. Taking the shortest black horizontal crossing of R, this somehow winds backwards and forwards almost all the way across R, containing segments that start and end in almost the same place but somehow never meet each other. Also, the shortest crossing of R is almost certainly at least 16 times longer than the shortest crossing of a slightly narrower rectangle. As the length of a crossing of R_s cannot scale as more than s^2, this is impossible.

All this sounds convincing; nevertheless, it is not so easy to deduce a contradiction. The problem is that, while the constant 16 in Claim 21 can be replaced by any larger *constant*, it cannot be replaced by a function of s tending to infinity: we can only apply (10) directly to a bounded number of rectangles. Otherwise, the $o(1)$ error probabilities accumulate.

To get around this, we use almost-independence of disjoint regions to 'square the error probability'. The idea is to consider the length (as a piecewise linear path) $L(R_s)$ of the shortest black horizontal crossing P of certain s by $2s$ rectangles R_s, when such a P exists. We take $L(R_s) = \infty$ if R_s has no black horizontal crossing. Even for widely separated rectangles, $L(R_s)$ and $L(R'_s)$ are not independent, so we modify the definition slightly to achieve independence.

As in Lemma 18, let $F(R_s) = F_r(R_s)$ be the event that every ball of radius $r = 2\sqrt{\log s}$ centred at a point in R_s contains at least one point of $\mathcal{P}$, and set

$$\widetilde{L}(R_s) = \begin{cases} L(R_s) & \text{if } F(R_s) \text{ holds,} \\ 0 & \text{otherwise.} \end{cases}$$

By Lemma 18, $\widetilde{L}(R_s) = L(R_s)$ whp. Furthermore, $\widetilde{L}(R_s)$ depends only on the restriction of $(\mathcal{P}^+, \mathcal{P}^-)$ to the r-neighbourhood of R_s. Hence, if R_s and R'_s are separated by a distance of at least $2r = o(s)$, the random variables $\widetilde{L}(R_s)$ and $\widetilde{L}(R'_s)$ are independent.

Roughly speaking, we wish to relate the distribution of $\widetilde{L}(R_s)$, which depends only on s, to the distribution of $\widetilde{L}(R_{s/2})$, using Claim 21. In fact, to leave a little elbow room, we relate $\widetilde{L}(R_s)$ to $\widetilde{L}(R_{0.47s})$, say. For $\eta > 0$ a (very small) constant, define $g(s)$ by

$$g(s) = \sup\{\, x : \mathbb{P}_p(\widetilde{L}(R_s) < x) \leq \eta \,\},$$

where R_s is any s by $2s$ rectangle. Recall that that $L(R_s) < \infty$ if and only if R_s has some black horizontal crossing, an event of probability $f_p(1/2, 2s) \geq f_p(1, s) \geq c_1 > 0$, from (8). As $\widetilde{L}(R_s) = L(R_s)$ whp, it follows that $0 < g(s) < \infty$ if η is chosen small enough, and then s large enough, which we shall assume from now on. Also, the supremum above is attained, so

$$\mathbb{P}_p(\widetilde{L}(R_s) < g(s)) \leq \eta. \tag{11}$$

Claim 21 tells us that, whp, every black horizontal crossing of R_s, including the shortest, contains 16 disjoint crossings P_i of slightly narrower rectangles. Using the fact that whp every crossing of a $0.96s$ by $2s$ rectangle is actually almost a crossing of a square, we can place a bounded number of $0.47s$ by $0.94s$ rectangles R_i, $1 \leq i \leq N$, such that, whp, each P_i includes crossings of some pair (R_j, R_k) of these rectangles, with R_j and R_k separated by a distance of at least $0.01s$. Here N is an absolute constant. For each such pair (R_j, R_k), the variables $\widetilde{L}(R_j)$ and $\widetilde{L}(R_k)$ are independent, so

$$\begin{aligned}\mathbb{P}_p\Big(\widetilde{L}(R_j) + \widetilde{L}(R_k) < g(0.47s)\Big) &\leq \mathbb{P}_p\Big(\widetilde{L}(R_j), \widetilde{L}(R_k) < g(0.47s)\Big) \\ &= \mathbb{P}_p\Big(\widetilde{L}(R_j) < g(0.47s)\Big)\mathbb{P}_p\Big(\widetilde{L}(R_k) < g(0.47s)\Big) \leq \eta^2,\end{aligned}$$

from (11). Hence, the probability that $\widetilde{L}(R_j) + \widetilde{L}(R_k) < g(0.47s)$ holds for *some* pair (R_j, R_k) separated by distance $0.01s$ is at most $N\eta^2$, and hence at most $\eta/2$ if we choose η small enough.

The proof of Theorem 19 is essentially complete: from the remarks above, whp every black horizontal crossing of R_s has length at least 16 times the minimum of $\widetilde{L}(R_j) + \widetilde{L}(R_k)$ over separated pairs (R_j, R_k). Thus,

$$\mathbb{P}_p\Big(\widetilde{L}(R_s) < 16g(0.47s)\Big) \leq \eta/2 + o(1) < \eta,$$

so $g(s) \geq 16g(0.47s)$, and $g(s)$ grows faster than s^3, say. As $g(s_0) > 0$ for some s_0, it follows that there are arbitrarily large s with $s^3 < g(s) < \infty$. But then, with probability bounded away from 1, the shortest black horizontal crossing of $R_s = [0, s] \times [0, 2s]$ has length at least s^3. This is impossible, since, whp, R_s meets only $O(s^2)$ Voronoi cells, and each has diameter at most $O(\log s)$. □

As noted in the original paper, the proof of Theorem 19 outlined above uses rather few properties of random Voronoi percolation: certain symmetries, the fact that horizontal and vertical crossings of a rectangle must meet, and an asymptotic independence property. For this reason, the proof carries over to many other contexts; see, for example, Bollobás and Riordan [2006b]. Van den Berg, Brouwer and Vágvölgyi [2006] proved a variant of this result for 'self-destructive percolation', a model of forest growth taking into account forest fires, which they used to prove results about the continuity of the percolation probability in this model.

Using Corollary 20 and Harris's Lemma, Theorem 19 easily implies that $\theta(1/2) = 0$, giving an alternative proof of Theorem 16. Of course, this result can be proved more cleanly by adapting Zhang's proof of Kesten's Theorem; see Zvavitch [1996]. However, Corollary 20 is a good starting point for the proof of the analogue of Kesten's Theorem, namely Theorem 17.

The main idea of the proof is to use some kind of sharp-threshold result to deduce that, for any $p > 1/2$ and any s_0, there is an $s > s_0$ with

$$f_p(3, s) \geq 0.99, \tag{12}$$

say. Before sketching the (rather lengthy) proof of (12), let us see how Theorem 17 follows. The argument, based on 1-independent bond percolation on $\mathbb{Z}^2$, is very close to an argument given in Chapter 3.

Given a $3s$ by s rectangle R_s with the long axis horizontal, let

$$G(R_s) = H(R_s) \cap V(S_1) \cap V(S_2),$$

where S_1 and S_2 are the two s by s 'end squares' of R_s; see Figure 16.

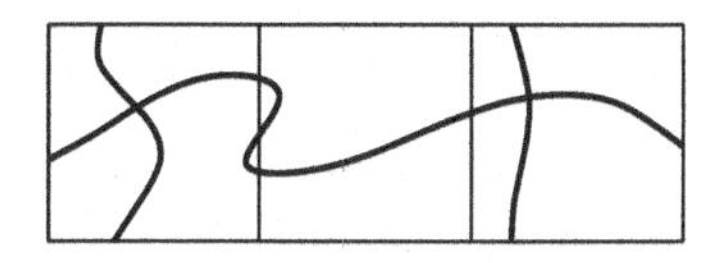

Figure 16. A $3s$ by s rectangle R_s such that $G(R_s)$ holds.

By translational and rotational symmetry,

$$\mathbb{P}_p\big(V(S_2)\big) = \mathbb{P}_p\big(V(S_1)\big) = \mathbb{P}_p\big(H(S_1)\big) = f_p(1, s) \geq f_p(3, s) \geq 0.99,$$

so $\mathbb{P}_p(G(R_s)) \geq 0.97$. Let

$$\widetilde{G}(R_s) = G(R_s) \cap F(R_s),$$

and define $\widetilde{G}(R'_s)$ similarly for a $3s$ by s rectangle R'_s with the long side vertical. From Lemma 18, $\mathbb{P}_p(F(R_s)) \to 1$ as $s \to \infty$. Thus, if s is large enough,

$$\mathbb{P}_p\big(\widetilde{G}(R_s)\big) \geq 0.9 \tag{13}$$

for every $3s$ by s rectangle R_s. By Lemma 18, the event $\widetilde{G}(R_s)$ depends only on the restriction of the Poisson processes $(\mathcal{P}^+, \mathcal{P}^-)$ to the r-neighbourhood of R_s. Choosing s large enough, we may assume that $r = 2\sqrt{\log s} < s/2$.

For each horizontal bond $e = ((a, b), (a + 1, b))$ of $\mathbb{Z}^2$, let

$$R_e = [2as, 2as + 3s] \times [2bs, 2bs + s]$$

be a $3s$ by s rectangle in $\mathbb{R}^2$. Similarly, for $e = ((a, b), (a, b + 1))$, set

$$R_e = [2as, 2as + s] \times [2bs, 2bs + 3s].$$

Let us define a bond percolation model $\widetilde{M}$ on $\mathbb{Z}^2$ by taking a bond e to be open if and only if $\widetilde{G}(R_e)$ holds. If e and f are vertex-disjoint bonds of $\mathbb{Z}^2$, then the rectangles R_e and R_f are separated by a distance of at least s. Hence, if S and T are sets of bonds of $\mathbb{Z}^2$ at graph distance at least 1, then the families $\{\widetilde{G}(R_e) : e \in S\}$ and $\{\widetilde{G}(R_e) : e \in T\}$ depend on the restrictions of the Poisson processes $(\mathcal{P}^+, \mathcal{P}^-)$ to the disjoint regions $\bigcup_{e\in S} R_e^{(r)}$ and $\bigcup_{e\in T} R_e^{(r)}$, and are thus independent. Hence, the probability measure associated to $\widetilde{M}$ is 1-independent. From (13), each bond is open in $\widetilde{M}$ with probability at least 0.9 so, from Lemma 15 of Chapter 3, with positive probability the origin is in an infinite open path in $\widetilde{M}$. Such a path guarantees an infinite black path meeting $[0, s]^2$ (see Figure 14 of Chapter 3), so

$$\mathbb{P}_p\big([0, s]^2 \text{ meets an unbounded black component}\big) > 0.$$

The increasing event that every point in $[0, s]^2$ is black has positive probability. If this event holds and $[0, s]^2$ meets an unbounded black component, then C_0 is unbounded. From Harris's Lemma, this happens with positive probability, so $\theta(p) > 0$. As $p > 1/2$ was arbitrary, and $\theta(1/2) = 0$, Theorem 17 follows.

Our aim for the rest of the section is to sketch the proof of (12), that for any $p > 1/2$ and any s_0 there is an $s > s_0$ with $f_p(3, s) \geq 0.99$; as we have seen, this implies Theorem 17. As a starting point, Corollary 20 is strong enough: one only needs $f_{1/2}(3,s) > c_0$ for *some* sufficiently large s. In a discrete setting, we could easily use the Friedgut–Kalai sharp-threshold result (Theorem 13 of Chapter 2) to deduce that, for this s, we have $f_p(3,s) > 0.99$. It is not unreasonable to expect it to be a simple matter to adapt this argument to random Voronoi percolation, by choosing a suitable discrete approximation, as in the proof of Theorem 3, say. However, as we shall see, there is a problem.

In Chapter 3, we presented various methods of deducing the statement for bond percolation on $\mathbb{Z}^2$ that corresponds to $f_p(3,s) \to 1$ for $p > 1/2$ fixed. One of these, the method based on the study of *symmetric* events, originated in the context of random Voronoi percolation. The others do not seem to adapt to this setting. As in Chapter 3, to define symmetric events we shall work in the *torus*, i.e., the quotient of $\mathbb{R}^2$ by a lattice.

More precisely, we shall work in the torus

$$\mathbb{T}^2 = \mathbb{T}^2_{10s} = \mathbb{R}^2/(10s\mathbb{Z})^2,$$

i.e., the surface obtained from the square $[0, 10s] \times [0, 10s]$ by identifying opposite sides. Here, s will be our scale parameter; we shall consider rectangles R in $\mathbb{T}^2$ with side-lengths ks, where $1 \le k \le 9$.

The notion of a random Voronoi tessellation makes perfect sense on the torus: we start with Poisson processes $\mathcal{P}^+$ and $\mathcal{P}^-$ of intensity p and $1-p$ on $\mathbb{T}^2$, which we may take to be the restriction of our processes on $\mathbb{R}^2$ to $[0, 10s]^2$. The definitions of the Voronoi cells and of the graph $G_{\mathcal{P}}$, $\mathcal{P} = \mathcal{P}^+ \cup \mathcal{P}^-$, are as before. We shall always consider s large. It is easy to check that whp every disc in $\mathbb{T}^2$ of radius $\sqrt{\log s}$, say, contains a point of $\mathcal{P}$, so the diameters of all Voronoi cells are at most $2\sqrt{\log s} = o(s)$. In particular, no cell 'wraps around' the torus.

The rectangles R we consider also do not come close to wrapping around the torus. Thus, events such as $H(R)$ have almost the same probability whether we regard R as a rectangle in $\mathbb{R}^2$ or in $\mathbb{T}^2$:

$$\mathbb{P}_p^{\mathbb{T}^2}(H(R)) = \mathbb{P}_p^{\mathbb{R}^2}(H(R)) + o(1)$$

as $s \to \infty$, where $\mathbb{P}_p^{\mathbb{T}^2}$ and $\mathbb{P}_p^{\mathbb{R}^2} = \mathbb{P}_p$ are the probability measures associated to Poisson processes $(\mathcal{P}^+, \mathcal{P}^-)$ on $\mathbb{T}^2 = \mathbb{T}^2_{10s}$ and on $\mathbb{R}^2$ respectively.

There is a natural notion of a 'symmetric' event with respect to the measure $\mathbb{P}_p^{\mathbb{T}^2}$, namely, an event that is preserved by translations of the torus. The is one of the two ingredients required for the application of the Friedgut–Kalai sharp-threshold result; the other is a discrete product space.

Let $\delta = \delta(s) > 0$ be such that $10s/\delta$ is an integer, and divide $\mathbb{T}^2 = \mathbb{T}^2_{10s}$ up into $(10s/\delta)^2$ squares S_i of side-length δ in the natural way. We shall approximate $(\mathcal{P}^+, \mathcal{P}^-)$ (defined on the torus) with a finite product measure as follows: a square S_i is *bad* if $|S_i \cap \mathcal{P}^-| > 0$, *neutral* if $|S_i \cap \mathcal{P}^-| = |S_i \cap \mathcal{P}^+| = 0$, and *good* if $|S_i \cap \mathcal{P}^-| = 0$ and $|S_i \cap \mathcal{P}^+| > 0$; see Figure 17. Thus, open points (points of $\mathcal{P}^+$) are good, and closed points

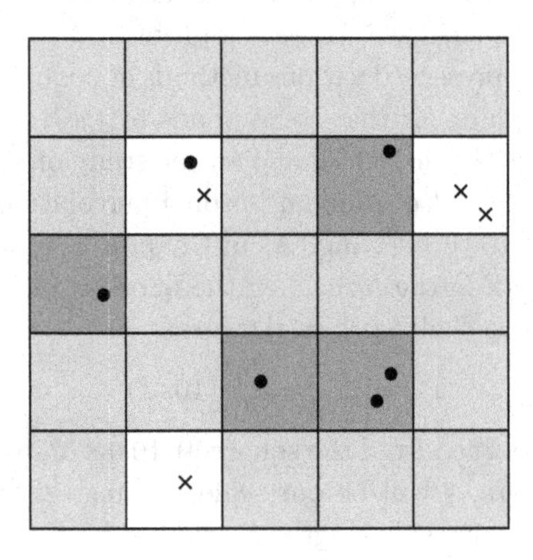

Figure 17. A small part of the decomposition of the torus into squares S_i. Points of $\mathcal{P}^+$ are shown by dots, those of $\mathcal{P}^-$ by crosses. A square is good (heavily shaded) if it meets $\mathcal{P}^+$ but not $\mathcal{P}^-$, neutral (lightly shaded) if it meets neither $\mathcal{P}^+$ nor $\mathcal{P}^-$, and bad (unshaded) if it meets $\mathcal{P}^-$. The 'crude state' of the torus is given by the shading of the squares. Usually, almost all squares are neutral.

are bad, and the presence of a bad point outweighs that of a good one; we shall see the reason for this choice later.

For each i, the probabilities that S_i is bad, neutral and good are

$$\begin{aligned} p_{\text{bad}} &= 1 - \exp\left(-\delta^2(1-p)\right) \sim \delta^2(1-p), \\ p_{\text{neutral}} &= \exp(-\delta^2) = 1 - O(\delta^2), \text{ and} \\ p_{\text{good}} &= \exp\left(-\delta^2(1-p)\right)\left(1 - \exp(-\delta^2 p)\right) \sim \delta^2 p, \end{aligned} \tag{14}$$

respectively, and these events are independent for different squares S_i. By the *crude state* of S_i we mean whether S_i is bad, neutral, or good; the *crude state* $CS = CS_\delta$ of the torus is given by the crude states of all $(10s/\delta)^2$ squares S_i.

It is clear that if δ is sufficiently small as a function of s, then CS essentially determines $(\mathcal{P}^+, \mathcal{P}^-)$, and thus $G_{\mathcal{P}}$ (defined on the torus).

Indeed, for s fixed, the σ-fields generated by $CS = CS_\delta$ are a filtration of that associated to $\mathbb{P}_p^{\mathbb{T}^2}$. In fact, for the discrete approximation to (essentially) encode the entire graph $G_{\mathcal{P}}$, it is enough to take $\delta = o(s^{-1})$. Indeed, it is easy to check that if δ is this small, then two things happen whp. First, no S_i contains more than one point of $\mathcal{P}$, so CS essentially determines $(\mathcal{P}^+, \mathcal{P}^-)$, up to shifting each point by a distance $\delta\sqrt{2}$. Second, there is no point of $\mathbb{T}^2$ such that the distances to the four closest points of $\mathcal{P}$ lie in an interval $[d, d + 2\sqrt{2}\delta]$. Thus, shifting points of $\mathcal{P}$ by at most $\sqrt{2}\delta$ does not change which Voronoi cells meet, and so does not affect the graph structure of $G_{\mathcal{P}}$.

As we shall see, we shall not be able to take δ this small, so the graph structure of $G_{\mathcal{P}}$ will be affected by the discrete approximation, although only 'slightly'.

The possible crude states CS of the torus correspond to the elements of $\Omega^n = \{-1, 0, 1\}^n$, where $n = (10s/\delta)^2$ is the number of squares S_i, and for $\omega = (\omega_i) \in \Omega^n$ we take ω_i to be -1, 0 or 1 if S_i is bad, neutral or good, respectively. Let us write $\mathbb{P}^n_{p_-,p_+}$ for the product probability measure on Ω^n in which each ω_i is -1, 0 or 1 with respective probabilities p_-, $1 - p_- - p_+$, and p_+. Then the discrete approximation to $\mathbb{P}_p^{\mathbb{T}^2}$ given by CS corresponds to the measure $\mathbb{P}^n_{p_{\text{bad}},p_{\text{good}}}$ on Ω^n. More precisely, if E is any event that depends only on the crude state of the torus, then

$$\mathbb{P}_p^{\mathbb{T}^2}(E) = \mathbb{P}^n_{p_{\text{bad}},p_{\text{good}}}(F), \tag{15}$$

where $F \subset \Omega^n$ is the corresponding event, and p_{bad}, p_{good} and p are related by (14).

It will clearly be convenient to use a form of the Friedgut–Kalai sharp-threshold result for powers of a 3-element probability space $\{-1, 0, 1\}$. In this context, an event $E \subset \Omega^n = \{-1, 0, 1\}^n$ is *increasing* if, whenever $\omega = (\omega_i) \in E$ and $\omega_i' \geq \omega_i$ for every i, then $\omega' = (\omega_i') \in E$. An event $E \subset \Omega^n$ is *symmetric* if it is invariant under the action on Ω^n of some group acting transitively on $[n] = \{1, 2, \ldots, n\}$.

From (14), both p_{bad} and p_{good} will be very small, so the sharp-threshold result we shall need is an equivalent of Theorem 14 of Chapter 2. As noted by Bollobás and Riordan [2006a], the proof given by Friedgut and Kalai [1996] extends immediately to give the following result.

Theorem 22. *There is an absolute constant c_3 such that, if $0 < q_- < p_- < 1/e$, $0 < p_+ < q_+ < 1/e$, $E \subset \{-1, 0, 1\}^n$ is symmetric and* *increasing, and* $\mathbb{P}^n_{p_-,p_+}(E) > \varepsilon$, *then* $\mathbb{P}^n_{q_-,q_+}(E) > 1 - \varepsilon$ *whenever*

$$\min\{q_+ - p_+, p_- - q_-\} \geq c_3 \log(1/\varepsilon) p_{\max} \log(1/p_{\max}) / \log n, \qquad (16)$$

where $p_{\max} = \max\{q_+, p_-\}$. □

When we come to apply this result, the form of (16) will matter; this is in sharp contrast to the situation for bond percolation on $\mathbb{Z}^2$. Indeed, when we used Theorem 13 of Chapter 2 to prove Kesten's Theorem in Chapter 3, the form of the corresponding bound was not important. All that mattered was that, with ε fixed and n tending to infinity, the increase in p required to raise $\mathbb{P}_p(E)$ from ε to $1 - \varepsilon$ tends to zero. Here, even though we are using a stronger result (with the extra factor $p_{\max} \log(1/p_{\max})$ working in our favour), there is a limit as to how small we can take δ while still getting a useful result from Theorem 22.

We shall compare $\mathbb{P}^{\mathbb{T}^2}_p(E)$ with $\mathbb{P}^{\mathbb{T}^2}_{1/2}(E)$ for a suitable event E, where $p > 1/2$ is fixed. From Corollary 20, the latter probability will be at least some small constant ε, and to prove (12) it will suffice to show that $\mathbb{P}^{\mathbb{T}^2}_p(E) \geq 1 - \varepsilon$. In doing this, we shall make a discrete approximation as above. From (14), when we apply Theorem 22, we shall have

$$q_+ - p_+,\ p_- - q_- \sim p\delta^2 - \delta^2/2 = \Theta(\delta^2).$$

Also, ε will be constant, $p_{\max} = \Theta(\delta^2)$, and $n = (10s/\delta)^2 = \Theta(s^2/\delta^2)$. Substituting these quantities into (16), we see that this condition will be satisfied if and only if $\delta \geq s^{-\gamma}$, for some constant $\gamma > 0$ depending on p. As p approaches $1/2$ from above, the constant γ tends to zero. Thus, we cannot afford to take a terribly fine discrete approximation; recall that $\delta = o(s^{-1})$ would be needed for the approximation process not to affect $G_{\mathcal{P}}$ at all.

With $\delta \sim s^{-\gamma}$, for γ a small positive constant, our discretization process will introduce 'defects' where, given the 'rounded' positions of the points of $\mathcal{P}$, we do not know which Voronoi cells actually meet. The density of these defects will be small (a negative power of s), so one might expect them not to matter. This turns out to be the case, but needs a lot of work to prove.

Benjamini and Schramm [1998] encountered the same problem in proving a certain 'conformal invariance' property of random Voronoi percolation in two and three dimensions. This is not conformal invariance in the sense of the conjecture of Aizenman and Langlands, Pouliot and Saint-Aubin [1994] discussed in Chapter 7. Instead, in two dimensions, what Benjamini and Schramm proved is essentially the following. Let

$D \subset \mathbb{R}^2$ be a (nice) domain, and let S_1 and S_2 be two segments of its boundary. Consider the Voronoi percolation associated to a Poisson process of intensity λ on R, using a certain metric ds to form the Voronoi cells, rather than the usual Euclidean metric. Then, as $\lambda \to \infty$, a fixed conformal (locally angle-preserving) change in the metric ds does not change the probability that there is a black path from S_1 to S_2 by more than $o(1)$.

Let us note that this is also a statement about defects: each Voronoi cell is very small, and, locally, the change in the metric is multiplication by a constant factor, so there will be very few 'defects' where different Voronoi cells meet for the two metrics: in two dimensions the expected number of defects is bounded as $\lambda \to \infty$; in three dimensions there are more, but the density is still very low. The result of Benjamini and Schramm is that these defects do not affect the crossing probability significantly; even though there are very few defects, the proof is far from simple.

In proving Theorem 17 we have a larger density of defects; on the other hand, we have the freedom to vary the probability p, replacing an arbitrary $p > 1/2$ by $(p + 1/2)/2$, say. Roughly speaking, we shall show that the effect of the defects is smaller than the effect of decreasing p in this way.

Let us now make some of the above ideas precise. Let R be a rectangle in $\mathbb{T}^2$. We wish to define an event E depending on the Poisson processes $(\mathcal{P}^+, \mathcal{P}^-)$ on $\mathbb{T}^2$ such that, whenever E holds, then $H(R)$ holds even if we move the points of $\mathcal{P}$ a little. Given $\eta > 0$, we say that a point $x \in \mathbb{T}^2$ is *η-robustly black* if there is a $z \in \mathcal{P}^+$ with $\mathrm{dist}(x, z') \geq \mathrm{dist}(x, z) + \eta$ for all $z' \in \mathcal{P}^-$; in other words, the nearest open point of $\mathcal{P}$ is at least a distance η closer than the nearest closed point. If we move all points of $\mathcal{P}$ at most a distance $\eta/2$, then any $x \in \mathbb{T}^2$ that was η-robustly black (and hence black) will still be black. We say that a path $P \subset \mathbb{T}^2$ is *η-robustly black* if every point of P is η-robustly black.

It turns out that if $0 < p_1 < p_2 < 1$ and $\gamma > 0$, then we can couple the probability measures $\mathbb{P}^{\mathbb{T}^2}_{p_1}$ and $\mathbb{P}^{\mathbb{T}^2}_{p_2}$ so that, whp, for every path P_1 that is black with respect to the first measure, there is a 'nearby' path P_2 that is η-robustly black with respect to the second measure, where $\eta = s^{-\gamma}$.

Theorem 23. *Let $0 < p_1 < p_2 < 1$ and $\gamma > 0$ be given. Let $\eta = \eta(s)$ be any function with $\eta(s) \leq s^{-\gamma}$. We may construct in the same* *probability space Poisson processes* $\mathcal{P}_1^+$, $\mathcal{P}_1^-$, $\mathcal{P}_2^+$ *and* $\mathcal{P}_2^-$ *on* $\mathbb{T}^2 = \mathbb{T}^2_{10s}$ *with intensities* p_1, $1-p_1$, p_2 *and* $1-p_2$, *respectively, such that* $\mathcal{P}_i^+$ *and* $\mathcal{P}_i^-$ *are independent for each* i, *and the following global event* E_{g} *holds whp as* $s \to \infty$: *for every piecewise-linear path* P_1 *that is black with respect to* $(\mathcal{P}_1^+, \mathcal{P}_1^-)$ *there is a piecewise-linear path* P_2 *that is* 4η*-robustly black with respect to* $(\mathcal{P}_2^+, \mathcal{P}_2^-)$, *with* $d_{\mathrm{H}}(P_1, P_2) \le (\log s)^2$. □

The proof of Theorem 23 is perhaps the hardest part of the proof of Theorem 17, and we shall say very little about it. Essentially, we must deal with certain 'potential defects', where the four closest points of $\mathcal{P}_1 = \mathcal{P}_1^+ \cup \mathcal{P}_1^-$ to some point $x \in \mathbb{T}^2$ are at almost the same distance (the distances differ by at most $s^{-\Theta(1)}$). We call such points of $\mathcal{P}_1$ *bad.* Roughly speaking, a *bad cluster* is a component in the graph on the bad points in which two bad points are adjacent if they are close enough to interact. The heart of the proof of Theorem 23 is a proof that, under the conditions of the theorem, whp the largest bad cluster has size $o(\log s)$. As the probability that a given point is bad is $s^{-\Theta(1)}$, one might expect the largest bad cluster to have size $O(1)$; it turns out, however, that there will be bad clusters of size $\Theta(\log s/\log\log s)$. Hence, the $o(\log s)$ bound needed for the proof of Theorem 23 is not far from best possible.

Using Theorems 22 and 23, it is not too hard to deduce (12) from Corollary 20, and so prove Theorem 17.

Proof of Theorem 17. Let $p > 1/2$ be fixed. As shown above, to prove that $\theta(p) > 0$, it suffices to show that, given s_0, there is an $s \ge s_0$ for which (12) holds.

Let γ be a positive constant to be chosen later. In fact, we shall set $\gamma = (p-1/2)/C$, where C is absolute constant. Let $s_1 \ge s_0$ be a large constant, such that all statements '... if s is large enough' in the rest of the proof hold for all $s \ge s_1$. By Corollary 20, there is an $s \ge s_1$ such that $f_{1/2}(9, s) \ge c_0$, where $c_0 > 0$ is an absolute constant; we fix such an s throughout the proof. Let R_1 be a fixed $9s$ by s rectangle in $\mathbb{R}^2$, so $\mathbb{P}_{1/2}(H(R_1)) \ge c_0$. Regarding R_1 as a rectangle in the torus $\mathbb{T}^2 = \mathbb{T}^2_{10s}$, let E_1 be the event that R_1 has a black horizontal crossing in the random Voronoi tessellation on the torus. As noted earlier, since R_1 does not come close to (within distance $O(\log s)$ of, say) wrapping around $\mathbb{T}^2$, we have

$$\mathbb{P}^{\mathbb{T}^2}_{1/2}(E_1) = \mathbb{P}^{\mathbb{R}^2}_{1/2}(H(R_1)) + o(1)$$

as $s \to \infty$. In particular, if s is sufficiently large, we have

$$\mathbb{P}_{1/2}^{\mathbb{T}^2}(E_1) \geq c_0/2, \tag{17}$$

say.

Let δ be chosen so that $10s/\delta$ is an integer, and $s^{-2\gamma} \leq \delta \leq s^{-\gamma}/4$; this is possible if s is large enough, which we may enforce by our choice of s_1. We shall first apply Theorem 23, and then 'discretize' by dividing the torus $\mathbb{T}^2$ into $(10s/\delta)^2$ small squares S_i of side-length δ, as above.

Set $p' = (p + 1/2)/2$, so $1/2 < p' < p$, and consider the coupled Poisson processes $(\mathcal{P}_1^+, \mathcal{P}_1^-)$ and $(\mathcal{P}_2^+, \mathcal{P}_2^-)$ whose existence is guaranteed by Theorem 23, applied with $p_1 = 1/2$, $p_2 = p'$, and $\eta = 4\delta \leq s^{-\gamma}$.

Let R_2 be an $8s$ by $2s$ rectangle obtained by moving the vertical sides of R_1 outwards by a distance $s/2$, and the horizontal sides inwards by the same distance; see Figure 18. Let E_2 be the event that there is a horizontal crossing of R_2 that is 4δ-robustly black.

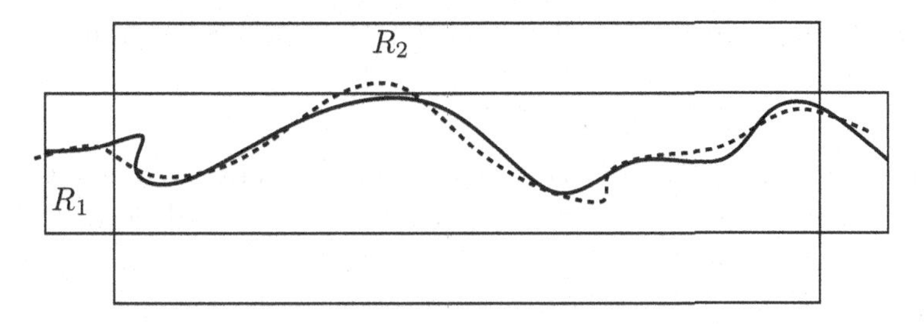

Figure 18. The $9s$ by s and $8s$ by $2s$ rectangles R_1 and R_2 (not to scale), together with a black path P_1 (solid curve) crossing R_1 horizontally, and a nearby robustly black path P_2 (dashed curve), part of which crosses R_2 horizontally.

Suppose that E_1 holds with respect to $(\mathcal{P}_1^+, \mathcal{P}_1^-)$; from (17), this event has probability at least $c_0/2$. Let $P_1 \subset R_1 \subset \mathbb{T}^2$ be a black path that crosses R_1 horizontally. If the global event E_{g} described in Theorem 23 holds, as it does whp, then there is a path P_2 with $d_{\mathrm{H}}(P_1, P_2) \leq (\log s)^2$ such that P_2 is 4δ-robustly black with respect to $(\mathcal{P}_2^+, \mathcal{P}_2^-)$. If s is sufficiently large that $(\log s)^2 < s/2$, then any such path P_2 contains a sub-path P_2' crossing R_2 horizontally, so E_2 holds with respect to $(\mathcal{P}_2^+, \mathcal{P}_2^-)$. If s is large enough, then E_{g} holds with probability at least $1 - c_0/4$, so

$$\mathbb{P}_{p'}^{\mathbb{T}^2}(E_2) \geq \mathbb{P}_{p'}^{\mathbb{T}^2}(E_1) - \mathbb{P}({E_{\mathrm{g}}}^c) \geq c_0/2 - c_0/4 \geq c_0/4.$$

Let E_3 be the event that *some* $8s$ by $2s$ rectangle in $\mathbb{T}^2$ has a 4δ-robustly black horizontal crossing. Then

$$\mathbb{P}_{p'}^{\mathbb{T}^2}(E_3) \geq \mathbb{P}_{p'}^{\mathbb{T}^2}(E_2) \geq c_0/4.$$

Let us divide $\mathbb{T}^2$ into $n = (10s/\delta)^2$ squares S_i as before, and define the crude state of each S_i and the crude state CS of the whole torus as above. Let E_3^δ be the event that the crude state CS of the torus is compatible with E_3. Since $\mathbb{P}(E_3) = \mathbb{E}(\mathbb{P}(E_3 \mid CS)) \leq \mathbb{P}(E_3^\delta)$, we have

$$\mathbb{P}_{p'}^{\mathbb{T}^2}(E_3^\delta) \geq c_0/4.$$

Let p_{bad} and p_{good} be defined by (14), and let p'_{bad} and p'_{good} be defined similarly, but with p' in place of p. Thus,

$$\begin{aligned} p'_{\mathrm{bad}} &= 1 - \exp\left(-\delta^2(1-p')\right) \sim \delta^2(1-p'), \text{ and} \\ p'_{\mathrm{good}} &= \exp\left(-\delta^2(1-p')\right)\left(1 - \exp(-\delta^2 p')\right) \sim \delta^2 p'. \end{aligned} \tag{18}$$

As the event E_3^δ depends only on CS, it may be viewed as an event F_3 in the finite product space $\Omega^n = \{-1, 0, 1\}^n$. In particular, from (15),

$$\mathbb{P}^n_{p'_{\mathrm{bad}}, p'_{\mathrm{good}}}(F_3) = \mathbb{P}_{p'}^{\mathbb{T}^2}(E_3^\delta) \geq c_0/4. \tag{19}$$

The event F_3 is symmetric, as E_3 and hence E_3^δ is preserved by a translation of the torus through $(i\delta, j\delta)$, $i, j \in \mathbb{Z}$. It is not hard to check that F_3 is also increasing.

From (14) and (18), we have

$$p'_{\mathrm{bad}} \sim (1-p')\delta^2,\ p'_{\mathrm{good}} \sim p'\delta^2,\ p_{\mathrm{bad}} \sim (1-p)\delta^2, \text{ and } p_{\mathrm{good}} \sim p\delta^2. \tag{20}$$

We shall apply Theorem 22 to $\{-1, 0, 1\}^n$, $n = (10s/\delta)^2$, with $p_- = p'_{\mathrm{bad}}$, $p_+ = p'_{\mathrm{good}}$, $q_- = p_{\mathrm{bad}}$ and $q_+ = p_{\mathrm{good}}$, and with $\varepsilon = \min\{c_0/4, 10^{-100}\}$, say. Note that $\varepsilon > 0$ is an absolute constant.

Let Δ be the quantity appearing on the right-hand side of (16) with this choice of parameters. From (20), all four of the quantities $p_\pm$ and $q_\pm$ are at most δ^2, if s is large enough. Hence, $p_{\max} \leq \delta^2$, and

$$\Delta = O\left(\delta^2 \log(1/\delta^2)/\log n\right),$$

where the implicit constant is absolute. By our choice of δ, we have $1/\delta \leq s^{2\gamma}$. On the other hand, $n = (10s/\delta)^2 \geq s^2$. Thus, $\log(1/\delta^2)/\log n \leq 2\gamma$. Hence, $\Delta \leq C\gamma\delta^2$ for some absolute constant C.

As C is an absolute constant, we may go back and choose γ small enough that $C\gamma \leq (p-p')/2$, so

$$\Delta \leq (p-p')\delta^2/2.$$

From (20), we have

$$q_+ - p_+, p_- - q_- \sim (p - p')\delta^2.$$

Thus, if s is large enough, $q_+ - p_+$ and $p_- - q_-$ are larger than Δ, i.e., the hypotheses of Theorem 22 are satisfied. Theorem 22 and (19) then imply

$$\mathbb{P}^n_{p_{\mathrm{bad}}, p_{\mathrm{good}}}(F_3) \geq 1 - \varepsilon \geq 1 - 10^{-100}.$$

Returning to the continuous setting, but considering the event E_3^δ, which depends only on CS, we have

$$\mathbb{P}_p^{\mathbb{T}^2}(E_3^\delta) = \mathbb{P}^n_{p_{\mathrm{bad}}, p_{\mathrm{good}}}(F_3) \geq 1 - 10^{-100}.$$

Suppose that CS is such that E_3^δ holds, and let $(\mathcal{P}^+, \mathcal{P}^-)$ be any configuration of the Poisson process consistent with this particular crude state of the torus. By the definition of E_3^δ, there is another realization consistent with the same crude state such that E_3 holds, i.e., such that there is a 4δ-robustly black path P crossing some $8s$ by $2s$ rectangle in $\mathbb{T}^2$. From the definition of bad, neutral and good squares, it follows that in the realization $(\mathcal{P}^+, \mathcal{P}^-)$, the path P is still black. Hence, writing E_4 for the event that *some* $8s$ by $2s$ rectangle in $\mathbb{T}^2$ has a black horizontal crossing, we have

$$\mathbb{P}_p^{\mathbb{T}^2}(E_4) \geq \mathbb{P}_p^{\mathbb{T}^2}(E_3^\delta) \geq 1 - 10^{-100}.$$

The rest of the argument is as in Chapter 3: we can cover $\mathbb{T}^2$ with $6s$ by $4s$ rectangles $R_1, \ldots, R_{25}$ so that, whenever E_4 holds, one of the R_i has a black horizontal crossing. Using the 'nth-root trick', it follows that

$$\mathbb{P}_p^{\mathbb{T}^2}(H(R_1)) \geq 1 - 10^{-4}.$$

Returning to the plane,

$$f_p(3/2, 4s) = \mathbb{P}_p^{\mathbb{R}^2}(H(R_1)) = \mathbb{P}_p^{\mathbb{T}^2}(H(R_1)) + o(1) \geq 0.999$$

if s is large enough. Appealing to (7), it is easy to deduce that $f_p(3, 4s) \geq 0.99$. As noted earlier, $\theta(p) > 0$ then follows easily by considering 1-independent percolation. Since $p > 1/2$ was arbitrary, and $\theta(1/2) = 0$ from Theorem 16, the result follows. □

The argument above was just a sketch of the proof of Theorem 17. In particular, we said almost nothing about the proof of Theorem 23. Although purely 'technical', such a result seems rather difficult to prove; this is actually a large part of the work in the original paper.

As shown by Bollobás and Riordan [2006a], for $p < 1/2$, the proof of Theorem 17 also gives exponential decay of the size of the open cluster C_0^G or corresponding black cluster C_0 containing the origin; this result follows from (12) by considering a suitable k-independent site percolation model on $\mathbb{Z}^2$.

Theorem 24. *Let $|C_0|$ denote the diameter of C_0, the area of C_0, or the number $|C_0^G|$ of Voronoi cells in C_0. Then, for any $p < 1/2$, there is a constant $c(p) > 0$ such that*

$$\mathbb{P}_p\big(|C_0| \geq n\big) \leq \exp(-c(p)n)$$

for every $n \geq 1$. □

As in Chapters 3 and 4, this implies that the Hammersley and Temperley critical probabilities coincide.

Theorem 25. *Let p_{T} be the Temperley critical probability for random Voronoi percolation in the plane above which $\mathbb{E}_p(|C_0|)$ diverges, where $|C_0|$ denotes the diameter of C_0, the area of C_0, or the number of Voronoi cells in C_0. Then $p_{\mathrm{T}} = p_{\mathrm{H}} = 1/2$.* □

We have seen that random Voronoi percolation can be very difficult to work with. Nevertheless, there is some hope that the conformal invariance conjecture of Aizenman and Langlands, Pouliot and Saint-Aubin [1994] discussed in Chapter 7 may be approachable for this model: the model has much more built-in symmetry than site or bond percolation on any lattice. Also, the result of Benjamini and Schramm [1998] provides a possible alternative method of attack; roughly speaking, they showed that for random Voronoi percolation, conformal invariance is equivalent to 'density invariance', i.e., the statement that, with suitable scaling, the crossing probabilities associated to two different inhomogeneous Poisson processes converge.

Bibliography

Abramowitz, M. and Stegun, I. A., Eds. (1966). *Handbook of Mathematical Functions, with Formulas, Graphs, and Mathematical Tables.* Dover, xiv+1046 pp.

Ahlswede R., Daykin D. E. (1978). An inequality for the weights of two families of sets, their unions and intersections. *Z. Wahrsch. Verw. Gebiete* **43**, 183–185.

Aizenman M. (1998). Scaling limit for the incipient spanning clusters. In *Mathematics of Multiscale Materials*, K. Golden, G. Grimmett, R. James, G. Milton, and P. Sen, eds. Springer.

Aizenman M., Barsky D. J. (1987). Sharpness of the phase transition in percolation models. *Comm. Math. Phys.* **108**, 489–526.

Aizenman M., Chayes J. T., Chayes L., Fröhlich J., Russo L. (1983). On a sharp transition from area law to perimeter law in a system of random surfaces. *Comm. Math. Phys.* **92**, 19–69.

Aizenman M., Delyon F., Souillard B. (1980). Lower bounds on the cluster size distribution. *J. Statist. Phys.* **23**, 267–280.

Aizenman M., Duplantier B., Aharony A. (1999). Path-crossing exponents and the external perimeter in 2d percolation. *Phys. Rev. Lett.* **83**, 1359–1362.

Aizenman M., Grimmett G. (1991). Strict monotonicity for critical points in percolation and ferromagnetic models. *J. Statist. Phys.* **63**, 817–835.

Aizenman M., Kesten H., Newman C. M. (1987). Uniqueness of the infinite cluster and continuity of connectivity functions for short and long range percolation. *Comm. Math. Phys.* **111**, 505–531.

Aizenman M., Newman C. M. (1984). Tree graph inequalities and critical behavior in percolation models. *J. Statist. Phys.* **36**, 107–143.

Aldous D. J. (1989). Stein's method in a two-dimensional coverage problem. *Statist. Probab. Lett.* **8**, 307–314.

Alexander K. S. (1996). The RSW theorem for continuum percolation and the CLT for Euclidean minimal spanning trees. *Ann. Appl. Probab.* **6**, 466–494.

Alon U., Drory A., Balberg I. (1990). Systematic derivation of percolation thresholds in continuum systems. *Phys. Rev. A* **42**, 4634–4638.

Ambartzumian R. V. (1990). *Factorization Calculus and Geometric Probability.* Cambridge University Press, xii+286 pp.

Appel M. J. B., Russo R. P. (1997). The minimum vertex degree of a graph on uniform points in $[0,1]^d$. *Adv. in Appl. Probab.* **29**, 582–594.

Appel M. J. B., Russo R. P. (2002). The connectivity of a graph on uniform points on $[0,1]^d$. *Statist. Probab. Lett.* **60**, 351–357.

Athreya S., Roy R., Sarkar A. (2004). On the coverage of space by random sets. *Adv. in Appl. Probab.* **36**, 1–18.

Baker D. R., Paul G., Sreenivasan S., Stanley H. E. (2002). Continuum percolation threshold for interpenetrating squares and cubes. *Phys. Rev. E* **66**, 046136, 5.

Balister P., Bollobás B., Quas A. (2005). Percolation in Voronoi tilings. *Random Structures Algorithms* **26**, 310–318.

Balister P., Bollobás B., Sarkar A., Walters M. (2005). Connectivity of random k-nearest-neighbour graphs. *Adv. in Appl. Probab.* **37**, 1–24.

Balister P., Bollobás B., Stacey A. (1993). Upper bounds for the critical probability of oriented percolation in two dimensions. *Proc. Roy. Soc. London Ser. A* **440**, 201–220.

Balister P., Bollobás B., Stacey A. (1994). Improved upper bounds for the critical probability of oriented percolation in two dimensions. *Random Structures Algorithms* **5**, 573–589.

Balister P., Bollobás B., Stacey A. (1999). Counting boundary paths for oriented percolation clusters. *Random Structures Algorithms* **14**, 1–28.

Balister P., Bollobás B., Walters M. (2004). Continuum percolation with steps in an annulus. *Ann. Appl. Probab.* **14**, 1869–1879.

Balister P., Bollobás B., Walters M. (2005). Continuum percolation with steps in the square or the disc. *Random Structures Algorithms* **26**, 392–403.

Beardon A. F. (1979). *Complex Analysis.* Wiley, xiii+239 pp.

Beffara V. (2004). Hausdorff dimensions for SLE_6. *Ann. Probab.* **32**, 2606–2629.

Beffara V. (2005). Cardy's formula on the triangular lattice, the easy way. Proceedings for a meeting held at the Fields Institute, September 2005. http://www.umpa.ens-lyon.fr/~vbeffara/files/Proceedings-Toronto.pdf.

Ben-Or M., Linial N. (1985). Collective coin flipping, robust voting schemes and minima of Banzhaf values. In *Proceedings of the 26th Annual IEEE Symposium on Foundations of Computer Science.* 408–416.

Ben-Or M., Linial N. (1990). Collective coin flipping. In *Randomness and Computation*, S. Micali, ed. Academic Press, 91–115.

Benjamini I., Schramm O. (1998). Conformal invariance of Voronoi percolation. *Comm. Math. Phys.* **197**, 75–107.

van den Berg J., Brouwer R., Vágvölgyi B. (2006). Continuity for self-destructive percolation in the plane. Preprint.

van den Berg J., Fiebig U. (1987). On a combinatorial conjecture concerning disjoint occurrences of events. *Ann. Probab.* **15**, 354–374.

van den Berg J., Keane M. (1984). On the continuity of the percolation probability function. In *Conference in Modern Analysis and Probability (New Haven, Conn., 1982).* American Mathematical Society, 61–65.

van den Berg J., Kesten H. (1985). Inequalities with applications to percolation and reliability. *J. Appl. Probab.* **22**, 556–569.

Bezdek K. (2002). Improving Rogers' upper bound for the density of unit ball packings via estimating the surface area of Voronoi cells from below in Euclidean d-space for all $d \geq 8$. *Discrete Comput. Geom.* **28**, 75–106.

Bezuidenhout C. E., Grimmett G. R., Kesten H. (1993). Strict inequality for critical values of Potts models and random-cluster processes. *Comm. Math. Phys.* **158**, 1–16.

Bollobás B. (1998). *Modern Graph Theory.* Springer, xiv+394 pp.

Bollobás B. (1999). *Linear Analysis*, Second edn. Cambridge University Press, xii+240 pp.

Bollobás B. (2001). *Random Graphs*, Second edn. Cambridge University Press, xviii+498 pp.

Bollobás B. (2006). *The Art of Mathematics: Coffee Time in Memphis.* Cambridge University Press.

Bollobás B., Kohayakawa Y. (1994). Percolation in high dimensions. *European J. Combin.* **15**, 113–125.

Bollobás B., Kohayakawa Y. (1995). A note on long-range percolation. In

Graph Theory, Combinatorics, and Algorithms, Vol. 1, 2 (Kalamazoo, MI, 1992). Wiley, 97–113.

Bollobás B., Riordan O. (2006a). The critical probability for random Voronoi percolation in the plane is 1/2. *Probab. Theory Related Fields.*

Bollobás B., Riordan O. (2006b). Sharp thresholds and percolation in the plane. *Random Structures Algorithms.*

Bollobás B., Riordan O. (2006c). A short proof of the Harris–Kesten Theorem. *Bull. London Math. Soc.* **38**, 470–484.

Bollobás B., Riordan O. (2006d). Percolation on dual lattices with two-fold symmetry. Preprint.

Bollobás B., Stacey A. (1997). Approximate upper bounds for the critical probability of oriented percolation in two dimensions based on rapidly mixing Markov chains. *J. Appl. Probab.* **34**, 859–867.

Bollobás B., Thomason A. (1985). Random graphs of small order. In *Random Graphs '83 (Poznań, 1983).* North-Holland, 47–97.

Borgs C., Chayes J. T., Kesten H., Spencer J. (1999). Uniform boundedness of critical crossing probabilities implies hyperscaling. *Random Structures Algorithms* **15**, 368–413.

Böröczky, Jr. K. (2004). *Finite Packing and Covering.* Cambridge University Press, xviii+380 pp.

Bourgain J., Kahn J., Kalai G., Katznelson Y., Linial N. (1992). The influence of variables in product spaces. *Israel J. Math.* **77**, 55–64.

Bräker H., Hsing T. (1998). On the area and perimeter of a random convex hull in a bounded convex set. *Probab. Theory Related Fields* **111**, 517–550.

Bräker H., Hsing T., Bingham N. H. (1998). On the Hausdorff distance between a convex set and an interior random convex hull. *Adv. in Appl. Probab.* **30**, 295–316.

Broadbent S. R., Hammersley J. M. (1957). Percolation processes. I. Crystals and mazes. *Proc. Cambridge Philos. Soc.* **53**, 629–641.

Burton R. M., Keane M. (1989). Density and uniqueness in percolation. *Comm. Math. Phys.* **121**, 501–505.

Cabo A. J., Groeneboom P. (1994). Limit theorems for functionals of convex hulls. *Probab. Theory Related Fields* **100**, 31–55.

Calka P. (2002a). The distributions of the smallest disks containing the Poisson–Voronoi typical cell and the Crofton cell in the plane. *Adv. in Appl. Probab.* **34**, 702–717.

Calka P. (2002b). La loi du plus petit disque contenant la cellule typique de Poisson–Voronoi. *C. R. Math. Acad. Sci. Paris* **334**, 325–330.

Calka P. (2003). An explicit expression for the distribution of the number of sides of the typical Poisson–Voronoi cell. *Adv. in Appl. Probab.* **35**, 863–870.

Calka P., Schreiber T. (2005). Limit theorems for the typical Poisson–Voronoi cell and the Crofton cell with a large inradius. *Ann. Probab.* **33**, 1625–1642.

Camia F., Newman C. M., Sidoravicius V. (2002). Cardy's formula for some dependent percolation models. *Bull. Braz. Math. Soc.* **33**, 147–156.

Camia F., Newman C. M., Sidoravicius V. (2004). A particular bit of universality: scaling limits of some dependent percolation models. *Comm. Math. Phys.* **246**, 311–332.

Cardy J. (2005). SLE for theoretical physicists. *Ann. Physics* **318**, 81–118.

Cardy J. L. (1992). Critical percolation in finite geometries. *J. Phys. A* **25**, L201–L206.

Chayes F. (1956). *Petrographic Modal Analysis.* Wiley, 113 pp.

Chayes J. T., Chayes L. (1986a). Inequality for the infinite-cluster density in Bernoulli percolation. *Phys. Rev. Lett.* **56**, 1619–1622.

Chayes J. T., Chayes L. (1986b). Percolation and random media. In *Phénomènes Critiques, Systèmes Aléatoires, Théories de Jauge, Part I, II (Les Houches, 1984).* North-Holland, 1001–1142.

Chayes J. T., Chayes L. (1987). On the upper critical dimension of Bernoulli percolation. *Comm. Math. Phys.* **113**, 27–48.

Chayes J. T., Puha A. L., Sweet T. (1999). Independent and dependent percolation. In *Probability Theory and Applications (Princeton, NJ, 1996).* American Mathematical Society, 49–166.

Chayes L., Lei H. K. (2006). Cardy's formula for certain models of the bond-triangular type. http://arxiv.org/abs/math-ph/0601023.

Chayes L., Schonmann R. H. (2000). Mixed percolation as a bridge between site and bond percolation. *Ann. Appl. Probab.* **10**, 1182–1196.

Coffman, Jr. E. G., Flatto L., Jelenković P. (2000). Interval packing: the vacant interval distribution. *Ann. Appl. Probab.* **10**, 240–257.

Conway A. R., Guttmann A. J. (1996). Square lattice self-avoiding walks and corrections to scaling. *Phys. Rev. Lett.* **77**, 5284–5287.

Cox J. T., Durrett R. (1983). Oriented percolation in dimensions $d \geq 4$: bounds and asymptotic formulas. *Math. Proc. Cambridge Philos. Soc.* **93**, 151–162.

de Bruijn N. G., Erdős P. (1952). Some linear and some quadratic recursion formulas. II. *Nederl. Akad. Wetensch. Proc. Ser. A.* **55** = *Indagationes Math.* **14**, 152–163.

Dean P. (1963). A new Monte Carlo method for percolation problems on a lattice. *Proc. Cambridge Philos. Soc.* **59**, 397–410.

Delesse A. (1848). Procédé méchanique pour déterminer la composition des roches. *Ann. des Mines (4th Ser.)* **13**, 379–388.

Dhar D. (1982). Diode-resistor percolation in two and three dimensions: I. Upper bounds on the critical probability. *J. Phys. A* **15**, 1849–1858.

Dhar D., Barma M. (1981). Monte Carlo simulation of directed percolation on a square lattice. *J. Phys. C* **14**, L1–L6.

Dirichlet G. L. (1850). Über die Reduktion der positiven quadratischen Formen mit drei unbestimmten ganzen Zahlen. *Journal für die Reine und Angewandte Mathematik* **40**, 209–227.

Domb C. (1959). Fluctuation phenomena and stochastic processes. *Nature* **184**, 509–512.

Domb C. (1972). A note on the series expansion method for clustering problems. *Biometrika* **59**, 209–211.

Domb C., Sykes M. F. (1961). Cluster size in random mixtures and percolation processes. *Phys. Rev.* **122**, 77–78.

Dubédat J. (2004). Critical percolation in annuli and SLE_6. *Comm. Math. Phys.* **245**, 627–637.

Dubson M., Garland J. (1985). Measurement of the conductivity exponent in two-dimensional percolating networks: square lattice versus random-void continuum. *Phys. Rev. B* **32**, 7621–7623.

Duren P. L. (1983). *Univalent Functions.* Springer, xiv+382 pp.

Durrett R. (1984). Oriented percolation in two dimensions. *Ann. Probab.* **12**, 999–1040.

Durrett R. (1985). Some general results concerning the critical exponents of percolation processes. *Z. Wahrsch. Verw. Gebiete* **69**, 421–437.

Durrett R. (1988). *Lecture Notes on Particle Systems and Percolation.* Wadsworth & Brooks/Cole Advanced Books & Software, viii+335 pp.

Dvoretzky A. (1956). On covering a circle by randomly placed arcs. *Proc. Nat. Acad. Sci. U.S.A.* **42**, 199–203.

Dvoretzky A., Robbins H. (1964). On the "parking" problem. *Magyar Tud. Akad. Mat. Kutató Int. Közl.* **9**, 209–225.

Elliott R. J., Heap B. R., Morgan D. J., Rushbrooke G. S. (1960). Equivalence of the critical concentrations in the Ising and Heisenberg models of ferromagnetism. *Phys. Rev. Lett.* **5**, 366–367.

Erdős P., Rényi A. (1960). On the evolution of random graphs. *Magyar Tud. Akad. Mat. Kutató Int. Közl.* **5**, 17–61.

Erdős P., Rényi A. (1961a). On the evolution of random graphs. *Bull. Inst. Internat. Statist.* **38**, 343–347.

Erdős P., Rényi A. (1961b). On the strength of connectedness of a random graph. *Acta Math. Acad. Sci. Hungar.* **12**, 261–267.

Eskin A., Mozes S., Oh H. (2005). On uniform exponential growth for linear groups. *Invent. Math.* **160**, 1–30.

Essam J. W., Guttmann A. J., De'Bell K. (1988). On two-dimensional directed percolation. *J. Phys. A* **21**, 3815–3832.

Falik D., Samorodnitsky A. (2005). Edge-isoperimetric inequalities and influences. Preprint.

Fejes Tóth L. (1953). *Lagerungen in der Ebene, auf der Kugel und im Raum.* Springer, x+197 pp.

Fejes Tóth L. (1972). *Lagerungen in der Ebene auf der Kugel und im Raum.* Springer, xi+238 pp. Zweite verbesserte und erweiterte Auflage.

Fekete M. (1923). Über die Verteilung der Wurzeln bei gewissen algebraischen Gleichungen mit ganzzahligen Koeffizienten. *Math. Z.* **17**, 228–249.

Finch S., Hueter I. (2004). Random convex hulls: a variance revisited. *Adv. in Appl. Probab.* **36**, 981–986.

Fisher M. E. (1961). Critical probabilities for cluster size and percolation problems. *J. Mathematical Phys.* **2**, 620–627.

Fisher M. E., Sykes M. F. (1959). Excluded-volume problem and the Ising model of ferromagnetism. *Phys. Rev.* **114**, 45–58.

Flatto L. (1973). A limit theorem for random coverings of a circle. *Israel J. Math.* **15**, 167–184.

Fortuin C. M., Kasteleyn P. W., Ginibre J. (1971). Correlation inequalities on some partially ordered sets. *Comm. Math. Phys.* **22**, 89–103.

Foss S. G., Zuyev S. A. (1996). On a Voronoi aggregative process related to a bivariate Poisson process. *Adv. in Appl. Probab.* **28**, 965–981.

Franceschetti M., Booth L., Cook M., Meester R., Bruck J. (2005). Continuum percolation with unreliable and spread-out connections. *J. Stat. Phys.* **118**, 721–734.

Freedman M. H. (1997). Percolation on the projective plane. *Math. Res. Lett.* **4**, 889–894.

Fremlin D. (1976). The clustering problem: Some Monte Carlo results. *J. Phys. (France)* **37**, 813–817.

Friedgut E. (2004). Influences in product spaces: KKL and BKKKL revisited. *Combin. Probab. Comput.* **13**, 17–29.

Friedgut E., Kalai G. (1996). Every monotone graph property has a sharp threshold. *Proc. Amer. Math. Soc.* **124**, 2993–3002.

Frisch H. L., Hammersley J. M. (1963). Percolation processes and related topics. *J. Soc. Indust. Appl. Math.* **11**, 894–918.

Gandolfi A. (1989). Uniqueness of the infinite cluster for stationary Gibbs states. *Ann. Probab.* **17**, 1403–1415.

Gandolfi A., Grimmett G., Russo L. (1988). On the uniqueness of the infinite cluster in the percolation model. *Comm. Math. Phys.* **114**, 549–552.

Gandolfi A., Keane M., Russo L. (1988). On the uniqueness of the infinite occupied cluster in dependent two-dimensional site percolation. *Ann. Probab.* **16**, 1147–1157.

Garboczi E., Thorpe M., DeVries M., Day A. (1991). Universal conductivity curve for a plane containing random holes. *Phys. Rev. A* **43**, 6473–6482.

Gaunt D. S., Ruskin H. (1978). Bond percolation processes in d-dimensions. *J. Phys. A* **11**, 1369–1380.

Gaunt D. S., Sykes M. F., Ruskin H. (1976). Percolation processes in d-dimensions. *J. Phys. A* **9**, 1899–1911.

Gawlinski E., Stanley H. (1981). Continuum percolation in two dimensions: Monte Carlo tests of scaling and universality for non-interacting discs. *J. Phys. A* **14**, L291–L299.

Gawlinski E. T., Redner S. (1983). Monte-Carlo renormalisation group for continuum percolation with excluded-volume interactions. *J. Phys. A* **146**, 1063–1071.

Gilbert E. N. (1959). Random graphs. *Ann. Math. Statist.* **30**, 1141–1144.

Gilbert E. N. (1961). Random plane networks. *J. Soc. Indust. Appl. Math.* **9**, 533–543.

Gilbert E. N. (1962). Random subdivisions of space into crystals. *Ann. Math. Statist.* **33**, 958–972.

González-Barrios J. M., Quiroz A. J. (2003). A clustering procedure based on the comparison between the k nearest neighbors graph and the minimal spanning tree. *Statist. Probab. Lett.* **62**, 23–34.

Gordon D. M. (1991). Percolation in high dimensions. *J. London Math. Soc.* **44**, 373–384.

Grassberger P. (1999). Conductivity exponent and backbone dimension in 2-d percolation. *Physica A* **262**, 251–263.

Gravner J., Griffeath D. (1997). Multitype threshold growth: convergence to Poisson–Voronoï tessellations. *Ann. Appl. Probab.* **7**, 615–647.

Griffiths R. B. (1967a). Correlations in Ising ferromagnets. I. *J. Mathematical Phys.* **8**, 478–483.

Griffiths R. B. (1967b). Correlations in Ising ferromagnets. II. External magnetic fields. *J. Mathematical Phys.* **8**, 484–489.

Grigorchuk R. I. (1983). On the Milnor problem of group growth. *Dokl. Akad. Nauk SSSR* **271**, 30–33.

Grigorchuk R. I. (1984). Degrees of growth of finitely generated groups and the theory of invariant means. *Izv. Akad. Nauk SSSR Ser. Mat.* **48**, 939–985. English translation: Math. USSR-Izv. 25 (1985), no. 2, 259–300.

Grimmett G. (1989). *Percolation.* Springer, xii+296 pp.

Grimmett G. (1994). Potts models and random-cluster processes with many-body interactions. *J. Statist. Phys.* **75**, 67–121.

Grimmett G. (1999). *Percolation*, Second edn. Springer, xiv+444 pp.

Grimmett G. (2004). The random-cluster model. In *Probability on Discrete Structures.* Springer, 73–123.

Grimmett G. R., Stacey A. M. (1998). Critical probabilities for site and bond percolation models. *Ann. Probab.* **26**, 1788–1812.

Groeneboom P. (1988). Limit theorems for convex hulls. *Probab. Theory Related Fields* **79**, 327–368.

Gromov M. (1981). *Structures Métriques pour les Variétés Riemanniennes.* CEDIC, iv+152 pp. Edited by J. Lafontaine and P. Pansu.

Grünbaum B., Shephard G. C. (1987). *Tilings and Patterns.* W. H. Freeman, xii+700 pp.

Guttmann A. J., Conway A. R. (2001). Square lattice self-avoiding walks and polygons. *Ann. Comb.* **5**, 319–345. Dedicated to the memory of Gian-Carlo Rota (Tianjin, 1999).

Haan S., Zwanzig R. (1977). Series expansions in a continuum percolation problem. *J. Phys. A* **10**, 1547–1555.

Hajek B. (1983). Adaptive transmission strategies and routing in mobile radio networks. *Proceedings of the Conference on Information Sciences and Systems*, 373–378.

Hall P. (1935). On representatives of subsets. *J. London Math. Soc.* **10**, 26–30.

Hall P. (1985a). Distribution of size, structure and number of vacant regions in a high-intensity mosaic. *Z. Wahrsch. Verw. Gebiete* **70**, 237–261.

Hall P. (1985b). On continuum percolation. *Ann. Probab.* **13**, 1250–1266.

Hall P. (1988). *Introduction to the Theory of Coverage Processes.* Wiley, xx+408 pp.

Hammersley J. M. (1957a). Percolation processes. II. The connective constant. *Proc. Cambridge Philos. Soc.* **53**, 642–645.

Hammersley J. M. (1957b). Percolation processes: Lower bounds for the critical probability. *Ann. Math. Statist.* **28**, 790–795.

Hammersley J. M. (1959). Bornes supérieures de la probabilité critique dans un processus de filtration. In *Le Calcul des Probabilités et ses Applications. Paris, 15-20 Juillet 1958.* Centre National de la Recherche Scientifique, 17–37.

Hammersley J. M. (1961a). Comparison of atom and bond percolation processes. *J. Mathematical Phys.* **2**, 728–733.

Hammersley J. M. (1961b). The number of polygons on a lattice. *Proc. Cambridge Philos. Soc.* **57**, 516–523.

Hammersley J. M. (1980). A generalization of McDiarmid's theorem for mixed Bernoulli percolation. *Math. Proc. Cambridge Philos. Soc.* **88**, 167–170.

Hammersley J. M., Morton K. W. (1954). Poor man's Monte Carlo. *J. Roy. Statist. Soc. Ser. B.* **16**, 23–38; discussion 61–75.

Hammond A. (2005). Critical exponents in percolation via lattice animals. *Electron. Comm. Probab.* **10**, 45–59 (electronic).

Hara T., Slade G. (1990). Mean-field critical behaviour for percolation in high dimensions. *Comm. Math. Phys.* **128**, 333–391.

Hara T., Slade G. (1995). The self-avoiding-walk and percolation critical points in high dimensions. *Combin. Probab. Comput.* **4**, 197–215.

Harris T. E. (1960). A lower bound for the critical probability in a certain percolation process. *Proc. Cambridge Philos. Soc.* **56**, 13–20.

Hatfield J. C. (1978). Ph. D. thesis, University of Minnesota.

Hayen A., Quine M. (2000). The proportion of triangles in a Poisson–Voronoi tessellation of the plane. *Adv. in Appl. Probab.* **32**, 67–74.

Henze N. (1983). Ein asymptotischer Satz über den maximalen Minimalabstand von unabhängigen Zufallsvektoren mit Anwendung auf einen Anpassungstest im $\mathbf{R}^p$ und auf der Kugel. *Metrika* **30**, 245–259.

Higuchi Y. (1982). Coexistence of the infinite (*) clusters: a remark on the square lattice site percolation. *Z. Wahrsch. Verw. Gebiete* **61**, 75–81.

Holley R., Liggett T. M. (1978). The survival of contact processes. *Ann. Probability* **6**, 198–206.

Holst L., Hüsler J. (1984). On the random coverage of the circle. *J. Appl. Probab.* **21**, 558–566.

Hou T., Li V. (1986). Transmission range control in multihop packet radio networks. *IEEE Transactions on Communications* **COM-34**, 38–44.

Hsing T. (1994). On the asymptotic distribution of the area outside a random convex hull in a disk. *Ann. Appl. Probab.* **4**, 478–493.

Hueter I. (1994). The convex hull of a normal sample. *Adv. in Appl. Probab.* **26**, 855–875.

Hueter I. (1999a). The convex hull of samples from self-similar distributions. *Adv. in Appl. Probab.* **31**, 34–47.

Hueter I. (1999b). Limit theorems for the convex hull of random points in higher dimensions. *Trans. Amer. Math. Soc.* **351**, 4337–4363.

Hughes B. D. (1995). *Random walks and random environments. Vol. 1. Random walks.* Oxford University Press, xxii+631 pp.

Hughes B. D. (1996). *Random walks and random environments. Vol. 2. Random environments.* Oxford University Press, xxiv+526 pp.

Huillet T. (2003). Random covering of the circle: the size of the connected components. *Adv. in Appl. Probab.* **35**, 563–582.

Janson S. (1986). Random coverings in several dimensions. *Acta Math.* **156**, 83–118.

Jensen I. (2004a). Enumeration of self-avoiding walks on the square lattice. *J. Phys. A* **37**, 5503–5524.

Jensen I. (2004b). Improved lower bounds on the connective constants for two-dimensional self-avoiding walks. *J. Phys. A* **37**, 11521–11529.

Jerauld G. R., Hatfield J. C., Scriven L. E., Davis H. T. (1984). Percolation and conduction on Voronoi and triangular networks: a case study in topological disorder. *J. Physics C: Solid State Physics* **17**, 1519–1529.

Jerauld G. R., Scriven L. E., Davis H. T. (1984). Percolation and conduction on the 3D voronoi and regular networks: a second case study in topological disorder. *J. Physics C: Solid State Physics* **17**, 3429–3439.

Johnson W. A., Mehl R. F. (1939). Reaction kinetics in processes of nucleation and growth. *Trans. A.I.M.M.E.* **135**, 416–458.

Jonasson J. (2001). Optimization of shape in continuum percolation. *Ann. Probab.* **29**, 624–635.

Kabatjanskiĭ G. A., Levenšteĭn V. I. (1978). Bounds for packings on the sphere and in space. *Problemy Peredači Informacii* **14**, 3–25.

Kager W., Nienhuis B. (2004). A guide to stochastic Löwner evolution and its applications. *J. Statist. Phys.* **115**, 1149–1229.

Kahn J., Kalai G., Linial N. (1988). The influence of variables on boolean functions. Proc. 29th Annual Symposium on Foundations of Computer Science, Computer Society Press.

Kelly D. G., Sherman S. (1968). General Griffiths' inequalities on correlations in Ising ferromagnets. *J. Mathematical Phys.* **9**, 466–484.

Kertész J., Vicsek T. (1980). Oriented bond percolation. *J. Phys. C* **13**, L343–L348.

Kesten H. (1963). On the number of self-avoiding walks. *J. Mathematical Phys.* **4**, 960–969.

Kesten H. (1964). On the number of self-avoiding walks. II. *J. Mathematical Phys.* **5**, 1128–1137.

Kesten H. (1980). The critical probability of bond percolation on the square lattice equals $\frac{1}{2}$. *Comm. Math. Phys.* **74**, 41–59.

Kesten H. (1981). Analyticity properties and power law estimates of functions in percolation theory. *J. Statist. Phys.* **25**, 717–756.

Kesten H. (1982). *Percolation Theory for Mathematicians.* Birkhäuser, iv+423 pp.

Kesten H. (1986). The incipient infinite cluster in two-dimensional percolation. *Probab. Theory Related Fields* **73**, 369–394.

Kesten H. (1987a). Percolation theory and first-passage percolation. *Ann. Probab.* **15**, 1231–1271.

Kesten H. (1987b). A scaling relation at criticality for 2D-percolation. In *Percolation Theory and Ergodic Theory of Infinite Particle Systems (Minneapolis, Minn., 1984–1985).* Springer, 203–212.

Kesten H. (1987c). Scaling relations for 2D-percolation. *Comm. Math. Phys.* **109**, 109–156.

Kesten H. (1988). Recent progress in rigorous percolation theory. *Astérisque*, 217–231. Colloque Paul Lévy sur les Processus Stochastiques (Palaiseau, 1987).

Kesten H. (1990). Asymptotics in high dimensions for percolation. In *Disorder in Physical Systems: A Volume in Honour of John Hammersley*, G. R. Grimmett and D. J. A. Welsh, eds. Oxford University Press, 219–240.

Kesten H. (1991). Asymptotics in high dimensions for the Fortuin–Kasteleyn random cluster model. In *Spatial Stochastic Processes: A Festschrift in Honor of Ted Harris on His Seventieth Birthday*, K. S. Alexander and J. C. Watkins, eds. Birkhäuser, 57–85.

Kesten H. (2003). First-passage percolation. In *From classical to modern probability*. Birkhäuser, 93–143.

Kesten H., Lee S. (1996). The central limit theorem for weighted minimal spanning trees on random points. *Ann. Appl. Probab.* **6**, 495–527.

Kesten H., Sidoravicius V., Zhang Y. (1998). Almost all words are seen in critical site percolation on the triangular lattice. *Electron. J. Probab.* **3**, no. 10, 75 pp. (electronic).

Kesten H., Zhang Y. (1987). Strict inequalities for some critical exponents in two-dimensional percolation. *J. Statist. Phys.* **46**, 1031–1055.

Kleban P., Zagier D. (2003). Crossing probabilities and modular forms. *J. Statist. Phys.* **113**, 431–454.

Kleinrock L., Silvester J. (1978). Optimum transmission radii for packet radio networks or why six is a magic number. IEEE Nat. Telecommun. Conf., 4.3.1–4.3.5.

Kleitman D. J. (1966). Families of non-disjoint subsets. *J. Combinatorial Theory* **1**, 153–155.

Kolmogorov A. N. (1956). *Foundations of the Theory of Probability.* Chelsea, viii+84 pp. English translation edited by Nathan Morrison, with an added bibliography by A. T. Bharucha-Reid; the translation is based on the original German monograph, which appeared in Ergebnisse der Mathematik in 1933, and also on the Russian translation by G. M. Bavli, published in 1936.

Langlands R., Pouliot P., Saint-Aubin Y. (1994). Conformal invariance in two-dimensional percolation. *Bull. Amer. Math. Soc.* **30**, 1–61.

Langlands R. P., Pichet C., Pouliot P., Saint-Aubin Y. (1992). On the universality of crossing probabilities in two-dimensional percolation. *J. Statist. Phys.* **67**, 553–574.

Lawler G. F. (2004). An introduction to the stochastic Loewner evolution. In *Random Walks and Geometry*, V. A. Kaimanovich, ed. de Gruyter, 261–293.

Lawler G. F. (2005). *Conformally Invariant Processes in the Plane.* American Mathematical Society, xii+242 pp.

Lawler G. F., Schramm O., Werner W. (2001a). The dimension of the planar Brownian frontier is 4/3. *Math. Res. Lett.* **8**, 13–23.

Lawler G. F., Schramm O., Werner W. (2001b). The dimension of the planar Brownian frontier is 4/3. *Math. Res. Lett.* **8**, 401–411. Printing error corrected.

Lawler G. F., Schramm O., Werner W. (2001c). Values of Brownian intersection exponents. I. Half-plane exponents. *Acta Math.* **187**, 237–273.

Lawler G. F., Schramm O., Werner W. (2001d). Values of Brownian intersection exponents. II. Plane exponents. *Acta Math.* **187**, 275–308.

Lawler G. F., Schramm O., Werner W. (2002a). Analyticity of intersection exponents for planar Brownian motion. *Acta Math.* **189**, 179–201.

Lawler G. F., Schramm O., Werner W. (2002b). One-arm exponent for critical 2D percolation. *Electron. J. Probab.* **7**, no. 2, 13 pp. (electronic).

Lawler G. F., Schramm O., Werner W. (2002c). Sharp estimates for Brownian non-intersection probabilities. In *In and out of Equilibrium (Mambucaba, 2000).* Birkhäuser, 113–131.

Lawler G. F., Schramm O., Werner W. (2002d). Values of Brownian intersection exponents. III. Two-sided exponents. *Ann. Inst. H. Poincaré Probab. Statist.* **38**, 109–123.

Lawler G. F., Schramm O., Werner W. (2003). Conformal restriction: the chordal case. *J. Amer. Math. Soc.* **16**, 917–955 (electronic).

Lawler G. F., Schramm O., Werner W. (2004). Conformal invariance of planar loop-erased random walks and uniform spanning trees. *Ann. Probab.* **32**, 939–995.

Liggett T. M. (1995). Survival of discrete time growth models, with applications to oriented percolation. *Ann. Appl. Probab.* **5**, 613–636.

Liggett T. M., Schonmann R. H., Stacey A. M. (1997). Domination by product measures. *Ann. Probab.* **25**, 71–95.

Lorenz B., Orgzall I., Heuer H.-O. (1993). Universality and cluster structures in continuum models of percolation with two different radius distributions. *J. Phys. A* **26**, 4711–4722.

Maehara H. (1988). A threshold for the size of random caps to cover a sphere. *Ann. Inst. Statist. Math.* **40**, 665–670.

Maehara H. (1990). On the intersection graph of random arcs on a circle. In *Random Graphs '87 (Poznań, 1987).* Wiley, 159–173.

Maehara H. (2004). On the intersection graph of random caps on a sphere. *European J. Combin.* **25**, 707–718.

Margulis G. A. (1974). Probabilistic characteristics of graphs with large connectivity. *Problemy Peredači Informacii* **10**, 101–108.

Matheron G. (1975). *Random Sets and Integral Geometry.* Wiley, xxiii+261 pp.

Meester R., Roy R. (1994). Uniqueness of unbounded occupied and vacant components in Boolean models. *Ann. Appl. Probab.* **4**, 933–951.

Meester R., Roy R. (1996). *Continuum Percolation.* Cambridge University Press, x+238 pp.

Meijering J. L. (1953). Interface area, edge length, and number of vertices in crystal aggregates with random nucleation. *Philips Research Reports* **8**, 270–290.

Menshikov M. V. (1986). Coincidence of critical points in percolation problems. *Dokl. Akad. Nauk SSSR* **288**, 1308–1311.

Menshikov M. V. (1987). Quantitative estimates and strong inequalities for the critical points of a graph and its subgraph. *Teor. Veroyatnost. i Primenen.* **32**, 599–602.

Menshikov M. V., Molchanov S. A., Sidorenko A. F. (1986). Percolation theory and some applications. In *Probability theory. Mathematical statistics. Theoretical cybernetics, Vol. 24 (Russian).* Akad. Nauk SSSR Vsesoyuz. Inst. Nauchn. i Tekhn. Inform., 53–110. Translated in J. Soviet Math. **42** (1988), no. 4, 1766–1810.

Milnor J. (1968). Problem 5603. *Amer. Math. Monthly* **75**, 685–686.

Molchanov I. (2005). *Theory of Random Sets.* Springer, xvi+488 pp.

Molchanov I., Scherbakov V. (2003). Coverage of the whole space. *Adv. in Appl. Probab.* **35**, 898–912.

Møller J. (1994). *Lectures on Random Voronoĭ Tessellations.* Springer, vi+134 pp.

Morrow G. J., Zhang Y. (2005). The sizes of the pioneering, lowest crossing and pivotal sites in critical percolation on the triangular lattice. *Ann. Appl. Probab.* **15**, 1832–1886.

Muchnik R., Pak I. (2001). Percolation on Grigorchuk groups. *Comm. Algebra* **29**, 661–671.

Newman C. M., Schulman L. S. (1981). Infinite clusters in percolation models. *J. Statist. Phys.* **26**, 613–628.

Newman M. E. J., Ziff R. M. (2000). Efficient Monte Carlo algorithm and high-precision results for percolation. *Phys. Rev. Lett.* **85**, 4104–4107.

Newman M. E. J., Ziff R. M. (2001). Fast Monte Carlo algorithm for site or bond percolation. *Phys. Rev. E* **64**, 016706, 16.

Ney P. E. (1962). A random interval filling problem. *Ann. Math. Statist.* **33**, 702–718.

Ni J., Chandler S. (1994). Connectivity properties of a random radio network. *Proceedings of the IEE – Communications* **141**, 289–296.

Ottavi H. (1979). Majorant et minorant du seuil de percolation de lien du kagomé. *J. de Physique* **40**, 233–237.

Parviainen R. (2005). Estimation of bond percolation thresholds on the Arhimedean lattices. Preprint available from http://www.ms.unimelb.edu.au/~robertp/.

Peierls R. (1936). On Ising's model of ferromagnetism. *Proc. Cambridge Philos. Soc.* **36**, 477–481.

Penrose M. D. (1993). On the spread-out limit for bond and continuum percolation. *Ann. Appl. Probab.* **3**, 253–276.

Penrose M. D. (1996). Continuum percolation and Euclidean minimal spanning trees in high dimensions. *Ann. Appl. Probab.* **6**, 528–544.

Penrose M. D. (1997). The longest edge of the random minimal spanning tree. *Ann. Appl. Probab.* **7**, 340–361.

Penrose M. D. (1999). On k-connectivity for a geometric random graph. *Random Structures Algorithms* **15**, 145–164.

Penrose M. D. (2003). *Random Geometric Graphs.* Oxford University Press, xiv+330 pp.

Penrose M. D., Pisztora A. (1996). Large deviations for discrete and continuous percolation. *Adv. in Appl. Probab.* **28**, 29–52.

Pike G. E., Seager C. H. (1974). Percolation and conductivity: A computer study. I. *Phys. Rev. B* **10**, 1421–1434.

Pönitz A., Tittmann P. (2000). Improved upper bounds for self-avoiding walks in $\mathbf{Z}^d$. *Electron. J. Combin.* **7**, Research Paper 21, 10 pp. (electronic).

Pyber L. (2004). Groups of intermediate subgroup growth and a problem of Grothendieck. *Duke Math. J.* **121**, 169–188.

Quintanilla J., Torquato S. (1999). Percolation for a model of statistically inhomogeneous random media. *J. Chem. Phys.* **111**, 5947–5954.

Quintanilla J., Torquato S., Ziff R. M. (2000). Efficient measurement of the percolation threshold for fully penetrable discs. *J. Phys. A* **33**, L399–L407.

Ráth B. (2005). Conformal invariance of critical percolation on the triangular lattice. Diploma Thesis, Institute of Mathematics, Budapest University of Technology and Economics. www.math.bme.hu/~rathb/rbperko.pdf.

Reimer D. (2000). Proof of the van den Berg–Kesten conjecture. *Combin. Probab. Comput.* **9**, 27–32.

Rényi A. (1958). On a one-dimensional problem concerning random space filling (in Hungarian). *Magyar Tud. Akad. Mat. Kutató Int. Közl.* **3**, 109–127.

Rényi A., Sulanke R. (1963). Über die konvexe Hülle von n zufällig gewählten Punkten. *Z. Wahrsch. Verw. Gebiete* **2**, 75–84.

Rényi A., Sulanke R. (1964). Über die konvexe Hülle von n zufällig gewählten Punkten. II. *Z. Wahrsch. Verw. Gebiete* **3**, 138–147.

Riordan O., Walters M. (2006). Rigorous confidence intervals for critical probabilities. Preprint.

Roberts F. D. K. (1967). A Monte Carlo solution of a two-dimensional unstructured cluster problem. *Biometrika* **54**, 625–628.

Rogers C. A. (1958). The packing of equal spheres. *Proc. London Math. Soc.* **8**, 609–620.

Rogers C. A. (1964). *Packing and Covering.* Cambridge University Press, viii+111 pp.

Rohde S., Schramm O. (2005). Basic properties of SLE. *Ann. of Math.* **161**, 883–924.

Rosso M. (1989). Concentration gradient approach to continuum percolation in two dimensions. *J. Phys. A* **22**, L131–L136.

Roy R. (1990). The Russo–Seymour–Welsh theorem and the equality of critical densities and the "dual" critical densities for continuum percolation on $\mathbf{R}^2$. *Ann. Probab.* **18**, 1563–1575.

Roy R. (1991). Percolation of Poisson sticks on the plane. *Probab. Theory Related Fields* **89**, 503–517.

Roy R., Tanemura H. (2002). Critical intensities of Boolean models with different underlying convex shapes. *Adv. in Appl. Probab.* **34**, 48–57.

Rudd W. G., Frisch H. L. (1970). Critical behavior in percolation processes. *Phys. Rev. B* **2**, 162–164.

Russo L. (1978). A note on percolation. *Z. Wahrsch. Verw. Gebiete* **43**, 39–48.

Russo L. (1981). On the critical percolation probabilities. *Z. Wahrsch. Verw. Gebiete* **56**, 229–237.

Russo L. (1982). An approximate zero-one law. *Z. Wahrsch. Verw. Gebiete* **61**, 129–139.

Saleur H., Duplantier B. (1987). Exact determination of the percolation hull exponent in two dimensions. *Phys. Rev. Lett.* **58**, 2325–2328.

Santaló L. A. (1976). *Integral Geometry and Geometric Probability.* Addison-Wesley, xvii+404 pp.

Schramm O. (2000). Scaling limits of loop-erased random walks and uniform spanning trees. *Israel J. Math.* **118**, 221–288.

Schramm O. (2001a). A percolation formula. *Electron. Comm. Probab.* **6**, 115–120 (electronic).

Schramm O. (2001b). Scaling limits of random processes and the outer boundary of planar Brownian motion. In *Current Developments in Mathematics, 2000.* International Press, 233–253.

Seager C. H., Pike G. E. (1974). Percolation and conductivity: A computer study. II. *Phys. Rev. B* **10**, 1435–1446.

Seymour P. D., Welsh D. J. A. (1978). Percolation probabilities on the square lattice. *Ann. Discrete Math.* **3**, 227–245.

Shepp L. A. (1972). Covering the circle with random arcs. *Israel J. Math.* **11**, 328–345.

Siegel A. F. (1979). Asymptotic coverage distributions on the circle. *Ann. Probab.* **7**, 651–661.

Siegel A. F., Holst L. (1982). Covering the circle with random arcs of random sizes. *J. Appl. Probab.* **19**, 373–381.

Silvester J. A. (1980). On the spatial capacity of packet radio networks. Department of Computer Science, UCLA, Engineering Report UCLA-ENG-8021.

Smirnov S. (2001a). Critical percolation in the plane: conformal invariance, Cardy's formula, scaling limits. *C. R. Acad. Sci. Paris Sér. I Math.* **333**, 239–244.

Smirnov S. (2001b). Critical percolation in the plane. I. Conformal invariance and Cardy's formula. II. Continuum scaling limit. http://www.math.kth.se/~stas/papers/.

Smirnov S., Werner W. (2001). Critical exponents for two-dimensional percolation. *Math. Res. Lett.* **8**, 729–744.

Smythe R. T., Wierman J. C. (1978). *First-passage Percolation on the Square Lattice.* Springer, viii+196 pp.

Solomon H., Weiner H. (1986). A review of the packing problem. *Comm. Statist. A—Theory Methods* **15**, 2571–2607.

Steele J. M., Shepp L. A., Eddy W. F. (1987). On the number of leaves of a Euclidean minimal spanning tree. *J. Appl. Probab.* **24**, 809–826.

Steele J. M., Tierney L. (1986). Boundary domination and the distribution of the largest nearest-neighbor link in higher dimensions. *J. Appl. Probab.* **23**, 524–528.

Steutel F. W. (1967). Random division of an interval. *Statistica Neerlandica* **21**, 231–244.

Stevens W. L. (1939). Solution to a geometrical problem in probability. *Ann. Eugenics* **9**, 315–320.

Stoyan D., Kendall W. S., Mecke J. (1987). *Stochastic Geometry and its Applications.* Wiley, 350 pp.

Stoyan D., Kendall W. S., Mecke J. (1995). *Stochastic Geometry and its Applications*, Second edn. Wiley, 456 pp.

Suding P. N., Ziff R. M. (1999). Site percolation thresholds for archimedean lattices. *Phys. Rev. E* **60**, 275–283.

Sykes M. F., Essam J. W. (1963). Some exact critical percolation probabilities for site and bond problems in two dimensions. *Phys. Rev. Lett.* **10**, 3–4.

Sykes M. F., Essam J. W. (1964). Exact critical percolation probabilities for site and bond problems in two dimensions. *J. Mathematical Phys.* **5**, 1117–1127.

Takagi H., Kleinrock L. (1984). Optimal transmission ranges for randomly distributed packet radio terminals. *IEEE Transactions on Communications* **COM-32**, 246–257.

Tasaki H. (1987). Hyperscaling inequalities for percolation. *Comm. Math. Phys.* **113**, 49–65.

Thiessen A. H., Alter J. C. (1911). Climatological data for July, 1911: District no. 10, Great Basin. *Monthly Weather Review* **31**, 1082–1089.

Vahidi-Asl M. Q., Wierman J. C. (1990). First-passage percolation on the Voronoĭ tessellation and Delaunay triangulation. In *Random Graphs '87 (Poznań, 1987)*. Wiley, 341–359.

Vahidi-Asl M. Q., Wierman J. C. (1992). A shape result for first-passage percolation on the Voronoĭ tessellation and Delaunay triangulation. In *Random Graphs, Vol. 2 (Poznań, 1989)*. Wiley, 247–262.

Vahidi-Asl M. Q., Wierman J. C. (1993). Upper and lower bounds for the route length of first-passage percolation in Voronoĭ tessellations. *Bull. Iranian Math. Soc.* **19**, 15–28.

Vicsek T., Kertész J. (1981). Monte Carlo renormalisation-group approach to percolation on a continuum: test of universality. *J. Phys. A* **14**, L31–L37.

Voronoi G. (1908). Nouvelles applications des paramètres continus à la théorie des formes quadratiques. *Journal für die Reine und Angewandte Mathematik* **133**, 97–178.

Werner W. (2004). Random planar curves and Schramm–Loewner evolutions. In *Lectures on Probability Theory and Statistics*. Lecture Notes in Math., vol. 1840. Springer, 107–195.

Werner W. (2005). Conformal restriction and related questions. *Probab. Surv.* **2**, 145–190 (electronic).

Wierman J. C. (1981). Bond percolation on honeycomb and triangular lattices. *Adv. in Appl. Probab.* **13**, 298–313.

Wierman J. C. (1984). A bond percolation critical probability determination based on the star-triangle transformation. *J. Phys. A* **17**, 1525–1530.

Wierman J. C. (1990). Bond percolation critical probability bounds for the Kagomé lattice by a substitution method. In *Disorder in Physical*

Systems: A Volume in Honour of John Hammersley, G. R. Grimmett and D. J. A. Welsh, eds. Oxford University Press, 349–360.

Wierman J. C. (1995). Substitution method critical probability bounds for the square lattice site percolation model. *Combin. Probab. Comput.* **4**, 181–188.

Wierman J. C. (2002). Bond percolation critical probability bounds for three Archimedean lattices. *Random Structures Algorithms* **20**, 507–518.

Wierman J. C. (2003a). Pairs of graphs with site and bond percolation critical probabilities in opposite orders. *Discrete Appl. Math.* **129**, 545–548.

Wierman J. C. (2003b). Upper and lower bounds for the Kagomé lattice bond percolation critical probability. *Combin. Probab. Comput.* **12**, 95–111.

Wierman J. C., Parviainen R. (2003). Ordering bond percolation critical probabilities.

Williams D. (1991). *Probability with Martingales.* Cambridge University Press, xvi+251 pp.

Wilson J. S. (2004). On exponential growth and uniformly exponential growth for groups. *Invent. Math.* **155**, 287–303.

Winterfeld P. H., Scriven L. E., Davis H. T. (1981). Percolation and conductivity of random two-dimensional composites. *J. Physics C* **14**, 2361–2376.

Wu F. Y. (1978). Percolation and the Potts model. *J. Statist. Phys.* **18**, 115–123.

Xue F., Kumar P. (2004). The number of neighbors needed for connectivity of wireless networks. *Wireless Networks* **10**, 169–181.

Yonezawa F., Sakamoto S., Aoki K., Nosé S., Hori M. (1988). Percolation in Penrose tiling and its dual in comparison with analysis for Kagomé, dice and square lattices. *J. Non-Crys. Solids* **106**, 262–269.

Zhan D. (2004). Stochastic Loewner evolution in doubly connected domains. *Probab. Theory Related Fields* **129**, 340–380.

Ziff R. M. (1995a). On Cardy's formula for the critical crossing probability in 2D percolation. *J. Phys. A* **28**, 1249–1255.

Ziff R. M. (1995b). Proof of crossing formula for 2D percolation. Addendum to: "On Cardy's formula for the critical crossing probability in 2D percolation" [J. Phys. A **28** (1995), no. 5, 1249–1255]. *J. Phys. A* **28**, 6479–6480.

Zvavitch A. (1996). The critical probability for Voronoi percolation. MSc. thesis, Weizmann Institute of Science.

Index

List of notation

$|A|$: cardinality, 4
$A \triangle B$: symmetric difference, 38
$A \,\square\, B$, $\mathcal{A} \,\square\, \mathcal{B}$: square/box product, 42
$A \subset B$: subset (equality allowed), 38
$A^{(\varepsilon)}$: closed ε-neighbourhood, 211
A_i: boundary arc of a (discrete) domain, 179, 193
A_i^+: outer boundary arc of a discrete domain, 193
$\mathrm{Bi}(n,p)$: binomial distribution, 29
$B_r(x)$: ball in a graph, 108
$B_r(z)$: ball in $\mathbb{R}^2$ or $\mathbb{C}$, 183
$B_r^+(x)$: out-ball in an oriented graph, 104
$\mathbb{C}$: complex numbers, 178
$C_x = \{y \in \Lambda : x \to y\}$: open cluster containing x, 4
C_x^+: open out-cluster, 25
$C_{\overrightarrow{\Lambda}}$: out-class graph, 84
$D_{r,\lambda}$: occupied set in $G_{r,\lambda}$, 242
$\mathbb{E}_p$ etc: expectation associated to $\mathbb{P}_p$ etc, 6
$E(\Lambda)$: edge set of a graph Λ, 1
$E_\delta^i(z)$: existence of open separating path, 201
$E_{r,\lambda}$: vacant set in $G_{r,\lambda}$, 243
$G(a)$: $G_{r,\lambda}$ with $\pi r^2 \lambda = a$, 244
$G_A(W)$: generalized Gilbert model, 243
G_δ: discrete domain, 191
$G_\delta^\pm$: discrete approximations to a domain, 212
$G_{n,d}$: finite random geometric graph with degree d, 255
$G_{r,\lambda}$: Gilbert model, 242
G_r: $G_{r,1}$, 242
$G_r(V_n)$, $G_r(\mathbb{T}_\ell^2, \lambda)$: finite random geometric graphs, 254
H: hexagonal lattice, 136
$H(R) = H_\mathrm{b}(R)$: R has a black horizontal crossing, 274
$H_{n,k}$: k-nearest graph, 257
I_k: there are k infinite open clusters, 118
$\mathcal{P} = \mathcal{P}_\lambda$: Poisson process, 241
$\mathcal{P}^+$: open points of $\mathcal{P}$, 268
$\mathcal{P}^-$: closed points of $\mathcal{P}$, 268
$\mathbb{P}_{\Lambda,p}^\mathrm{b}$: probability measure associated to bond percolation, 2
$\mathbb{P}_{\Lambda,p}^\mathrm{s}$: probability measure associated to site percolation, 2
Q_p^n, $Q_\mathbf{p}^n$: weighted hypercube, 37
$\overrightarrow{Q}$: quadrant of $\overrightarrow{\mathbb{Z}}^2$, 168
$R_n(x) = \{x^+ \xrightarrow{n}\}$: there is an open path from $S_1^+(x)$ to $S_n^+(x)$, 87
SLE_κ: Schramm–Loewner evolution, 235
$S_r(x)$: sphere in a graph, 79
$S_r^+(x)$: out-sphere in an oriented graph, 86
T: triangular lattice, 129
$\mathbb{T}_\ell^2$, $\mathbb{T}^2$: torus, 254
$\mathbb{T}_n^2$, $\mathbb{T}^2$: discrete torus, 65
$V(\Lambda)$: vertex set of a graph Λ, 1
$V_\mathrm{b}(R)$, $V_\mathrm{w}(R)$: R has a black/white vertical crossing, 275
V_x: Voronoi cell of x, 263
$\overrightarrow{\mathbb{Z}}^d$: natural orientation of $\mathbb{Z}^d$, 30

$d(x)$: degree of a vertex/site, 21
$d(x,y)$: graph distance, 61
$d(x,y)$ in an oriented graph: distance from x to y, 86
$\mathrm{dist}(x,y)$: Euclidean distance, 211
d_H: Hausdorff distance, 211
$f_\delta^i(z)$: probability of $E_\delta^i(z)$, 207

Printed in the United States
By Bookmasters